Managing the Health Risks of Climate Change

Jeffrey Kurebwa
Bindura University of Science Education, Zimbabwe

David Makwerere
Bindura University of Science Education, Zimbabwe

IGI Global
Scientific Publishing
Publishing Tomorrow's Research Today

Vice President of Editorial — Melissa Wagner
Managing Editor of Acquisitions — Mikaela Felty
Managing Editor of Book Development — Jocelynn Hessler
Production Manager — Mike Brehm
Cover Design — Phillip Shickler

Published in the United States of America by
IGI Global Scientific Publishing
701 East Chocolate Avenue
Hershey, PA, 17033, USA
Tel: 717-533-8845
Fax: 717-533-8661
Website: https://www.igi-global.com E-mail: cust@igi-global.com

Library of Congress Cataloging-in-Publication Data

Names: Kurebwa, Jeffrey, 1982- editor. | Makwerere, David, 1977- editor.
Title: Managing the health risks of climate change / edited by Jeffrey Kurebwa, David Makwerere.
Description: Hershey, PA : IGI Global Scientific Publishing, [2025] | Includes bibliographical references and index. | Summary: "This edited book seeks to provide new evidence on managing the health risks of climate change"-- Provided by publisher.
Identifiers: LCCN 2024052916 (print) | LCCN 2024052917 (ebook) | ISBN 9798369341834 (hardcover) | ISBN 9798369341841 (ebook)
Subjects: LCSH: Climatic changes--Health aspects. | Medical climatology. | Environmental health. | Human beings--Effect of climate on.
Classification: LCC RA793 .M35 2025 (print) | LCC RA793 (ebook) | DDC 613/.11--dc23/eng/20250124
LC record available at https://lccn.loc.gov/2024052916
LC ebook record available at https://lccn.loc.gov/2024052917

British Cataloguing in Publication Data
A Cataloguing in Publication record for this book is available from the British Library.

Table of Contents

Detailed Table of Contents

In today's world, climate change is a potential threat to women's health and well-being. The climate crisis impacts women's health through various sources. Women are increasingly vulnerable to physical, emotional, social, and psychological health outcomes. The United Nations (UN) has acknowledged that current changes in temperature and atmosphere due to climate change have a significant effect on various aspects of women's health. In this light, the purpose of this book chapter is to highlight the importance of the biological, psychological, and social impact of climate change on women's health. By thoroughly examining these studies, we aim to give a comprehensive conclusion of the complex relationship between climate change and women's health and well-being. Results underline the necessity of fostering an interdisciplinary approach to policy development to focus on women's overall health and development. In conclusion, this study successfully enhances our understanding of women's health and drives meaningful action in research and real-world applications.

The impacts of climate change pose serious threats to public health and healthcare systems globally. Extreme weather events, rising temperatures, shifting disease patterns, and other climate-driven changes are straining the ability of communities to protect human health and well-being. This book chapter explores the critical need to build health-resilient systems that can withstand and adapt to the challenges of a changing climate. Pathways through which climate change affects human health,

Preface

This edited book seeks to provide new evidence on managing the health risks of climate change. Climate change poses a significant threat to human health, impacting the physical environment and influencing various facets of both natural and human systems, including social and economic conditions as well as the operation of health systems. This phenomenon acts as a threat multiplier, jeopardizing the health advancements achieved over the past thirty years. The changing climate has led to an increase in the frequency and severity of weather-related events such as storms, extreme heat, floods, droughts, and wildfires. These climatic hazards have both direct and indirect repercussions on health, heightening the risks of mortality, non-communicable diseases, the emergence and spread of infectious diseases, and health crises. Furthermore, climate change is a direct contributor to humanitarian emergencies, manifesting through heatwaves, cyclones, wildfires, floods, tropical storms, and hurricanes (Intergovernmental Panel on Climate Change (IPCC) (2021).

The scale, frequency, and intensity of climate-related challenges are on the rise. According to research conducted by the World Bank (2024), approximately 3.6 billion individuals worldwide reside in regions that are particularly vulnerable to the impacts of climate change. Projections indicate that between 2030 and 2050, climate change could lead to an additional 250,000 fatalities annually due to factors such as undernutrition, malaria, diarrhea, and heat stress. Furthermore, the direct health-related costs, not accounting for expenses in sectors that influence health like agriculture and water sanitation, are anticipated to reach between US$ 2–4 billion each year by 2030 (World Bank, 2024). Regions with inadequate health infrastructure, predominantly in African nations, will struggle the most to manage these challenges without external support for preparation and response efforts.

The consensus within the scientific community is that climate change is occurring, primarily driven by human activities, and poses significant risks to human health (Field et al., 2014). The health impacts associated with climate variability and change are varied, potentially influencing a broad spectrum of health outcomes. Alterations in weather patterns can modify the frequency and severity of morbidity and mortality linked to extreme weather events, as well as changes in the levels of ozone, particulate matter, and aeroallergens (Smith et al., 2014). Additionally, shifts in climate and weather may create conditions that alter the geographic distribution, seasonal patterns, and incidence of certain infectious diseases in various regions, exemplified by the potential for malaria to spread into higher altitude areas in parts of Sub-Saharan Africa. Variations in water supply and agricultural output may have significant implications for undernutrition, especially in certain regions of Africa and Asia (Lloyd, Kovats, & Chalabi 2011). While climate change may lead to improved health outcomes in specific areas, the net effect is expected to be harmful to health and well-being, particularly in low- and lower-middle-income nations that face greater challenges related to climate-sensitive health issues (Smith et al., 2014).

The shifting climate is contributing to a rise in heat-related health issues and fatalities, altering the dynamics of infectious disease spread, which heightens the risk of severe outbreaks and pandemics (Costello, 2018). Additionally, it is adversely affecting maternal and child health, while exacerbating the health consequences associated with extreme weather phenomena like floods, droughts, wildfires, and storms. This climate crisis places considerable pressure on healthcare systems, as it not only escalates the need for medical services but also hampers the systems' capacity to effectively address these growing challenges (Ebi, 2021). The ongoing climate crisis is significantly undermining access to essential human necessities, including food security, safe drinking water, sanitation, and clean air. Recent data from the World Bank indicates that rising temperatures could result in at least 21 million more fatalities by the year 2050, attributable to five major health threats: extreme heat, stunting, diarrhea, malaria, and dengue (World Bank, 2024).

The impact of climate change on human health is profound, presenting a considerable risk that could worsen various health conditions. With rising global temperatures and an increase in the frequency and intensity of extreme weather events, individuals and policymakers must implement proactive strategies to mitigate these emerging health threats (Martin, 2021). A major health issue linked to climate change is the proliferation of infectious diseases. Warmer climates enable disease-carrying organisms like mosquitoes and ticks to expand into new areas, thereby exposing communities to illnesses that were previously unfamiliar to them. The World Health Organization (WHO) (2023) has cautioned that the prevalence of vector-borne diseases, including malaria, dengue fever, and Zika virus, is expected to rise as a consequence of climate change. To reduce this risk, it is essential to enhance public health systems, advance disease monitoring frameworks, and adopt effective vector management strategies. Additionally, climate change plays a significant role in the rising occurrence and severity of extreme weather phenomena, including heatwaves, droughts, and floods. These events can adversely affect human health both directly and indirectly, resulting in injuries, illnesses, and fatalities. For instance, heatwaves can worsen existing health issues, particularly among at-risk groups such as the elderly and individuals with chronic respiratory or cardiovascular conditions. Strategies to combat these challenges involve the development of improved early warning systems, bolstering emergency response capabilities, and implementing urban planning initiatives that foster infrastructure resilient to heat (Hessi, 2020).

Climate change poses significant threats to food and water security, which can adversely affect nutrition and contribute to the proliferation of waterborne diseases. Variations in precipitation and prolonged droughts can hinder agricultural output, resulting in food scarcity and increased malnutrition. Furthermore, flooding and the pollution of water supplies heighten the likelihood of waterborne illnesses (Watts et al., 2018). To address these challenges, it is essential to implement sustainable agricultural methods, enhance access to clean water and sanitation, and invest in the infrastructure necessary for water treatment and distribution. The escalating health risks are not being counterbalanced by adequate investments in research and the fortification of health systems. In 2020, a significant 63% of countries lacked a robust implementation of national emergency frameworks (Romanello et al., 2021). Out of 91 nations that participated in a World Health Organization survey regarding climate change and health, merely 47 indicated the existence of a national adaptation plan, with a lack of sufficient resources identified as the primary obstacle to advancement. Alarmingly, only 0.3% of the US $5.1 billion allocated for global climate change adaptation from 2018 to 2020 was channeled toward health systems, thereby leaving at-risk populations and regions without the essential human and financial resources necessary to preserve lives (Ebi, 2021). The current scenario regarding the health impacts of climate change and the responses to it presents a rather grim outlook; however, this situation can be significantly enhanced through meaningful progress in mitigating the severity of climate change and in developing health systems that are both climate-resilient and environmentally sustainable. A thorough and diverse strategy is essential for effectively addressing the health risks associated with climate change. This strategy should focus on bolstering public health systems, improving disease monitoring and response mechanisms, fostering infrastructure that can withstand climate impacts, and encouraging individuals and communities to embrace healthy and sustainable practices. Policymakers, healthcare experts, and researchers need to work together in crafting and executing effective measures aimed at safeguarding human health amid the challenges posed by a changing climate (Ebi, 2021).

In the first chapter, the authors conduct a comprehensive review of the bio-psychosocial effects of climate change on women's health and overall well-being. They contend that climate change poses a significant threat to women's health in contemporary society. The repercussions of the climate crisis manifest in multiple ways, rendering women increasingly susceptible to adverse physical, emotional, social, and psychological health outcomes. The United Nations has recognized that the ongoing alterations in temperature and atmospheric conditions due to climate change profoundly influence various dimensions of women's health. Consequently, this chapter aims to underscore the critical importance of understanding the biological, psychological, and social ramifications of climate change on women's health. Through an in-depth analysis of these studies, the objective is to provide a detailed conclusion regarding the intricate connection between climate change and the health and well-being of women. The findings emphasize the importance of adopting an interdisciplinary strategy in policy formulation that prioritizes women's holistic health and development. Ultimately, this research significantly contributes to our comprehension of women's health and encourages impactful initiatives in both academic research and practical applications.

The second chapter emphasizes that climate change threatens public health and healthcare infrastructures worldwide. The increasing frequency of extreme weather events, elevated temperatures, and alterations in disease distribution are overwhelming communities' capacities to safeguard health and well-being. This chapter highlights the urgent need for developing health-resilient systems capable of enduring and adapting to the adversities of a changing climate. How climate change influences human health encompasses both direct effects, such as those from heatwaves, wildfires, and natural disasters, and indirect effects, including challenges to food and water security, the proliferation of infectious diseases, and the displacement of populations. This chapter of the book offers actionable insights for policymakers, public health officials, healthcare managers, and community leaders on developing the adaptive capacity essential for safeguarding human health amid climate change. By implementing proactive measures to strengthen health resilience, communities can mitigate their susceptibility to climate-related challenges and promote equitable access to high-quality healthcare services as climate impacts become more severe.

Chapter three highlights that eco-anxiety involves not only worries about personal survival but also the welfare of future generations. It identifies media portrayals, societal interpretations, and individual experiences as significant factors that exacerbate eco-anxiety, thereby enriching the understanding of its underlying causes. By acknowledging both the detrimental and beneficial aspects of eco-anxiety, the chapter underscores the critical need to address personal distress within the larger framework of climate action. Investigating the effects of eco-anxiety, also referred to as climate distress, on mental health is crucial, given its widespread prevalence and the substantial negative impact it has on individuals' daily functioning. The rise in eco-anxiety is increasingly evident as public concern over climate change grows, leading to significant and alarming environmental transformations.

Chapter four aims to elucidate the effects of climate change on individuals with disabilities. It delves deeper into the notions of climate change, climate adaptation, ableism, and the various models about persons with disabilities. The chapter highlights the significant challenges that climate change poses to this demographic and offers actionable recommendations for different stakeholders to mitigate the adverse effects of climate change on individuals with disabilities. It is essential to guarantee the active involvement of individuals with disabilities and their representative organizations throughout the entire climate change policymaking and planning process at every level. This includes investing in the enhancement of inclusive strategies within key sectors that have significant potential for fostering climate-resilient and inclusive communities. Additionally, it is important to advocate for a comprehensive and equitable programmatic approach to climate resilience that weaves together social, economic, health, and climate factors, ultimately yielding extensive advantages for all marginalized groups.

In the fifth chapter, the authors delve into the effects of natural disasters on the mental well-being of children and adolescents residing in Joshimath, Uttarakhand, India. The presence of such disasters has been linked to heightened death anxiety and various psychological disorders among the youth. Specifically, these events contribute to an increase in post-traumatic stress disorder (PTSD) cases within this demographic. To support their analysis, the chapter presents a study involving a sample of 280 individuals aged between 12 and 18 years, comprising 116 boys and 164 girls, with an average age of 14.52 years and a standard deviation of 1.70. The participants were selected from various schools that experienced the adverse effects of the natural disaster. 1. The findings emphasize the necessity for policymakers to create strategies and resources aimed at mitigating the impact of natural disasters on the mental well-being of children and adolescents.

Chapter six examines the significance of air conditioning technology in fostering healthier living conditions in the context of climate change. It addresses how climate change affects indoor air quality (IAQ) and the subsequent effects on human health. The chapter advocates for a comprehensive strategy that encompasses ventilation, source control, air purification, and moisture management. Additionally, it recommends adaptation measures for climate change, including the implementation of advanced air conditioning systems and early warning mechanisms. The text underscores the dangers linked to conventional cooling methods, especially for at-risk populations. It also highlights technological innovations such as High-Efficiency Particulate Air (HEPA) filters and solar-powered systems as means to enhance IAQ while reducing environmental harm. Furthermore, it explores sustainable construction practices, photocatalytic oxidation, and the use of green roofs and walls. The chapter recognizes the difficulties associated with data privacy and the complexities of integration; however, it highlights the unification of air conditioning technology, adaptation to climate change, and sustainable design as essential components for creating healthier indoor environments.

Chapter Seven posits that the escalating worries regarding climate change and its effects on mental health have become a critical issue deserving of focus. Studies indicate that people living in areas prone to severe weather phenomena are especially at risk of facing mental health difficulties due to climate-induced stressors. The erratic weather conditions, persistent droughts, and food shortages have intensified pre-existing mental health issues and have led to an increase in new instances of anxiety, depression, and trauma within these communities. Examining the connection between climate change and mental health reveals the intricate ways in which climate-related stressors impact the psychological well-being of individuals, families, and communities. Understanding this complex relationship is essential for creating effective interventions and policies aimed at alleviating the negative consequences of climate change on mental health.

In the eighth chapter, the authors examine the concept of digital transformation, emphasizing its role in leveraging digital technologies for the advancement of society and the healthcare sector. Healthcare systems need to embrace these technologies to create innovative solutions that enhance the delivery of care and tackle various medical challenges. This transformation involves significant shifts related to the internet and digital tools, as well as their integration with emerging therapies and best practices aimed at improving health management. Effectively managing the vast amounts of data collected can lead to better patient outcomes and lower service costs. Furthermore, the impact of digital technologies on medical education will prompt experts to develop new training methodologies, thereby opening up fresh opportunities for practitioners in their respective fields. This chapter examines the evolution of healthcare through digital transformation, emphasizing the role of Augmented Reality (AR) and Virtual Reality (VR) technologies that are revolutionizing the delivery of healthcare services, training methodologies, and patient engagement worldwide.

Chapter nine concludes the discussion by examining how climate change influences household food security, utilizing Integral Theory as a framework for analysis. It delves into the four essential pillars of food security: availability, access, utilization, and stability. Additionally, the chapter explores various determinants of food security alongside the climatic factors that impact it. The findings suggest that enhancing agricultural practices and adopting strategies such as diversification, conservation, and climate-smart approaches are crucial for bolstering food security. Furthermore, the study advocates for advancements in agricultural extension services, improvements in storage technologies, the establishment of food banks, and the incorporation of spiritual practices to foster food security.

Jeffrey Kurebwa
Bindura University of Science Education, Zimbabwe

David Makwerere
Bindura University of Science Education, Zimbabwe

References

Chiabai, A., Balakrishnan, S., Sarangi, G., & Nischal, S. (2010). Human Health. In *Costing Adaptation: Preparing for Climate Change in India*. A. Markandya., & A. Mishra (Eds.). New Delhi: TERI Press.

Costello, A. (2018). The Lancet Countdown on Health and Climate Change: From 25 Years of Inaction to a Global Transformation for Public Health. *Lancet*, 391(10120), 581–630. DOI: 10.1016/S0140-6736(17)32464-9 PMID: 29096948

Ebi, K. L., Boyer, C., Ogden, N., Paz, S., Berry, P., Campbell-Lendrum, D., Hess, J. J., & Woodward, A. (2021). Burning embers: Synthesis of the health risks of climate change. *Environmental Research Letters*, 16(4), 044042. DOI: 10.1088/1748-9326/abeadd

Field, C. B., Barros, V. R., Dokken, D. J., Mach, K. J., & Mastrandrea, M. D. (Eds.). (2014). *Impacts, Adaptation, and Vulnerability. Part A: Global and Sectoral Aspects. Contribution of Working Group II to the Fifth Assessment Report of the Intergovernmental Panel on Climate Change*. Cambridge University Press.

Hess, J. J., Ranadive, N., Boyer, C., Aleksandrowicz, L., Anenberg, S. C., Aunan, K., Belesova, K., Bell, M. L., Bickersteth, S., Bowen, K., Burden, M., Campbell-Lendrum, D., Carlton, E., Cissé, G., Cohen, F., Dai, H., Dangour, A. D., Dasgupta, P., Frumkin, H., & Ebi, K. L. (2020). Guidelines for modeling and reporting health effects of climate change mitigation actions. *Environmental Health Perspectives*, 128(11), 115001. DOI: 10.1289/EHP6745 PMID: 33170741

Intergovernmental Panel on Climate Change (IPCC). (2021). [*The Physical Science Basis. Contribution of Working Group I to the Sixth Assessment Report of the Intergovernmental Panel on Climate Change*. www.ipcc.org]. *Climatic Change*, 2021, •••.

Lloyd, S. J., Kovats, R. S., & Chalabi, Z. (2011). Climate Change, Crop Yields, and Undernutrition: Development of a Model to Quantify the Impact of Climate Scenarios on Child Undernutrition. *Environmental Health Perspectives*, 119(12), 1817–1823. DOI: 10.1289/ehp.1003311 PMID: 21844000

Martin, M. A., Sendra, O. A., Bastos, A., Bauer, N., Bertram, C., Blenckner, T., Bowen, K., Brando, P. M., Rudolph, T. B., Büchs, M., Bustamante, M., Chen, D., Cleugh, H., Dasgupta, P., Denton, F., Donges, J. F., Donkor, F. K., Duan, H., Duarte, C. M., & Woodcock, J. (2021). Ten new insights in climate science 2021: A horizon scan. *Glob. Sustain.*, 4, E25. DOI: 10.1017/sus.2021.25

Romanello, M., McGushin, A., Di Napoli, C., Drummond, P., Hughes, N., Jamart, L., Kennard, H., Lampard, P., Solano Rodriguez, B., Arnell, N., Ayeb-Karlsson, S., Belesova, K., Cai, W., Campbell-Lendrum, D., Capstick, S., Chambers, J., Chu, L., Ciampi, L., Dalin, C., & Hamilton, I. (2021). The 2021 Report of the Lancet Countdown on Health and Climate Change: Code Red for a Healthy Future. *Lancet*, 398(10311), 1619–1662. DOI: 10.1016/S0140-6736(21)01787-6 PMID: 34687662

Smith, K. R., Woodward, A., Campbell-Lendrum, D., Chadee, D. D., & Honda, Y. (2014). Human Health: Impacts, Adaptation, and Co-Benefits. In C.B, Field, V.R, Barros., D.J, Dokken., K.J. Mach., & M.D. Mastrandrea (Eds.). *Climate Change 2014: Impacts, Adaptation, and Vulnerability; Part A: Global and Sectoral Aspects. Contribution of Working Group II to the Fifth Assessment Report of the Intergovernmental Panel on Climate Change*. New York: Cambridge University Press.

Watts, N., Amann, M., Ayeb-Karlsson, S., Belesova, K., Bouley, T., Boykoff, M., et al. (2018). The Lancet Countdown on health and climate change: from 25 years of inaction to a Global Transformation for Public Health. *The Lancet,* 10;391 (10120):581– 630.

World Bank. (2024). *Health and Climate Change*. www.worldbank.org

World Health Organization (WHO). (2018). *Climate change and Health*. www.who.int

Introduction

Climate change is increasingly acknowledged as one of the most urgent global health issues of our era. Its effects are complex, influencing both the environment and the health and well-being of populations across the globe. With rising temperatures and a growing frequency of extreme weather events, the health risks linked to climate change are expected to increase, endangering millions of lives and placing significant pressure on health systems that are frequently unprepared to cope with such challenges. The health consequences of climate change are extensive and diverse, affecting a wide array of physical and mental health conditions. These consequences include illnesses related to heat exposure, respiratory problems arising from deteriorating air quality, diseases transmitted by vectors, and mental health issues related to climate anxiety and displacement.

Vulnerable groups, such as children, the elderly, and individuals with pre-existing health issues, face heightened risks, underscoring the critical necessity for effective management strategies (Willets & Haines, 2023). Climate change has emerged as a paramount concern on the global health agenda, endangering individuals and communities through various exposure pathways that affect a wide range of physical and mental health conditions, as well as social determinants of health. The potential displacement of millions due to climate change threatens to destabilize social structures in numerous ways. Additionally, climatic stressors have significant implications for youth development and the well-being of future generations. The last eight years have recorded the highest temperatures, and as global temperatures continue to rise, the adverse health impacts associated with climate change are expected to escalate.

In light of the mounting evidence, the current approaches to risk management and climate mitigation are insufficiently aligned with the pressing nature of the issue, resulting in a continued separation between the domains of climate change and global health security. There are notable deficiencies in addressing the health consequences of climate change across all three pillars of climate action: mitigation, adaptation, and loss and damage (Willets & Haines, 2023). Although parties to the UN Framework Convention on Climate Change (UNFCCC, have pledged to reduce the negative health impacts of climate change through national policies, there has been a lack of emphasis on health considerations in international climate negotiations. Despite an expanding array of research that connects climate change to various health outcomes, there remains a considerable gap in the incorporation of health factors within climate action strategies.

Nationally determined contributions (NDCs) established under the UN Framework Convention on Climate Change (UNFCCC) frequently emphasize health considerations but often overlook the health advantages associated with mitigation and adaptation efforts. This gap highlights the necessity for a more integrated strategy that aligns climate initiatives with public health goals. In response to these issues, this book seeks to deliver an in-depth examination of the health threats linked to climate change while offering practical strategies for risk management. It will investigate the contributions of diverse stakeholders, such as governmental bodies, health organizations, and local communities, in building resilience and improving health outcomes amid the challenges posed by climate change.

By integrating climate action with health policy, we can enhance our preparedness for the health consequences associated with a changing climate, thereby protecting the well-being of both present and future generations. Climate change poses a significant risk to human health, impacting not only the physical environment but also various dimensions of both natural and human systems, which encompass social and economic conditions as well as the operation of health care systems. Consequently, it acts as a threat multiplier, jeopardizing and potentially reversing the health advancements achieved over the past decades. As climate conditions evolve, we are witnessing an increase in the frequency and severity of weather-related events, such as storms, extreme heat, floods, droughts, and wildfires.

Weather and climate-related hazards have both direct and indirect effects on health, heightening the likelihood of fatalities, the prevalence of non-communicable diseases, the emergence and transmission of infectious diseases, and various health emergencies. The repercussions of climate change extend to our health workforce and infrastructure, diminishing the ability to achieve universal health coverage. More fundamentally, climate-related shocks and escalating pressures, such as alterations in temperature and precipitation, droughts, floods, and rising sea levels, compromise the environmental and social factors that are essential for both physical and mental well-being. Climate change influences every facet of health, encompassing clean air, water, and soil, as well as food systems and livelihoods. A continued delay in addressing climate change will exacerbate health risks, jeopardize decades of advancements in global health, and violate our shared commitments to uphold the human right to health for everyone (Thacker et al. 2021).

The Sixth IPCC (2023) report has determined that the risks associated with climate change are manifesting more rapidly and will escalate in severity sooner than earlier projections indicated, complicating adaptation efforts in the face of rising global temperatures. The IPCC (2023) report highlights that approximately 3.6 billion individuals currently reside in regions that are particularly vulnerable to the effects of climate change. Notably, low-income nations and small island developing states, which contribute insignificantly to global emissions, are experiencing the most severe health consequences. In areas deemed vulnerable, the mortality rate resulting from extreme weather events over the past decade has been 15 times greater than that in regions considered less vulnerable (WHO, 2024).

Climate change is affecting health in numerous ways, contributing to mortality and morbidity through the rising frequency of extreme weather events, including heatwaves, storms, and floods. It disrupts food systems, increases the prevalence of zoonotic diseases, and exacerbates food, water, and vector-borne illnesses, while also impacting mental health (Prina et al. 2024). In addition, climate change threatens various social determinants essential for maintaining good health, such as economic stability, equity, and access to healthcare and social support. The health risks associated with climate change disproportionately affect the most vulnerable and marginalized groups, including women, children, ethnic minorities, impoverished communities, migrants or displaced individuals, the elderly, and those with pre-existing health conditions (Conway et al. 2023)

It is clear that climate change has a significant impact on human health; however, accurately assessing the extent and consequences of various health risks associated with climate change remains a complex task. Nevertheless, advancements in scientific research are increasingly enabling us to link rising rates of illness and death to global warming, as well as to more precisely evaluate the associated risks and the magnitude of these health challenges. According to data from the World Health Organization (2023), approximately 2 billion individuals do not have access to safe drinking water, and around 600 million people experience foodborne illnesses each year, with children under the age of five accounting for 30% of the fatalities related to these illnesses. The stressors associated with climate change further amplify the risks of waterborne and foodborne diseases. In 2020, an estimated 770 million people were affected by hunger, with the majority located in Africa and Asia (Proulx et al. 2024). The effects of climate change compromise food availability, quality, and diversity, thereby worsening food security and nutrition crises.

Changes in temperature and precipitation are contributing to the increased transmission of vector-borne diseases. In the absence of effective preventive measures, the annual death toll from these diseases, which currently exceeds 700,000, is likely to escalate (IPCC, 2018). Climate change not only triggers immediate mental health challenges, such as anxiety and post-traumatic stress disorder but also leads to long-term psychological issues stemming from factors like displacement and the breakdown of social cohesion. Recent studies indicate that 37% of fatalities related to heat exposure can be attributed to anthropogenic climate change. Furthermore, there has been a 70% increase in heat-related deaths among individuals aged 65 and older over the past twenty years (World Health Organisation, 2024). In 2020, an additional 98 million people faced food insecurity compared to the average from 1981 to 2010. The World Health Organization (2024) has conservatively estimated that climate change could result in an additional 250,000 deaths each year by the 2030s due to its effects on diseases such as malaria and the risks associated with coastal flooding. Nevertheless, challenges in modeling persist, particularly in accurately assessing risks related to drought and migration pressures.

The climate crisis poses a significant threat to the advancements made over the past five decades in areas such as development, global health, and poverty alleviation, while also exacerbating existing health disparities both among and within populations. It poses serious challenges to the achievement of Universal Health Coverage by intensifying the current disease burden and heightening barriers to accessing essential health services, particularly during critical times of need. Approximately 930 million individuals, representing about 12% of the global population, allocate at least 10% of their household income to healthcare expenses (WHO, 2024). As the most vulnerable populations remain largely uninsured, health-related shocks and stresses are already driving around 100 million people into poverty annually, a situation that is further aggravated by the effects of climate change (WHO, 2024).

In the short to medium term, the health consequences of climate change will primarily hinge on the vulnerability of populations, their capacity to withstand the current pace of climate change, and the degree and speed of their adaptation efforts. Over the long term, the impacts will increasingly rely on the extent of proactive measures taken now to mitigate emissions and prevent crossing critical temperature thresholds and potential irreversible tipping points. Although no individual is entirely immune to these threats, it is the individuals whose health is most adversely affected by the climate crisis—those who have contributed the least to its origins and who possess the least means to safeguard themselves and their families—who are most at risk. This includes populations in low-income and marginalized nations and communities (IPCC, 2018).

Addressing the health impacts of climate change highlights the necessity for equity: those who contribute most to emissions should shoulder the greatest costs associated with mitigation and adaptation, thereby prioritizing health equity and the needs of vulnerable populations. As the global climate crisis intensifies, its harmful effects on human health and well-being are expected to escalate correspondingly. No individual is immune to its consequences; however, certain groups—particularly women, children, the elderly, ethnic minorities, individuals with pre-existing health conditions, and those living in poverty—are disproportionately affected. The shifting climate is leading to an increase in heat-related illnesses and fatalities, altering the dynamics of infectious disease spread, which raises the likelihood of severe outbreaks and pandemics, while also exacerbating maternal and child health issues. Furthermore, extreme weather events such as floods, droughts, wildfires, and storms are intensifying health challenges (IPCC, 2018). The strain on health systems is considerable, as the demand for health services rises even as the capacity to meet this demand diminishes. Additionally, the climate crisis is rapidly undermining access to essential human needs, including food security, safe drinking water, sanitation, and clean air. According to recent data from the World Bank (2024), a warmer climate could result in at least 21 million additional deaths by 2050 due to five significant health risks: extreme heat, stunting, diarrhea, malaria, and dengue.

Measures to Strengthen Health Systems

To enhance health systems' ability to anticipate, identify, prepare for, and react to climate-related risks and disasters, it is essential to implement measures such as developing surveillance and early-warning systems that are informed by climate data, expanding the capacity of the health workforce in the area of climate and health, and ensuring that healthcare infrastructure is resilient to climate impacts.

Health systems should be assisted in their transition towards low-carbon, high-quality service delivery. This can be achieved by implementing clean, renewable energy sources for infrastructure and transportation fleets, as well as utilizing low-carbon medicines and medical equipment. This transition is essential, as the health sector is not only impacted by climate change but also plays a significant role in exacerbating the crisis, accounting for approximately 5 percent of global greenhouse gas (GHG) emissions (WHO, 2024).

Efforts to combat climate change and its health implications must focus on addressing the underlying causes by fostering collaboration across various sectors. This includes enhancing initiatives related to One Health, improving water, sanitation, and hygiene (WASH) practices, and promoting energy efficiency.

Demographic groups, including the elderly, children, and individuals with pre-existing health conditions, exhibit heightened susceptibility to the effects of climate change. Furthermore, low-income communities and indigenous populations are likely to face disproportionate challenges due to their restricted access to healthcare services and increased exposure to environmental risks. The encouraging aspect is that, notwithstanding the severe repercussions, there exist numerous viable and efficient strategies to mitigate greenhouse gas emissions and adjust to anthropogenic climate change, all of which are currently accessible, as stated by scientists in the IPCC (2018) report. The IPCC report advocates for the integration of health considerations into policies related to food, infrastructure, social protection, and water. This includes enhancing public health initiatives, bolstering the resilience of health systems, minimizing the vulnerability of water and sanitation systems to flooding, and advancing surveillance and early warning mechanisms, as well as the development of vaccines (WHO, 2008).

The Link between Climate Change and Health

Rising temperatures and alterations in rainfall patterns are facilitating the broader dissemination of diseases such as malaria, dengue, West Nile virus, and Lyme disease, as the vectors responsible for these illnesses, including mosquitoes and other insects, are expanding their habitats. For example, shifts in temperature and precipitation across Africa may lead to a potential increase in yellow fever fatalities by as much as 25% by the year 2050 (WHO, 2024). Evidence indicates that malaria-carrying mosquitoes have already begun to infiltrate more temperate regions, having extended their range by approximately 6.5 meters annually over the past century and moving away from the Equator by about 4.7 kilometers each year. In response to this growing threat, a malaria vaccine is currently being deployed in high-risk areas, while a dengue vaccine exists, albeit with limited global accessibility (WHO, 2024).

Climate change significantly impacts the accessibility and quality of water and food resources. This situation can result in increased instances of dehydration and the proliferation of waterborne illnesses, as the availability of clean water diminishes—currently, approximately half of the global population experiences severe water scarcity for at least one month each year (WHO, 2024). Additionally, global warming adversely affects agricultural productivity, potentially leading to food shortages and rising prices, which in turn exacerbates malnutrition. Undernourishment compromises the immune system, heightening individuals' vulnerability to diseases, thereby underscoring the critical need for routine immunization against essential childhood illnesses and the capacity to effectively address outbreaks of vaccine-preventable diseases such as cholera.

Rising temperatures can contribute to a higher formation of ground-level ozone and exacerbate pollution from particulate matter. This issue is not confined to specific regions often regarded as polluted; in fact, a report from the United Nations Environment Programme (UNEP, 2019) indicated that 99% of the global population resided in areas that failed to comply with the air quality standards set by the World Health Organization (WHO) in 2021.

Climate change is resulting in an increase in the frequency and intensity of extreme weather phenomena, including severe heatwaves, wildfires, and flooding. These events not only create humanitarian crises and result in loss of life but also contribute to displacement and migration, particularly affecting low-lying coastal regions and areas vulnerable to such extreme conditions. The movement of populations can heighten the risk of infectious diseases due to overcrowding and inadequate sanitation. Many migrant groups often lack access to healthcare services, particularly to essential, location-based health interventions like routine immunizations and maternal health programs, which leaves them without critical medical care and increases their susceptibility to various diseases.

Global warming is adversely impacting our mental well-being in numerous ways. The occurrence of extreme weather events leads to the displacement of populations, which can significantly elevate stress levels and adversely affect mental health as individuals continuously seek essential resources such as food, water, shelter, and healthcare access. According to the IPCC (2018), elevated temperatures, particularly during heat waves, contribute to various detrimental effects on mental health, including increased rates of depression and anxiety, with established correlations between rising temperatures and suicide attempts. Research shows that for each 1°C rise in the monthly average temperature, there is a corresponding 2.2% increase in mental health-related fatalities. Furthermore, as noted by the IPCC (2018), air pollution plays a critical role in influencing mental health, with rising instances of depression and anxiety correlating with deteriorating air quality. In the United States, short-term exposure to pollutants such as PM2.5 and NO2 has been linked to a heightened risk of acute psychiatric hospital admissions. Similarly, in the United Kingdom, elevated air pollution levels have been associated with an increased demand for mental health services among individuals suffering from psychotic or mood disorders (IPCC, 2018).

High temperatures are linked to negative birth outcomes, notably preterm births, and stillbirths, in addition to contributing to hypertension and gestational diabetes during pregnancy. Furthermore, heatwaves can impair cognitive function, which adversely impacts the learning capabilities of children and adolescents, while also elevating the risk of heart attacks and respiratory issues among the elderly population (WHO, 2024). Ambient air pollution elevates the risk of hypertension in pregnant women, contributes to low birth weight, and is associated with preterm deliveries, as well as adverse effects on the development of the fetal brain and lungs. Additionally, it heightens the susceptibility to respiratory illnesses in both children and the elderly, who are also at an increased risk for cancer, cardiovascular diseases, and pneumonia (WHO, 2024).

Climate-related natural disasters exert considerable effects on both mental and physical health. Events such as flooding and drought diminish the availability of safe drinking water and food resources, leading to a rise in diarrheal diseases and malnutrition. Additionally, wildfires have been linked to a higher incidence of respiratory illnesses and an increase in cardiovascular mortality rates among the elderly population. Climate change has a universal impact; however, the resulting displacements and disruptions pose significant challenges for individuals who rely on consistent access to healthcare and social services. Vulnerable groups, including infants, the elderly, and pregnant women, are particularly at risk due to physiological factors such as challenges in regulating body temperature, susceptibility to dehydration, and compromised immune systems (WHO, 2024). These populations also experience heightened effects from the secondary consequences of climate change and associated disasters, including shortages of food and water, as well as increases in vector-borne and waterborne diseases. By examining the health repercussions of various climate-related hazards on specific demographics, researchers seek to assist governments and organizations in identifying risks and formulating appropriate responses. According to Friel (2021), there is currently a lack of climate adaptation strategies specifically designed to meet the unique needs of women, infants, children, adolescents, and older adults who may face mobility and cognitive challenges. It is essential to prepare childcare, social care, and educational systems for the challenges posed by extreme weather and rising temperatures, while also involving individuals of all ages in climate action, discussions, and planning efforts.

Climate change has significantly transformed weather patterns on both global and local scales, leading to rising sea levels, a trend that is expected to persist. There is a notable increase in average temperatures, precipitation levels, and other climatic factors such as humidity. Furthermore, the variability of weather is intensifying, resulting in a higher frequency of heat waves, many of which are more severe and prolonged, alongside an increase in floods and droughts. These climatic shifts are contributing to a rise in injuries, illnesses, and fatalities associated with various health issues sensitive to climate conditions. The health risks in the future will be influenced not only by the dangers posed by a changing climate but also by the vulnerability of individuals and communities to these threats, as well as the preparedness and responsiveness of health systems to manage the associated risks effectively (Evans, 2021). Such risks encompass fatalities and injuries from extreme weather events, the spread of infectious diseases, and challenges related to food and water security. The distribution of these risks is uneven, leading to the creation of new inequities and the exacerbation of existing ones. Projections indicate that most of these risks will escalate with each increment of warming. Adopting an equity-focused approach to transition from incremental to transformational resilience could mitigate vulnerability and enhance sustainability for all; however, significant additional funding is essential to enable proactive and effective measures within the health system (Proulx, 2024).

Climate change has a profound impact on health, particularly affecting vulnerable groups such as pregnant women, newborns, infants, children, at-risk adults, and the elderly, who are more susceptible to the consequences of extreme or prolonged heat. Additionally, it can adversely influence health indirectly by disrupting the fundamental determinants of well-being. For instance, diminished agricultural yields can lead to food scarcity, resulting in inadequate nutrition and reduced overall health and vitality. Furthermore, climate change may expand the habitats of disease-carrying organisms, such as certain mosquito species that thrive at higher elevations, thereby increasing the prevalence of diseases like malaria, West Nile virus, and Zika in previously unaffected areas (WHO, 2022). The frequency of climate-related disasters is on the rise globally, impacting regions including small island nations. These direct and indirect health effects accumulate over time and throughout an individual's life. The repercussions of climate change can be intensified by pre-existing inequalities and the interplay with other crises, yet they can also be alleviated through effective governance that leverages collective intelligence for the public good, implements evidence-based policies, fosters community engagement, and involves all sectors of society.

Good governance and the integration of health considerations into all policies are grounded in the essential belief that every individual possesses the right to health as an integral aspect of human rights (WHO, 2008). Additionally, governments worldwide have pledged to pursue the Sustainable Development Goals (United Nations General Assembly, 2015), with the majority endorsing the Paris Agreement (United Nations, 2015). These global policy frameworks emphasize several key points: first, that the fundamental needs of all individuals, including access to food, shelter, and healthcare, must be fulfilled without any form of discrimination; second, that the relationship between climate change and health is deeply interconnected, with effects that transcend national boundaries; and third, that strategic policy decisions can significantly influence efforts to anticipate, mitigate, and adapt to climate change when enacted at national, regional, and global levels. Moreover, research suggests that effective policies and initiatives must also tackle the social, economic, commercial, and political factors that drive climate change (Friel, 2023).

References

Conway, F., Portela, A., Filippi, V., Chou, D., & Kovats, S. (2024). Climate change, air pollution and maternal and newborn health: An overview of reviews of health outcomes. *Journal of Global Health*, 14, 14. DOI: 10.7189/jogh.14.04128 PMID: 38785109

Evans, S. (2021). Which countries are historically responsible for climate change? www.carbonbrief.org

Friel, S. (2023). Climate change mitigation: Tackling the commercial determinants of planetary health inequity. *Lancet*, 402(10419), 2269–2271. DOI: 10.1016/S0140-6736(23)02512-6 PMID: 37977168

Intergovernmental Panel on Climate Change (IPCC). (2018). *Global Warming of 1.5°C*.www.ipcc.ch

Mycoo, M., Wairiu, M., Campbell, D., Duvat, V., Golbuu, Y., Maharaj, S., ... (2022). Small Islands. In: H.O. Pörtner., D.C. Roberts., M. Tignor., E.S. Poloczanska., K. Mintenbeck., A. Alegria., M. Craig., S. Langsdorf., S. Löschke., V. Möller., A. Okem., & B. Rama. (Eds). *Climate Change 2022: Impacts, Adaptation and Vulnerability. Contribution of Working Group II to the Sixth Assessment Report of the Intergovernmental Panel on Climate Change*. Cambridge, UK: Cambridge University Press; 2022. p. 2043–2121.

Nazril, I. S., & Winkel, J. (2017). *Climate Change and Social Inequality. Department of Economic and Social Affairs Working Paper No. 152* ST/ESA/2017/DWP/152. 2017. www.un.org

Prina, M., Khan, N., & Akhter Kan, S. (2024). Healthy ageing and climate change: An assessment of the impact of climate hazards on older people. *Global Journal of Health, 14*.

Proulx, K., Daelmans, B., Baltag, V., & Banati, P. (2024). Climate change impacts on child and adolescent health and wellbeing: A narrative review. *Journal of Global Health*, •••, 4. PMID: 38781568

Romanello, M., di Napoli, C., Green, C., Kennard, H., Lampard, P., & Scamman, D. (2023). The 2023 report of the Lancet Countdown on health and climate change: The imperative for a health-centred response in a world facing irreversible harms. *Lancet*, 402(10419), 2346–2394. DOI: 10.1016/S0140-6736(23)01859-7 PMID: 37977174

Thacker, S., Adshead, D., Fantini, C., Palmer, R., Ghosal, R., & Adeoti, T.. (2021). *Infrastructure for climate action*. United Nations Office for Project Services. Denmark.

United Nations. (2015). *Paris Agreement*. United Nations.

United Nations. (2022). *The human right to a clean, healthy and sustainable environment: resolution/adopted by the General Assembly*.www.un.org

United Nations General Assembly. (2015). *Transforming our World: The 2030 Agenda for Sustainable Development*. United Nations General Assembly.

Willets, E., & Haines, A. (2023). *Managing the Health Risks of Climate Change*. www.cfr.org

World Bank Group. (2024). *Health and Climate Change*.www.worldbank.org

World Health Organisation. (2024). *Climate Change*. www.who.int

World Health Organization. (2008). *The Right to Health*. www.who.int

World Health Organization. (2022). *Mental Health and Climate Change: Policy Brief*.www.who.int

Chapter 1
A Systematic Review on Biopsychosocial Impact of Climate Change on Women's Health and Well-Being

Shreya Pharswan
https://orcid.org/0009-0000-6607-588X
Banaras Hindu University, Varanasi, India

Urvashi Pipaliya
https://orcid.org/0009-0009-1391-3531
Central University of Haryana, India

Ravindra Singh
https://orcid.org/0000-0003-3062-1812
Magadh University, Bodh Gaya, India

ABSTRACT

In today's world, climate change is a potential threat to women's health and well-being. The climate crisis impacts women's health through various sources. Women are increasingly vulnerable to physical, emotional, social, and psychological health outcomes. The United Nations (UN) has acknowledged that current changes in temperature and atmosphere due to climate change have a significant effect on various aspects of women's health. In this light, the purpose of this book chapter is to highlight the importance of the biological, psychological, and social impact of climate change on women's health. By thoroughly examining these studies, we aim

DOI: 10.4018/979-8-3693-4183-4.ch001

to give a comprehensive conclusion of the complex relationship between climate change and women's health and well-being. Results underline the necessity of fostering an interdisciplinary approach to policy development to focus on women's overall health and development. In conclusion, this study successfully enhances our understanding of women's health and drives meaningful action in research and real-world applications.

1. INTRODUCTION

Recent incidents have attracted our attention towards climate change and it's impact on various domains of human lives. Climate change poses a substantial threat not only to the environment but also to the overall health and well-being and economic opportunities of individuals and groups. The United Nations Framework Convention on Climate Change (UNFCCC) defines the term 'climate change' as direct or indirect consequences of human activities. In addition to the variabilities observed in natural climates over comparable periods, climate change also alters the composition of global temperatures. The evolving climatic conditions are influenced by both anthropogenic (e.g., burning of fossil fuels) and natural phenomena such as tectonic movements and volcanic eruptions (Masika, 2002). These factors have triggered a number of environmental changes, including melting polar ice, rising sea levels, intensified heat waves, floods, water scarcity, and declining biodiversity. These changes have resulted in rising global temperatures and unstable shifting of weather patterns (Desai & Mandal, 2021), which in turn increasingly affect human lives and their overall well-being (Dankelman et al., 2008). Well-being is a concept often coupled with 'health' as in the term 'health and well-being' (Knight & McNaught, 2011), 'quality of life', 'human development', and 'welfare'. Contemporary discussions on well-being often start from the definition that 'health is not merely the absence of diseases, but a state of complete well-being' (World Health Organization, 1946). This expanded definition includes the concept of well-being. The WHO defines well-being as a positive state experienced by individuals and societies that is determined by economic, environmental, and social conditions.

The interaction between climate change, human health, and well-being has been a widely discussed matter in the last few decades. The nature and extent of the events affecting human lives due to climate change are largely dependent upon one's social, political, and economic status. Climate change unites communities to take action against a common threat or create social instability in the face of increasingly scarce resources (Fritze et al., 2008). Additionally, the effects of climate change are not gender-neutral (International Union for Conservation of Nature, 2015). In other words, they disproportionately affect men and women in various ways. Women are often at a higher risk due to biological, political, and cultural factors (Kumar, 2016; World Health Organization, 2014). Women's vulnerabilities have become especially apparent in south-Asian regions like India, where long-established gender inequalities persist (Desai & Mandal, 2021). Goh (2021) highlighted that the challenges faced by poor rural women that increases their vulnerability to climate-change stressors include restricted access to healthcare, education, and economic opportunities. Additionally, women are often overlooked as potential contributors to climate change strategies (Dankelman et al., 2008) due to which they are excluded from discussions and decision-making regarding adaptive strategies to counter climate-induced events.

The climatic burden on women indiscriminately influences every domain of their lives, including biological, psychological, emotional, cognitive, and social areas. Moreover, the impacts of climate change can adversely affect women at any stage of life, from birth and puberty to pregnancy and old age. The disproportionate consequences of climate variability ultimately have a long-term or a life-long impact on women's overall well-being. Therefore, the United Nations Framework Convention on Climate Change (2011) acknowledged that systematic inquiry and comprehensive analysis is required to understand the complex link between climate change, gender disparities, and their consequences on human health and well-being. The World Health Organization (WHO) also recognizes the multifaceted nature of challenges faced by women due to climate change. Subsequently, the World Health Organization (WHO) has placed the "health impacts of climate and environmental change" as a priority, acknowledging the need to prioritize the well-being of women, children, and adolescents at the heart of global health and development (World Health Organization, 2017). These goals are aligns with the United Nations' Sustainable Development Goals (SDGs), that focus on poverty reduction, gender equality, sustainability, and climate action (Sorensen et al., 2018).

The frequency with increased intensity of disasters is continuously increasing around the globe. The United Nations Intergovernmental Panel on Climate Change (IPCC) emphasizes on the extensive impacts (E.g. heatwaves, floods, droughts, cyclones, and the erosion) of climate change on ecology and human lives highlighting (Dankelman et al., 2008; Intergovernmental Panel on Climate Change, 2014). The annual number of disasters has increased from approximately 75 to 400 since 1975 (United Nations International Strategy for Disaster Reduction, 2008), leading to increased casualties, deaths, and destruction of resources.

Figure 1. The various impacts of climate change

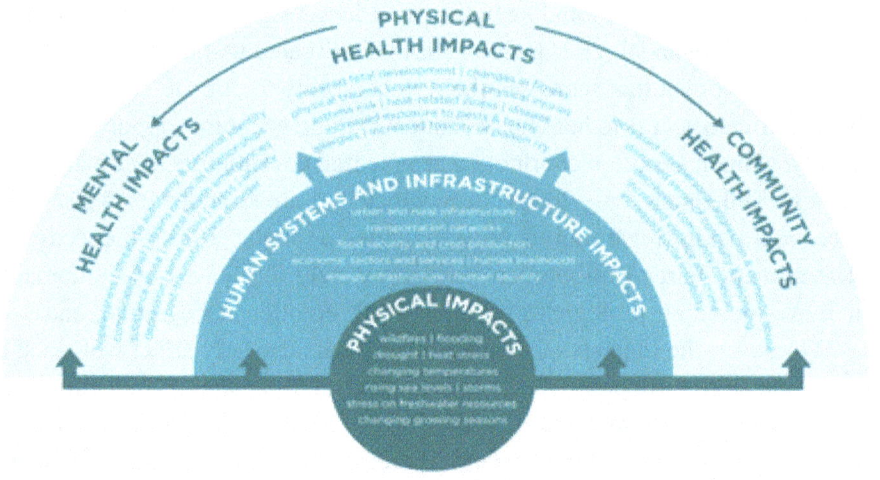

Source: *Linking the physical and psychological impacts of climate change. Clayton, S., Manning, C., & Hodge, C. (2014). Beyond storms & droughts: The psychological impact of climate change. American Psychological Association and EcoAmerica, (June), 51.*

The association among climate change and biopsychosocial well-being of women can be illustrated by the example of anxious mothers. Pregnant women are more vulnerable group may potentially experience adverse outcomes such as miscarriage, premature birth etc, (Bekkar et al., 2020; Tan et al., 2017; Patel et al., 2020). Moreover, trauma from extreme weather conditions (heat and cold) increases the risk of experiencing prenatal maternal stress- PNMS (King et al., 2012) PTSD, and depression (Crimmins et al., 2016). These factors further exacerbate the burdens on pregnant women by limiting or declining socio-cultural opportunities for community participation, access to resources (Sorenson et al., 2018), and even social status and entitlements (Ahmad and Fajber, 2009). The integration of all these components exerts detrimental consequences on the overall well-being of expectant mothers.

Local or community-level examples have proven women's active involvement in post-disaster recovery strategies as is evident from case studies from Bangladesh, Ghana, and Senegal. In addition to managing most of the household, they contribute to rebuilding houses, securing livestock, treating other family members, securing an income, and managing childcare and education (Dankelman et al., 2008). In rural areas, women have also demonstrated initiatives in contributing to production activities such as gathering seeds, harvesting, poultry rearing, food processing, garment making, and handicrafts. Rural women, particularly from underprivileged and low-income communities also engage in traditionally male-dominated activities such as paid labor in construction, earthwork, and field-based agricultural work (Dankelman et al., 2008). Nevertheless, their role and contributions to climate-related solutions are still overlooked. This book chapter aims to highlight the impact of climate change, on women health. The systematic review focuses on three primary areas: biological, psychological, and social aspects of women's lives.

2. METHODOLOGY

This article is a systematic review based on previous literature on the pervasiveness of biological, psychological, and social factors of climate change that impact women's health and well-being. The information was taken from open-access sources such as JSOTR, Google Scholar, PubMed, Inderscience, Science Direct, ProQuest, PsychINFO, Medline, and Web of Science, and credible newspaper articles (Published between 1992 to 2024). The review includes numerous studies and credible information sources of government organizations (such as the World Health Organization, Intergovernmental Panel on Climate Change, and UNICEF) that examine the health and well-being of women, and emphasize how climate change affects women's biological, psychological, and social health. The advanced search option was adopted for the systematic review from web sources with keywords viz. 'Climate Change', 'Natural Disasters', 'Well-Being', 'Mental Health', 'Stress', 'Depression', 'Hormones', 'Social Relations', 'Biopsychosocial', 'Physical Health,' 'Social Impact,' 'Eco-Anxiety,' 'Sustainable Development Goals (SDG's)' 'Women's Mental Health', and 'Women's Health'. The chapter highlighted a particular focus on research carried out involving women.

Search Strategy and Inclusion and Exclusion Criteria

We have adopted the PRISMA guideline checklist to select and reject the preferred articles for our review purposes. We have considered the English language for reaching the manuscripts, and an attempt was made to include all countries and populations across the world. Apart from Google Scholar and other various database searches, an additional manual search of the reference lists was performed to include the relevant studies. Citations mining was done through the searches of the above-mentioned online databases. Further they all are imported into Endnote X9, to remove duplicate manuscripts. At the primary stage the database search identified 460 research papers. After removing 149 duplicates, 311 articles remained, which were subsequently screened for relevance by the theme, titles, and abstracts. Inclusion criteria were peer-reviewed articles addressing the biological, psychological, or social impacts of climate change on women's health reported by the biopsychosocial impact of climate change. This process resulted in 110 being deemed irrelevant and excluded, and 201 were retained for further screening. Further, two reviewers have carefully assessed the worthiness of included articles at both stages and according to feedback 56 were removed from the review process resulting in a total of 145 articles. Moreover, few uncertainties or disagreements were resolved through discussion, and with this additional 58 articles were excluded. Finally, after all above attempts a total of 87 article were reviewed to conclude our findings (See Figure. 1).

Figure 2. Adapted prisma guideline checklist diagram for systematic reviews

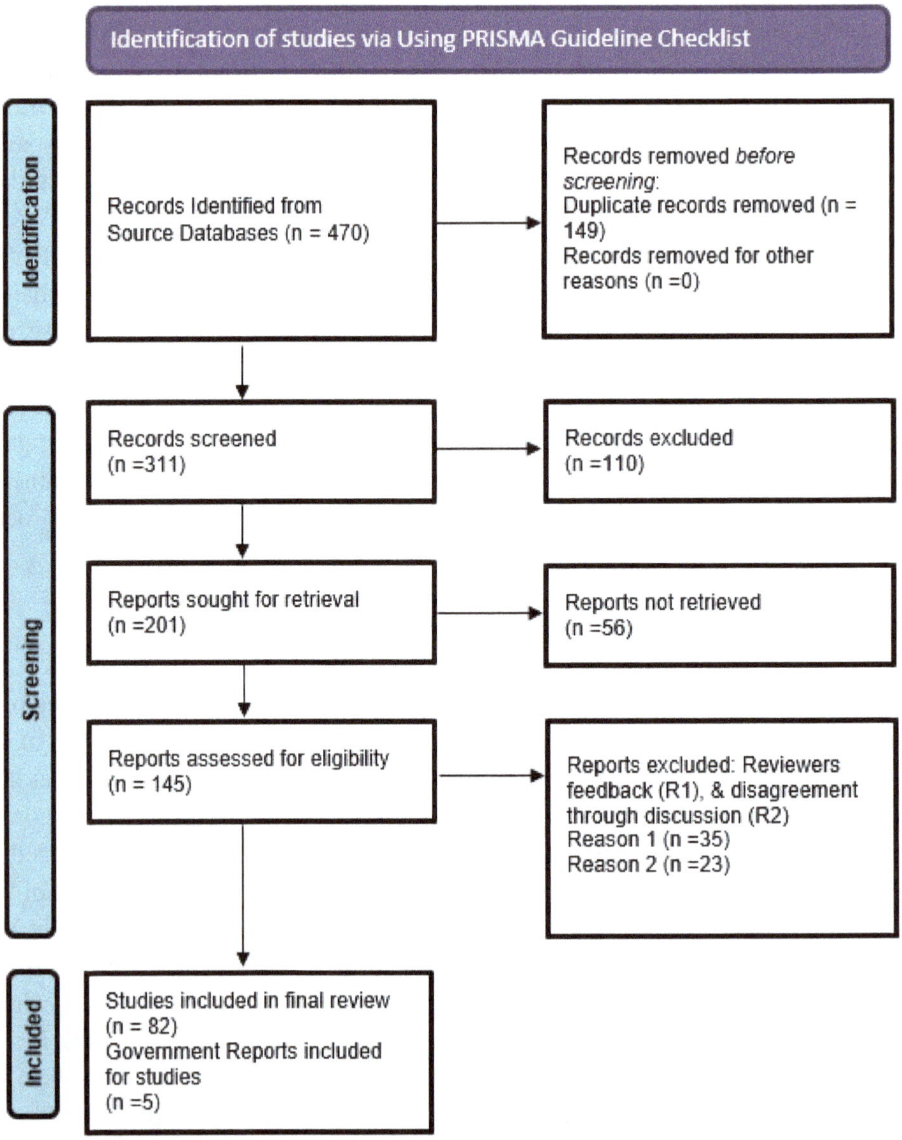

3. REVIEW OF LITERATURE

3.1 Biological impact Climate Change on Women's Health

Research on the biological influence of climate change has focused largely on the direct physical health impacts, particularly, in the context of:

- Impacts of poor air quality and air pollution
- Water scarcity and sanitation issues
- Heightened temperatures and prolonged exposures to heat
- Vector-borne diseases
- Extreme weather events leading to fatal injuries and death
- Reproductive and maternal health

The effects of climate change impact women and girls disproportionately based on geographical and socioeconomic contexts. Variables like heightened susceptibility to heat, inadequate air, eroding water quality, and the onslaught of extreme weather phenomena disrupt the physiological well-being of women (Crimmins et al., 2016). These disruptions involve disturbances in hormonal equilibrium, fertility-linked disorders, and a heightened vulnerability to various vector-borne diseases (Girardi & Bremer, 2022). In the Indian context, several vector-borne diseases are endemic such as malaria, dengue, chikungunya, and Japanese encephalitis (Sorensen et al., 2018). Studies have determined significant correlations between climatic shifts and disease prevalence. Laneri et al. (2010) reported that climate change data can be useful for forecasting malaria epidemics in deserts and semi-arid regions of India. Additionally, the unavailability of clean water due to climate change adversely affects sanitation systems and contributes to various illnesses (Patel et al., 2020).

Air pollution can trigger severe reproductive defects during gametogenesis, thereby impacting fertility (Canelón & Boland, 2020). Moreover, declining outdoor quality and harmful pollutants in the air are also responsible for aggravating a heightened risk of asthma and declining cardiopulmonary health (Tiotiu et al., 2020; Fann et al., 2016). Heightened levels of air pollution expose pregnant women to elevated risks of miscarriage, gestational diabetes, preterm birth, and stillbirths (Bekkar et al., 2020; Tan et al., 2017). Furthermore, climate change exposes women to harmful endocrine-disrupting chemicals (EDCs) that can be found in polluted air, water, and soil. These chemicals have been proven to have adverse effects on ovarian reserves in women, contributing to infertility and premature menopause (Patel et al., 2015).

The climate impacts result in adverse reproductive health outcomes among women such as early menarche, early menopause, infertility, and complications during pregnancy and childbirth (Girardi & Bremer, 2022). Changes in the timing of menarche due to climate change also impact women's health (Canelón & Boland, 2020). These shifts in timing results in heightened disease burden, severe mental health conditions, fertility-related issues, cardiovascular diseases, and bone health disorders (Canelón & Boland, 2020). Climate-induced disasters often result in destroying sanitation systems in the affected areas causing heightened complications for women and adolescent girls (Dankelman et al., 2008). In a case study from Bangladesh, it was reported that after a cyclone, many women refrained from using the toilet during the day and subsequently suffered from urinary tract infections (Mirza, 2002).

The difficulties faced by pregnant women due to the deteriorating climatic conditions arise from inadequate access to clean water, nutrition, sanitation, and shelter. (Patel et al., 2020). These conditions not only impact the health and well-being of pregnant women but also pose significant threats to fetal development. Importantly, pregnancy itself heightens vulnerability, thereby increasing the susceptibility of pregnant women to the adverse effects of climatic stressors (Patel et al., 2020). A report based on the aftermath of Hurricane Katrina denoted that exposure to harmful contaminants, lack of healthcare access, and psychological stress during the disaster increased cases of adverse reproductive outcomes among women (Callaghan et al., 2007). High temperatures and particularly prolonged exposure to heat, are linked to adverse pregnancy outcomes, including stillbirths (Basu et al., 2016; Ha et al., 2017), congenital birth defects (Van Zutphen et al., 2012), preterm deliveries and low birth weight (Strand et al., 2011). This susceptibility transcends maternal age and ethnicity, suggesting a universal risk posed by climatic shifts. Emerging evidence highlights the reality of how climate change, environmental pollution, and disasters disproportionately affect women's health in developing regions. A one-month study conducted by Kabir et al. (2016) in the South-West part of Bangladesh, particularly the two most vulnerable coastal districts, namely, Barguna and Khulna found that disasters such as cyclones accentuate existing gender-based disparities.

A review conducted by Van Den Berg et al. (2005) demonstrated that in comparison to the general population, medically unidentified physical symptoms (MUPS) are more common among people affected by catastrophes and among survivors of disasters. The risk variables associated with MUPS included PTSD symptoms, (Van Den Berg et al., 2005). Evidence suggests that the mortality rate due to disasters is higher for women as compared to men (Aguilar, 2004; Rofi et al., 2006). This gender difference is even larger when the affected women have low socio-economic status (World Health Organization, 2014). For instance, a 1991 cyclone in Bangladesh resulted in 140,000 deaths, out of which 90% of victims were females (Aguilar, 2004). Similarly, during the Asian Tsunami, the female mortality rate was reported to be 1.9 times greater than males (Rofi et al., 2006). In 2008, cyclone Nargis resulted in 130,000 deaths, out of which 61% of victims were women (World Health Organization, 2014).

3.2 Psychological Impact on Well-being

The detrimental impact of climate-related events influences one's mental health, thereby worsening their overall well-being and quality of life. A substantial portion of the population experiences neuropsychiatric disorders, particularly depressive disorders, brain syndromes, and dementias, with a notable difference between genders (Dimitrov, 2019). An alarming 41.9 percent of these disorders affect women, as opposed to 29.3 percent among men (World Health Organisation, 2001). This disproportionate impact highlights the vulnerability of women, particularly in scenarios marked by internal conflicts, disasters, and displacements. (Dimitrov, 2019).

The adverse effects of climate events extend beyond the physical burden, imposing a psychological and emotional toll, particularly on women. In instances of displacement and destitution resulting from disasters, evidence indicates a rise in cases of depression and suicides among women, as established support networks and societies crumble under the weight of calamity (Asian Development Bank, United Nations and World Bank, 2005). For example, impacts of natural disasters include acute stress and anxiety with PTSD (Bandla et al., 2019; Galea, Nandi, & Vlahov, 2005); somatic disorders (Van Den Berg et al., 2005); and other problems such as drug abuse, higher rates of suicide, along with elevated risk of child abuse (Fritze et al., 2008). Environmental disasters can lead women to experience emotional issues (E.g. anger, frustration, irritability, disbelief, and death anxiety (Dimitrov, 2019). The psychological distress was reported in drought-affected areas and flood-stricken regions in women (Coêlho, 2004).

A review conducted by Clayton (2021) indicated that acute climatic events threaten social and economic structure triggers migration, job loss, food and water scarcity. Thereby, it has a direct impact on mental health and well-being in women. Furthermore, it was also found that chronic climatic changes such as drought, air pollution, and heightened temperatures lead to suicidal ideation, burnout, violence, anger, psychotic episodes, and hospitalization (Clayton, 2021). Galea, Nandi, & Vlahov (2005) reported that the likelihood of PTSD after natural disasters was more common among women as compared to men. Additionally, a case study based on the Rigolet community of the Nunatsiavut region of Canada reported that environmental changes resulted in stress, resurfacing previous traumas, and increasing the possibility of drug and alcohol abuse and suicidal ideation (Cunsolo Willox et al., 2013).

The dimension of mental health also encompasses other areas including social, and cultural dimensions that affect mental well-being (Shooshtari, 2018). For instance, pre-existing social hierarchies and vulnerabilities disproportionate the burden of the adverse consequences of climate change in women (Arora-Jonsson, 2011). Furthermore, the extent of experiencing disaster-induced traumas depend on one's social relationships. A study conducted by Sezgin & Punamäki (2016) in Eastern Anatolia, Turkey reported that earthquake trauma was associated with deteriorated social relationships. The gender-specific psychological impacts have also been reported by the studies conducted in the aftermath of climate-induced disasters. A post-earthquake study in Pakistan revealed a higher prevalence of post-traumatic stress disorders among women compared to men. The study reported the prevalence rate of 55.2% post-traumatic stress symptoms among women and 33.4% among men highlighting the differentiated impact of climate-induced events on mental health of men and women. (Naeema et al., 2011). Another major finding in the study conducted by Naeema et al. (2011) also emphasized on the importance of family structure in the recovery from post-traumatic stress disorder among men and women. A post-cyclone survey conducted in Bangladesh found that women were at a higher risk of experiencing depression and PTSD after surviving cyclone Mora (Mamun et al., 2019). Another post-disaster survey from Bangladesh reported that exposure to floods within the 12 months before the survey rounds was associated with high rates of depression, anxiety, and co-occurring depression and anxiety among women (Wahid et al., 2023).

The growing awareness of climate change is also associated with the phenomenon of "climate anxiety" or "eco-anxiety", which has become a significant psychological concern worldwide in the last few decades. Eco-anxiety refers to a number of negative emotions associated with the realization of climate change (Coffey et al., 2021). Although a certain level of anxiety can potentially act as an adaptive function to confront climate threats, excessive levels of climate anxiety act as a stressor affecting mental health. Eco-anxiety results in increasing the mental burden, particularly among people who are more vulnerable to multiple stressors and less resilient (Clayton, 2020). Women are more prone to worry and exhibit higher levels of eco-anxiety as compared to men (Boluda et al., 2022), which further increases their susceptibility to other mental health conditions.

3.3 Social Determinants and Vulnerabilities

Social norms play a vital role in shaping women's vulnerabilities due to the impacts of climate events (Ahmad, 2012). Climate-induced disasters heighten women's susceptibility to sexual harassment, trafficking, and prostitution due to displacement and poverty (Patel et al., 2020). The challenges posed by climate change often restrict women's responses primarily due to the existing gender norms. Women are often excluded from decision-making processes at home and in communities due to patriarchal norms. Dimitrov (2019) also highlighted the financial vulnerabilities faced by women due to difficulties in finding employment opportunities, limited access to resources and materials, and poor compensation. Climate-induced disasters and subsequent unstable environmental conditions have further heightened the differentiated impact, leading to a distressing rise in incidents of domestic violence (Stork et al., 2015). This shift in gender roles and responsibilities is often faced with backlash and increased violence against women (Stork et al., 2015), particularly in patriarchal communities.

The influence of social identities such as caste and religion further heighten women's vulnerability during and after disasters (Patel et al., 2020). Factors such as limiting women's access to healthcare, cooling facilities, and proper clothing for cooling further increase their susceptibility to heat-related risks (Sorenson et al., 2018). Additionally, access to early-warning information in case of disasters is often “gendered” (Ahmad et al., 2009). This information is typically obtained through communication channels like radio, television, cell phones or in certain pilot programs which are majorly utilized by men as compared to women (Ahmad et al., 2009). A study by Tobin-Gurley et al. (2010) in the aftermath of Hurricane Katrina reported on the recovery-related challenges faced by single-displaced mothers. Women's vulnerability to climate change also depends on their socioeconomic status and social entitlements (Ahmad and Fajber, 2009). Cultural norms can restrict women's movement during disasters, as was seen during the Asian Tsunami, where the female casualty rate was higher than men (United Nations International Strategy for Disaster Reduction, 2008). A survey conducted after the Asian Tsunami reported that the female mortality rate was 1.9 times greater than males (Rofi et al., 2006). This report highlighted the cause of this phenomenon as women's prioritization of communal well-being over their own safety.

Migration, a consequence of climate impacts, affects women and men differently. Moving from one place to another emphasizes the ecological vulnerability experienced by women when adjusting to a new environment, primarily due to environmental degradation (Dimitrov, 2019). For instance, a study from Bangladesh reported that women face distinct challenges during migration due to cyclones, losing access to livelihood options, and encountering inadequate shelter and sanitation facilities (Kartiki, 2011). A study from India has shown that climate change in the Sunderbans has resulted in a decline in its two main occupations- agriculture and fishing (Karmakar, 2021). This resulted in reduced income and migration of men to other states. Consequently, women are left to care for their families, provide social security, and earn money through odd jobs, doubling their burden on top of the challenges posed by climate change (Karmakar, 2021).

Climate change is a pressing global concern that has garnered significant attention. However, its relationship with violence against women remains underexplored. This concern is gaining attention due to its link with exacerbating Sexual and Gender-Based

Violence (Desai & Mandal, 2021). While climate change affects entire populations, it uniquely exposes women and girls to heightened vulnerabilities, particularly due to their gender (Institute of Development Studies, 2008). In times of crises, particularly conflicts, and disasters, women confront heightened risks of SGBV due to family and societal breakdowns, often accompanied by forced displacement (United Nations High Commissioner for Refugees, 2003). This leaves them susceptible to physical, sexual, and psychological harm, as well as denial of resources and services. A case study from Bangladesh reported that young women and adolescent girls were sexually harassed at an alarming rate after a natural disaster because of separation or isolation from other family members and lack of attention or observation (Alam & Rahman, 2014). Another study by Ross et al. (2024) reported that climate variability has a direct effect on intimate-partner violence including physical, sexual, and psychological forms of violence. The study also revealed that this direct association is mediated by water insecurity due to climate events.

The harmful effects of climate-induced migration differ in terms of human capital and income for women and men. While men are more likely to migrate in search of employment, women often remain to care for households (Goh., 2012). Such male outmigration can burden women, leading to increased workloads and exposure to abuse. Moreover, women encounter challenges in post-disaster economic recovery, as available employment opportunities in reconstruction fields are traditionally male-dominated (Tobin-Gurley et al., 2010). Climate change presents differential impacts on various segments of society, with a heightened influence on women's diverse experiences (Masika, 2002). Gender divisions in labor also emerge as a significant factor affecting women's capacity to cope with environmental shifts. The sociocultural explanation emphasizing the role of women as a caregiver also highlights their heightened exposure to biomass (Sorenson et al., 2018). Continues contact to biomass exposes women to air pollution linked with a variety of health risks, and reproductive issues (Chafe et al., 2014). In India, a major population lacks access to inefficient energy sources like quality fuel biomass (World Bank, 2018). A study reported that the Indian women in drought-affected areas of would usually skip meals or eat two meals a day, significantly affecting their overall well-being (Patel et al., 2020). In the context of South Asia, women confront amplified impacts of climate change due to factors such as poverty, gender inequality, insecure land rights, and limited access to education (Yadav and Lal, 2018). These challenges place poor, pregnant, older, and agriculture-dependent women at elevated risk (World Bank, 2019).

Contrasting with other research, a study conducted by Alam and Rahman (2014) highlighted some benefits that women experience post-disaster, such as social inclusion due to gender equality in jobs (due to lack of manpower), economic opportunities, and resources. Rather, it highlights the transforming effect on women that provides them with opportunities to contribute to post-disaster recovery. Enarson (2000) also highlighted the opportunities that arise for women after disasters to challenge and uplift their societal status.

Table 1. Summery Table of Review of Literature of the Previous Study

Author (s)	Population	Methods	Main Objective	Major Findings
Basu et al. (2019)	California	Stratified case-crossover design	Examining the impact of increasing temperatures on stillbirths.	Higher temperatures and prolonged exposure to heat are associated with increased risks of stillbirths.
Girardi & Bremer (2022)	Multiple countries	Literature Review	Examining the effects of climate-related events and ongoing environmental changes on women's reproductive health.	Climate-related disruptions are linked with disturbances in hormonal equilibrium, fertility-linked disorders, and a heightened vulnerability to various vector-borne diseases.
Kabir (2016)	Bangladesh	Epidemiological field study	Inquiring the consequences of cyclones Sidr and Aila on the affected population.	Women were particularly vulnerable to the disaster. Inadequate nutrition and sanitation contributed to irregular menstruation, leucorrhoea, anemia, and infertility.
Canelón & Boland (2020)	Multiple countries	Systematic Literature review	Examining the impact of climate-related events on the timing of menarche and other aspects of women's health.	Extreme weather events disrupt hormonal rhythms, thereby impacting the timing of menarche. This increases the disease burden including mental health conditions, fertility-related issues, cardiovascular diseases, and bone health disorders.
Naeema et al. (2011)	Pakistan	Cross-sectional survey	Finding the prevalence of PTSD among men and women after in an earthquake.	The prevalence rate of PTSD was reported to be 55.2% among women and 33.4% among men.
Coêlho, A.E.L (2004)	Brazil	Cross-sectional survey	Comparing the psychological responses of individuals living a drought prone area with those living drought-free city of the same state.	Women living in the drought prone area were found to have increased anxiety levels and higher emotional distress than women and men in the drought-free area.

continued on following page

Table 1. Continued

Author (s)	Population	Methods	Main Objective	Major Findings
Sezgin & Punamäki (2016)	Eastern Anatolia, Turkey	Cross-sectional survey	Examining the association between earthquake trauma and women's social relationships.	Women with poor marital and neighborhood relationships experienced higher levels of distress and reported greater symptoms of post-traumatic stress disorder.
Wahid et al. (2023)	Bangladesh	Longitudinal panel survey	Analyzing the impact of climate-induced disasters and other stressors on the mental health of affected individuals.	Women who experienced floods within the 12 months before the survey rounds were diagnosed with higher rates of depression, anxiety, and co-occurring depression and anxiety.
Tobin-Gurley et al. (2010)	Colorado, United States	Qualitative, in-depth interviews	Exploring the recovery-related challenges faced by displaced single mothers following Hurricane Katrina.	The difficulties faced by single women during post-hurricane recovery included mistreatment, lack of awareness about available resources, breakdown of informal support networks, and bureaucratic problems in accessing assistance.
Fatema et al. (2023)	Bangladesh	Qualitative, in-depth interviews	Exploring the gender-based vulnerabilities faced by women during natural disasters.	The risks faced by women included poor housing and shelter conditions, risk of abuse, migration and difficulties finding a safe place, social shame, hygiene and sanitation issues, and foregoing food to feed their husbands and children.
Alam & Rahman (2014)	Bangladesh	Mixed method (quantitative and qualitative data collection)	Exploring the diverse impacts of climate-induced disasters on women's health and development.	Disasters have a particularly negative impact on women by disrupting families, food availability, job security and income, health, and safety. Women and girls experience harassment at an alarming rate because of separation or isolation from other family members.

continued on following page

Table 1. Continued

Author (s)	Population	Methods	Main Objective	Major Findings
Karmakar (2021)	Africa	Quantitative study	Investigating the impact of climate-related events on intimate partner violence.	Climate variability is directly associated with intimate-partner violence which can occur in the form of physical, sexual, and psychological violence. Additionally, this direct association is mediated by water insecurity due to climate events.
Barkin et al. (2022)	Multiple countries	A systematic review	Exploring the impact of climate events on women's mental health in their perinatal period and acknowledging the variations in climate as a potential threat to them.	Exposure to EWEs is a trauma that can exacerbate the already difficult adjustment to motherhood and cause signs of poor mood such as anxiety and depression.
Boluda et al. (2022)	Multiple countries	A systematic review	Analyzing existing literature on eco-anxiety to understand its impacts on mental health.	Eco-anxiety was associated with a number of psychological functions including, depression, anxiety, stress, PTSD, functional impairments and insomnia.
Patel et al. (2020)	South Asia	A systematic literature review	Exploring the impacts of climate change on women and their resilience against climate-induced events.	Climate change significantly affect the socio-economic status of women and has disproportionate effects on various domains of life.
Bekkar et al. (2020)	United States of America	A systematic literature review	Investigate prenatal exposure to fine particulate matter, ozone, heat, and the association of these factors with pregnancy outcomes.	Pregnant women who are exposed to higher air pollution levels and heat are at a greater risk for adverse pregnancy outcomes
Clayton (2021)	Multiple countries	Review study	Analyzing the current and future effects of climate-related events on mental health outcomes.	Extreme weather and inconsistent changes due to climate change have a detrimental impact on mental health leading to disorders like depression, anxiety, and PTSD.

continued on following page

Table 1. Continued

Author (s)	Population	Methods	Main Objective	Major Findings
Coffey et al. (2021)	Multiple countries	Scoping review	Exploring the operational function and characteristics of eco-anxiety through past literature.	Individuals connected to nature are highly vulnerable to eco-anxiety.
Goh (2013)	Developing countries	A systematic literature review	Assessing the gendered impact of climate-related events on human well-being and resources.	In terms of resources and well-being, women are more susceptible to the negative impacts of climate change particularly due to the existing social and cultural norms regarding their gender.
Obradovich et al. (2018)	United States of America	Empirical study	Providing empirical evidence of climate change impacts on mental health outcomes.	Exposure to higher temperatures, amplified rainfall, global warming, and intense hurricanes results in worsening psychological health.
Hayes et al. (2018)	Multiple Countries	A systematic review study	Analyzing the current and potential risks and impacts of climate change on mental health and recommending priority actions for the addressing this issue.	There is still a lot of scope for the advancement of research in this field. The resulting consequences of climate change on mental health are also accelerating with various direct, indirect, and overarching effects.

4. MAJOR FINDINGS OF THE STUDY

The present book chapter attempts to uncover the multifaceted nature of climate change and its varied impacts on women's health and well-being.

Key Climate Change Incidents Affecting Women's Health and Well-Being

Natural calamities such as hurricanes, floods, and wildfires have a significantly negative impact on women's mental health (Norris et al., 2002). When Women encounter any climate disaster, they are more likely to develop post-traumatic stress disorder (PTSD). These impacts on their mental health are sustained by average coping mechanisms and low social support networks. It was also reported by the studies that women are disproportionately affected by flooding, which results in acute anxiety and depression (Tunstall et al., 2006). Extreme and sudden exposure to stressors with prior psychiatric problems, with weak coping mechanisms, are likely to increase adverse outcomes in the female gender, especially middle age, women having ethnic minority status (Norris et al., 2002). In Addition, women exposed to wildfires report a higher amount of stress compared to men. Moreover, extreme weather conditions like excessive temperature or cold can negatively affect women's mental health (Norris et al., 2002). Zhao et al, (2021) stated that the hospitalization rate due to psychological health is much higher in women with age 65 and above. On the other side, heavy cold spells can trigger stress and traumatic episodes. Whereas high humidity and poor air quality are correlated with increased anxiety and depression and negatively associated mental health in women (Vandentorren et al., 2006; Pun et al., 2017; Obradovich et al. (2018).

Women's reproductive health is more exposed to environmental toxins agents, hazardous chemicals, and infectious diseases. Studies have reported a negative correlation between increasing temperatures and low birth weight along with a high rate of miscarriage (Bekkar et al., 2020). Malnutrition during pregnancy can lead to adverse birth outcomes and long-term health issues for both mothers and children (Bekkar et al., 2020). Women with higher body fat percentage, and hormonal issues are vulnerable to heat-related illnesses (Zhao et al., 2021). The loss of homes and loved ones can lead to intense feelings of grief, fear, flashbacks uncontrollable thoughts, and helplessness which significantly impair QOL and well-being. The uncertainty about the future, and financial problems, contribute to these poor mental health circumstances (Hayes et al., 2018). To cope with irresistible traumatic episodes, a woman may turn to substance use (Hayes et al., 2018).

Women, often taking care of house meal preparation are involved in water collection, where they face increased workloads and health risks associated with poor water quality (WHO, 2014). They are often primary caregivers and have limited access to resources and decision-making, making it harder for them to recover from climate-related disasters (Neumayer, & Plümper, 2007). This economic inequality exacerbates the health impacts of climate-related shocks in women (Alston, 2013). Natural disasters can lead to an increase in gender-based violence, including domestic violence and sexual exploitation which exacerbate their trauma. Climate-induced displacement and resource scarcity can increase the risk of gender-based violence (Alston, 2013).

The relevant quantitative data reveals the association between EWE (Extreme Weather Events) exposure and unfavorable psychological outcomes (Barkin et al., 2022). Greater loss and intense experiences during EWEs put a person more at risk for mood disorders including anxiety and depression (Barkin et al., 2022). While other concepts such as "eco-anxiety" and "climate despair" have gained some attention and popularity, there is still little to no attention on how climate change impacts mental health of maternal women (Barkin et al., 2022). It has been found that perinatal mood and anxicty disorders (PMADs) are the most common disorders during childbirth and forward (Wisner et al., 2013), moreover, it's an immense term that includes various serious problems such as perinatal obsessive-compulsive disorder, posttraumatic stress disorder, and postpartum psychosis (Center for Women's Mood Disorders, 2022). Among all these perinatal anxiety and depression are the most common ones with 10% to 20% (Van Niel & Payne, 2020) and 16% to 17% (Fairbrother et al., 2016) respectively. However, Harville et al. (2010) noticed that the personal resilience of an individual works as a protective barrier against disastrous loss and negative mental health outcomes.

Evidence has indicated that long-term exposure to extreme heat increases the risks related to delivery in pregnant women (Barreca et al., 2019). Acceleration of childbirth due to adverse heat exposure leads to shorter gestation period, which is further linked to adverse health and cognitive outcomes of the child (Barreca et al., 2019). The term "eco-anxiety" refers to the mindset that links anxiety related to views of the detrimental effects of climate change with climate change itself (Coffey et al., 2021). In the case of underdeveloped coping skills eco-anxiety works as a stressor to mental health (Boluda et al., 2022). Studies have connected Eco-Anxiety with decreased self-rated mental health, unwillingness to have children, symptoms of sadness, anxiety, and PTSD, as well as functional impairment, stress, insomnia, and other adverse mental health outcomes (Boluda et al., 2022). However, it has been observed that engaging in pro-environment behavior helps in subsiding the symptoms of MDD induced by climate anxiety (Boluda et al., 2022). Findings have also highlighted that this impact can be especially seen in women, poorer countries, and the younger generation (Boluda et al., 2022). Research have highlighted that women experience higher levels of stress and anxiety than men because they exhibit more behavioral symptoms of post-traumatic stress disorder (PTSD) after a disaster (Doherty et al.,2011). Sundbladd et al., (2007) while investigating the emotional and cognitive risk assessments discovered that women were more concerned about the repercussions, even though men and women did not differ in their estimations of the likelihood that climate change would have major implications.

By momentary and spatial association of an individual's mental health with environmental difficulty exposure, it has been found that exposure to fleeting weather conditions, increasing global warming, and intense hurricanes are related to substandard mental health (Obradovich et al., 2018). An average maximum temperature < 30 degrees Celsius increases the probability of mental health issues by 1% and months with precipitation of more than 25 days in a month amplify mental health problems by 2% (Obradovich et al., 2018). This association between temperature < 30 degrees Celsius and mental health issues is observed 1.6 times more in women than in men. Bekker et al., 2020, found an association between PM2.5, ozone, and heat exposure with severe obstetrical results such as stillbirth, pre-term birth as well as low birth weight.

Statistical data has shown that asthmatic mothers have 52% more chance of having a preterm delivery. It has also been observed that if a mother is exposed to air pollution during their third trimester it increases their chance of stillbirth by 42%. Other factors that cause complications during pregnancy include a rise in the temperature, proximity to the solid waste ground, and an increase in temperature by 1% before the delivery. There is a risk of underdiagnosis of the actual effects of climate-related events on mental health as well as stigmatizing the "normal" reaction to it which results in the undertreatment or aid to mental health problems (Hayes et al., 2018). It has been observed that climate change has also resulted in some positive changes including feelings of compassion, altruism, post-traumatic growth, or even increased acceptance of climate change and engagement with climate mitigation (Hayes et al., 2018). The key benefit of understanding this relationship is having in-depth knowledge of the risks and impacts of climate change which helps in the development of proper policy, mental health response, and a better resilient system (Hayes et al., 2018).

5. DISCUSSION

Unique Vulnerabilities of Women in Climate Change

Climate change is a major challenge for our overall health in the day-to-day world. It disproportionately affects human lives in various ways. Men and women differ in the physical and mental aspects, behaviors, and societal roles and responsibilities. Consequently, the impact of environmental calamities on them varies according to these differences. Therefore, climate change impacts are not gender-neutral. The biopsychosocial impact of climate change on women encompasses its effects on physical health, psychological health, and social dimensions. The repercussions of climate change and its various aspects (rising temperatures, unstable weather patterns, disasters, etc.) disrupt the physiological well-being of women resulting in increased vulnerabilities to diseases, infections, fertility-linked disorders, and several complications during pregnancy (Bekkar et al.,2020; Tan et al.,2017; Girardi & Bremer, 2022). Psychologically, these outcomes further enhance their susceptibility to experience stress (Galea, Nandi, & Vlahov, 2005), depression, anxiety, trauma (Bandla et al., 2019), and overall mental well-being. Additionally, the conditions of disasters, internal conflicts, and displacements together with social, cultural, and gender norms (Patel et al., 2020). restrict women's access to healthcare (Sorenson et al., 2018), education, and other resources. These factors place a disproportionate burden on women leading to economic vulnerabilities, sexual and gender-based violence (Desai & Mandal, 2021)., and increased workload (Goh., 2012) due to their caregiving roles.

Further, the present book chapter identifies crucial health challenges and attempts to evaluate the direct and indirect effects of climate change on women's physical, mental health, and social health. The findings of the study highlight the requirement for pro-gender-sensitive intervention policies to mitigate the adverse impacts of climate change on women. As we have witnessed biological vulnerabilities can aggravate mental health, while social inequalities further compound these issues (UN Women, 2018). Thus, a holistic approach that interplays these factors would be required with new knowledge and skills to mitigate vulnerabilities and handle climate-related risks. Therefore, investing in quality education for combating climate change plays a key role in achieving the Millennium Development Goals (Anderson, 2010). Furthermore, these plans also aim to ensure women's access to funds that provide them with financial support to access technology that helps in mitigation strategies. This further benefits the environment by promoting cleaner technology adoption and promoting gender equity by expanding women's technological opportunities (Anderson, 2010). Mitigating the adverse consequences of climate change on women requires easy access to healthcare and resources, an enhanced support system, quality education, and access to technology. Environmental changes and calamities can either enhance or constrain adaptive capacities due to the pre-existing gender roles and norms (World Health Organization, 2014). Therefore, adaptation strategies need to consider women's and men's relative and different capabilities, resilience levels, and vulnerabilities. Moreover, the disproportionate impact of climate change on women necessitates the need to uplift and involve women in social and political spheres and ensure their equal participation.

The climate-related impact on women, particularly in underprivileged and low-income communities emphasizes the need for inclusive and gender-sensitive approaches to climate solutions. Climate change and environmental events not only impact the health and well-being of affected individuals and communities but also have long-term effects on future generations. The review highlights the need to combine and coordinate efforts across various scientific disciplines to address the multifaceted challenges arising due to climate events. For instance, the integration of social sciences with natural sciences is necessary to comprehend the relationship between climate change and general human health. Climate change affects women and men in three spheres- biological, psychological, and social. In this respect, a systematic approach provides a detailed insight into the dynamics of climate and its biopsychosocial impact on men and women. Moreover, it emphasizes how local communities perceive, understand, and respond to climate change.

6. CONCLUSION

This review highlights the existence of the complex relationship between climate-driven biopsychosocial impacts and women's health. While considering the influence of climate change on women's health and well-being, it is important to address all three dimensions: biological, psychological, and social, ensuring none of them are overlooked. Women are often at a disadvantage when it comes to climate issues. This book chapter also suggests some practical strategies, and policy recommendations to mitigate these challenges in women. In Conclusion, we can infer from the above review that women may face specific health challenges due to climate change such as reproductive health, mental, and social health. Additionally, the disruption of social networks and support systems, is crucial for women's health and well-being, as they often rely on such networks for both emotional and social support.

7. FUTURE RESEARCH DIRECTIONS

Future research should focus on the adaptation of Climate-related strategies in the case of women since they often bear unequal burdens. Collaboration with various stakeholders would be required to address climate change and its impacts on women's lives. Additionally, Upcoming research should explore the long-term biopsychosocial impacts of climate change on women, different geographical regions, and cultural and socioeconomic status. Furthermore, longitudinal studies can provide cumulative results of the impact of climate change on women's health and well-being. Access to health care, social support, psycho-education, and better economic opportunities can do the job (Sorensen et al., 2018).

8. LIMITATION OF THE STUDY

The current study is not without its limitations. The drawback in the current study stems from the difficulty in comparing the differential impacts experienced by men and women. Additionally, much of the literature consists of qualitative case studies with small sample sizes that focus on a specific region. Although important, this narrow focus inhibits generalisability which results in reducing the awareness among health personnel and global decision-makers, especially about the unique (reproductive) challenges facing women and girls. Lastly, most of the studies on climate change and women's health have focused on rural women owing to their dependency on natural resources, lack of education, decision-making, and social rules. Fewer studies have been done on how well-educated women are affected by climate change who may have access to more resources, opportunities, and abilities to deal with the problems of climate change. However, this does not mean that they are resistant to the health risks of climate change which affects women of all ages, socioeconomic status, and regions. In fact, well-educated women can play a pivotal role as leaders, advocates, and entrepreneurs in promoting sustainable development and effective responses to climate change.

REFERENCES

Aguilar, L., Araujo, A., & Quesada-Aguilar, A. (2007). Gender and climate change. *IUCN (International Union for the Conservation of Nature) Fact Sheet,* http://www. genderandenvironment. org/admin/admin_biblioteca/documentos/Fact*sheet%20ClimateChange. pdf.*

Ahmad, N. (2012). Gender and Climate Change in Bangladesh: The Role of Institutions in Reducing Gender Gaps in Adaptation Program. *Social Development Papers of World Bank, 126*(67820). https://hdl.handle.net/10986/27416

Ahmed, S., & Fajber, E. (2009). Engendering adaptation to climate variability in Gujarat, India. *Gender and Development,* 17(1), 33–50. DOI: 10.1080/13552070802696896

Alam, K., & Rahman, M. H. (2014). Women in natural disasters: A case study from the southern coastal region of Bangladesh. *International Journal of Disaster Risk Reduction,* 8, 68–82. DOI: 10.1016/j.ijdrr.2014.01.003

Alston, M. (2014). Gender mainstreaming and climate change. *Women's Studies International Forum, 47*(Part B), 287–294. DOI: 10.1016/j.wsif.2013.01.016

Anderson, A. (2010). *Combating climate change through quality education.* Brookings Global Economy and Development., https://www.brookings.edu/articles/combating-climate-change-through-quality-education

Arora-Jonsson, S. (2011). Virtue and vulnerability: Discourses on women, gender, and climate change. *Global Environmental Change,* 21(2), 744–751. DOI: 10.1016/j.gloenvcha.2011.01.005

Asian Development Bank, United Nations, and World Bank. (2005). Maldives Tsunami: impact and recovery. Joint needs Assessment by World Bank–ADB–UN System, Male.

Bandla, S., Nappinnai, N. R., & Gopalasamy, S. (2019). Psychiatric morbidity in December 2015 flood-affected population in Tamil Nadu, India. *The International Journal of Social Psychiatry,* 65(4), 338–344. DOI: 10.1177/0020764019846166 PMID: 31068043

Barkin, J. L., Philipsborn, R. P., Curry, C. L., Upadhyay, S., Geller, P. A., Pardon, M., Dimmock, J., Bridges, C. C., Sikes, C. A., Kondracki, A. J., & Buoli, M. (2024). Climate change is an emerging threat to perinatal mental health. *Journal of the American Psychiatric Nurses Association,* 30(3), 683–689. DOI: 10.1177/10783903221139831 PMID: 36482670

Barreca, A., & Schaller, J. (2019). The impact of high ambient temperatures on delivery timing and gestational lengths. *Nature Climate Change*, 10(1), 77–82. DOI: 10.1038/s41558-019-0632-4

Basu, R., Sarovar, V., & Malig, B. J. (2016). Association between high ambient temperature and risk of stillbirth in California. *American Journal of Epidemiology*, 183(10), 894–901. DOI: 10.1093/aje/kwv295 PMID: 27037268

Bekkar, B., Pacheco, S., Basu, R., Basu, R., & Denicola, N. (2020). Association of air pollution and heat exposure with preterm birth, low birth weight, and stillbirth in the US: A systematic review. *JAMA Network Open*, 3(6), e208243. Advance online publication. DOI: 10.1001/jamanetworkopen.2020.8243 PMID: 32556259

Boluda-Verdú, I., Senent-Valero, M., Casas-Escolano, M., Matijasevich, A., & Pastor-Valero, M. (2022). Fear for the future: Eco-anxiety and health implications, a systematic review. *Journal of Environmental Psychology*, 84, 101904. DOI: 10.1016/j.jenvp.2022.101904

Callaghan, W. M., Rasmussen, S. A., Jamieson, D. J., Ventura, S. J., Farr, S. L., Sutton, P. D., & Posner, S. F. (2007). Health concerns of women and infants in times of natural disasters: Lessons learned from Hurricane Katrina. *Maternal and Child Health Journal*, 11(4), 307–311. DOI: 10.1007/s10995-007-0177-4 PMID: 17253147

Canelón, S. P., & Boland, M. R. (2020). A systematic literature review of factors affecting the timing of Menarche: The potential for climate change to impact women's health. *International Journal of Environmental Research and Public Health*, 17(5), 1703. DOI: 10.3390/ijerph17051703 PMID: 32150950

Center for Women's Mood Disorders. (2022). Perinatal mood and anxiety disorders. https://www.med.unc.edu/psych/ wmd/resources/mood-disorders/perinatal/

Chafe, Z. A., Brauer, M., Klimont, Z., Van Dingenen, R., Mehta, S., Rao, S., Riahi, K., Dentener, F., & Smith, K. R. (2014). Household cooking with solid fuels contributes to ambient PM2.5 air pollution and the burden of disease. *Environmental Health Perspectives*, 122(12), 1314–1320. DOI: 10.1289/ehp.1206340 PMID: 25192243

Clayton, S. (2020). Climate anxiety: Psychological responses to climate change. *Journal of Anxiety Disorders*, 74, 102263. DOI: 10.1016/j.janxdis.2020.102263 PMID: 32623280

Clayton, S. (2021). Climate Change and Mental Health. *Current Environmental Health Reports*, 8(1), 1–6. DOI: 10.1007/s40572-020-00303-3 PMID: 33389625

Coêlho, A. E. L. (2004). Psychological responses to drought in Northeastern Brazil. *Interamerican Journal of Psychology*, 38(1), 95–103. DOI: 10.30849/rip/ijp.v38i1.620

Coffey, Y., Bhullar, N., Durkin, J., Islam, M. S., & Usher, K. (2021). Understanding Eco-anxiety: A systematic scoping review of current literature and identified knowledge gaps. *The Journal of Climate Change and Health*, 3, 10037. DOI: 10.1016/j.joclim.2021.100047

Crimmins, A., Balbus, J., Gamble, J. L., Beard, C. B., Bell, J. E., Dodgen, D., ... (2016). The impacts of climate change on human health in the United States: A scientific assessment (p. 312). Washington, DC: US Global change research program. DOI: 10.7930/J0R49NQX

Cunsolo Willox, A., Harper, S. L., Ford, J. D., Edge, V. L., Landman, K., Houle, K., Blake, S., & Wolfrey, C. (2013). Climate change and mental health: An exploratory case study from Rigolet, Nunatsiavut, Canada. *Climatic Change*, 121(2), 255–270. DOI: 10.1007/s10584-013-0875-4

Dankelman, I., Alam, K., Ahmed, W. B., Gueye, Y. D., Fatema, N., & Mensah-Kutin, R. (2008). Gender, climate change, and human security: Lessons from Bangladesh, Ghana, and Senegal. WEDO with ABANTU for Development in Ghana, Action Aid Bangladesh, and ENDA Senegal. http://www.wedo.org/wp-content/uploads/hsn-study-final-may-20-2008.pdf

Desai, B. H., & Mandal, M. (2021). Role of climate change in exacerbating sexual and gender-based violence against women: A new challenge for international law. *Environmental Policy and Law*, 51(3), 137–157. DOI: 10.3233/EPL-210055

Dimitrov, B. G. (2019). Effects of climate change on women. *Res. Rev. Int. J. Multidis*, 4(5), 201–215.

Doherty, T. J., & Clayton, S. (2011). The psychological impacts of global climate change. *The American Psychologist*, 66(4), 265–276. DOI: 10.1037/a0023141 PMID: 21553952

Enarson, E. (2000). *Gender and Natural Disasters. IPCRR Working Paper no. 1.* International Labour Organization., Retrieved from https://www.ilo.org/publications/gender-and-natural-disasters

Fairbrother, N., Janssen, P., Antony, M. M., Tucker, E., & Young, A. H. (2016). Perinatal anxiety disorder prevalence and incidence. *Journal of Affective Disorders*, 20(2), 148–155. DOI: 10.1016/j.jad.2015.12.082 PMID: 27131505

Fann, N., Brennan, T., Dolwick, P., Gamble, J. L., Ilacqua, V., & Kolb, L.. (2016). Air Quality Impacts. In *The impacts of climate change on human health in the United States: A scientific assessment* (pp. 69–98). Global Change Research Program., health2016.globalchange.gov/

Fatema, S. R., East, L., Islam, S., & Usher, K. (2023). Gender-based vulnerabilities for women during natural disasters in Bangladesh. *Frontiers in Communication*, 8, 1–8, 1180406. DOI: 10.3389/fcomm.2023.1180406

Fritze, J. G., Blashki, G. A., Burke, S., & Wiseman, J. (2008). Hope, despair, and transformation: Climate change and the promotion of mental health and wellbeing. *International Journal of Mental Health Systems*, 2(1), 1–10. DOI: 10.1186/1752-4458-2-13 PMID: 18799005

Galea, S., Nandi, A., & Vlahov, D. (2005). The epidemiology of posttraumatic stress disorder after disasters. *Epidemiologic Reviews*, 27(1), 78–91. DOI: 10.1093/epirev/mxi003 PMID: 15958429

Girardi, G., & Bremer, A. A.Girardi & Bremer. (2022). Effects of Climate and Environmental Changes on Women's Reproductive Health. *Journal of Women's Health*, 31(6), 755–757. DOI: 10.1089/jwh.2021.0631 PMID: 35708571

Goh, A. (2012). A literature review of the gender- differentiated impacts of climate change in developing countries. *CAPRi, 106*, 1–38. Retrieved from http://www.indiaenvironmentportal.org.in/files/file/a literature review of the gender-differentiated impacts of climate change.pdf

Ha, S., Liu, D., Zhu, Y., Kim, S. S., Sherman, S., Grantz, K. L., & Mendola, P. (2017). Ambient temperature and stillbirth: A multi-center retrospective cohort study. *Environmental Health Perspectives*, 125(6), 067011. DOI: 10.1289/EHP945 PMID: 28650842

Harville, E. W., Xiong, X., & Buekens, P. (2010). Disasters and perinatal health: A systematic review. *Obstetrical & Gynecological Survey*, 65(11), 713–728. DOI: 10.1097/OGX.0b013e31820eddbe PMID: 21375788

Hayes, K., Blashki, G., Wiseman, J., Burke, S., & Reifels, L. (2018). Climate change and mental health: Risks, impacts and priority actions. *International Journal of Mental Health Systems*, 12(1), 1–12. DOI: 10.1186/s13033-018-0210-6 PMID: 29881451

Institute of Development Studies. (2008). *Gender and climate change: Mapping the linkages*. BRIDGE, Institute of Development Studies. https://www.adequations.org/IMG/pdf/GenderAndClimateChange.pdf

Intergovernmental Panel on Climate Change. (2014). Summary for policymakers. In: *Climate Change 2014: Impacts, Adaptation, and Vulnerability. Part A: Global and Sectoral Aspects*. Contribution of Working Group II to the Fifth Assessment Report of the Intergovernmental Panel on Climate Change [Field, C.B., V.R. Barros, D.J. Dokken, K.J. Mach, M.D. Mastrandrea, T.E. Bilir, M. Chatterjee, K.L. Ebi, Y.O. Estrada, R.C. Genova, B. Girma, E.S. Kissel, A.N. Levy, S. MacCracken, P.R. Mastrandrea, and L. L. White (eds.)]. Cambridge University Press, Cambridge, United Kingdom and New York, NY, USA, 1-32.

International Union for Conservation of Nature. (2015). *Gender and climate change: IUCN issues in brief.* https://www.iucn.org/resources/issues-brief/gender-and-climate-change

Kabir, R., Khan, H. T. A., Ball, E., & Caldwell, K. (2016). Climate change impact: The experience of the coastal areas of Bangladesh affected by Cyclones Sidr and Aila. *Journal of Environmental and Public Health*, 2016, 1–9. Advance online publication. DOI: 10.1155/2016/9654753 PMID: 27867400

Karmakar, D. (2021). Gender and Climate Change: The Condition of Women in the Indian Sundarbans. *Scholars Journal of Arts. Humanities and Social Sciences*, 9(1), 1–5. DOI: 10.36347/sjahss.2021.v09i01.001

Kartiki, K. (2011). Climate change and migration: A case study from rural Bangladesh. *Gender and Development*, 19(1), 23–38. DOI: 10.1080/13552074.2011.554017

King, S., Dancause, K., Turcotte-Tremblay, A. M., Veru, F., & Laplante, D. P. (2012). Using natural disasters to study the effects of prenatal maternal stress on child health and development. *Birth Defects Research. Part C, Embryo Today*, 96(4), 273–288. DOI: 10.1002/bdrc.21026 PMID: 24203917

Knight, A., & McNaught, A. (Eds.). (2011). *Understanding Well-being: an introduction for students and practitioners of health and social care*. Lantern.

Kumar, V. (2016). Gender-responsive climate change strategies for sustainable development. *Productivity*, 57(2), 182.

Laneri, K., Bhadra, A., Ionides, E. L., Bouma, M., Dhiman, R. C., Yadav, R. S., & Pascual, M. (2010). Forcing versus feedback: Epidemic malaria and monsoon rains in northwest India. *PLoS Computational Biology*, 6(9), e1000898. DOI: 10.1371/journal.pcbi.1000898 PMID: 20824122

Mamun, M. A., Huq, N., Papia, Z. F., Tasfina, S., & Gozal, D. (2019). Prevalence of depression among Bangladeshi village women subsequent to a natural disaster: A pilot study. *Psychiatry Research*, 276, 124–128. DOI: 10.1016/j.psychres.2019.05.007 PMID: 31077883

Masika, R. (2002). *Gender, development and climate change*. Oxfam GB., DOI: 10.3362/9780855987374

Mirza, M. M. Q. (2002). Global warming and changes in the probability of occurrence of floods in Bangladesh and implications. *Global Environmental Change*, 12(2), 127–138. DOI: 10.1016/S0959-3780(02)00002-X

Naeema, F., Ayub, M., Masood, K., Gul, H., Farrukh, A., Saheen, A., Waheed, W., & Chaudhry, H. R. (2011). Prevalence and psychosocial risk factors of PTSD: 18 months after Kashmir earthquake in Pakistan. *Journal of Affective Disorders*, 130(1-2), 268–274. DOI: 10.1016/j.jad.2010.10.035 PMID: 21071091

Neumayer, E., & Plümper, T. (2007). The gendered nature of natural disasters: The impact of catastrophic events on the gender gap in life expectancy, 1981–2002. *Annals of the Association of American Geographers*, 97(3), 551–566. DOI: 10.1111/j.1467-8306.2007.00563.x

Norris, F. H., Friedman, M. J., Watson, P. J., Byrne, C. M., Diaz, E., & Kaniasty, K. (2002). 60,000 disaster victims speak: Part I. An empirical review of the empirical literature, 1981-2001. *Psychiatry*, 65(3), 207–239. DOI: 10.1521/psyc.65.3.207.20173 PMID: 12405079

Obradovich, N., Migliorini, R., Paulus, M. P., & Rahwan, I. (2018). Empirical evidence of mental health risks posed by climate change. *Proceedings of the National Academy of Sciences of the United States of America*, 115(43), 10953–10958. DOI: 10.1073/pnas.1801528115 PMID: 30297424

Patel, S., Zhou, C., Rattan, S., & Flaws, J. A. (2015). Effects of endocrine-disrupting chemicals on the ovary. *Biology of Reproduction*, 93(1), 20–21. DOI: 10.1095/biolreprod.115.130336 PMID: 26063868

Patel, S. K., Agrawal, G., Mathew, B., Patel, S., Mohanty, B., & Singh, A. (2020). Climate change and women in South Asia: A review and future policy implications. *World Journal of Science. Technology and Sustainable Development*, 17(2), 145–166. DOI: 10.1108/WJSTSD-10-2018-0059

Pun, V. C., Manjourides, J., & Suh, H. (2017). Association of ambient air pollution with depressive and anxiety symptoms in older adults: Results from the NSHAP Study. *Environmental Health Perspectives*, 125(3), 342–348. DOI: 10.1289/EHP494 PMID: 27517877

Rofi, A., Doocy, S., & Robinson, C. (2006). Tsunami mortality and displacement in Aceh province, Indonesia. *Disasters*, 30(3), 340–350. DOI: 10.1111/j.0361-3666.2005.00324.x PMID: 16911432

Ross, A., Mack, E. A., Marcantonio, R., Miller-Graff, L., Pearson, A. L., Smith, A. C., & Zimmer, A. (2024). A mediation analysis of the linkages between climate variability, water insecurity, and interpersonal violence. *Climate and Development*, 16(2), 134–148. DOI: 10.1080/17565529.2023.2186746

Sezgin, A. U., & Punamäki, R. L. (2016). Perceived changes in social relations after earthquake trauma among eastern Anatolian women: Associated factors and mental health consequences. *Stress and Health*, 32(4), 355–366. DOI: 10.1002/smi.2629 PMID: 25516368

Shooshtari, S., Abedi, M., Bahrami, M., & Samouei, R. (2018). The mental health needs of women in natural disasters: A qualitative study with a preventive approach. *Journal of Family Medicine and Primary Care*, 7(4), 678. DOI: 10.4103/jfmpc.jfmpc_333_17 PMID: 30234037

Sorensen, C., Saunik, S., Sehgal, M., Tewary, A., Govindan, M., Lemery, J., & Balbus, J. (2018). Climate change and women's health: Impacts and opportunities in India. *GeoHealth*, 2(10), 283–297. DOI: 10.1029/2018GH000163 PMID: 32159002

Stork, A., Travis, C., & Halle, S. (2015). Gender-Sensitivity in natural resource management in Côte d'Ivoire and Sudan. *Peace Review*, 27(2), 147–155. DOI: 10.1080/10402659.2015.1037617

Strand, L. B., Barnett, A. G., & Tong, S. (2011). The influence of season and ambient temperature on birth outcomes: A review of the epidemiological literature. *Environmental Research*, 111(3), 451–462. DOI: 10.1016/j.envres.2011.01.023 PMID: 21333980

Sundblad, E.-L., Biel, A., & Gärling, T. (2007). Cognitive and affective risk judgements related to climate change. *Journal of Environmental Psychology*, 27(2), 97–106. DOI: 10.1016/j.jenvp.2007.01.003

Tan, Y., Yang, R., Zhao, J., Cao, Z., Chen, Y., & Zhang, B. (2017). The associations between air pollution and adverse pregnancy outcomes in China. *Advances in Experimental Medicine and Biology*, 1017, 181–214. DOI: 10.1007/978-981-10-5657-4_8 PMID: 29177963

Tiotiu, A. I., Novakova, P., Nedeva, D., Chong-Neto, H. J., Novakova, S., Steiropoulos, P., & Kowal, K. (2020). Impact of air pollution on asthma outcomes. *International Journal of Environmental Research and Public Health*, 17(17), 6212. Advance online publication. DOI: 10.3390/ijerph17176212 PMID: 32867076

Tobin-Gurley, J., Peek, L., & Loomis, J. (2010). Displaced single mothers in the aftermath of Hurricane Katrina: Resource needs and resource acquisition. *International Journal of Mass Emergencies and Disasters*, 28(2), 170–206. DOI: 10.1177/028072701002800202

Tunstall, S., Tapsell, S., Green, C., Floyd, P., & George, C. (2006). The health effects of flooding: Social research results from England and Wales. *Journal of Water and Health*, 4(3), 365–380. DOI: 10.2166/wh.2006.031 PMID: 17036844

United Nations Framework Convention on Climate Change. (2011). *Reducing Vulnerability to Climate Change, Climate Variability and Extremes, Land Degradation and Loss of Biodiversity: Environmental and Developmental Challenges and Opportunities.*https://unfccc.int/resource/docs/publications/pub_nwp_reducing_vulnerability.pdf

United Nations High Commissioner for Refugees. (2003). *Sexual and Gender-Based Violence against Refugees, Returnees, and Internally Displaced Persons: Guidance for Prevention and Response.*https://www.unhcr.org/sites/default/files/legacy-pdf/3f696bcc4.pdf

United Nations International Strategy for Disaster Reduction. (2008). *Gender perspectives: Integrating disaster risk reduction into climate change adaptation*. United Nations International Strategy for Disaster Reduction. https://www.unisdr.org/files/3391_GenderPerspectivesIntegrating DRRCC Adaptation.pdf

Van Den Berg, B., Grievink, L., Yzermans, J., & Lebret, E. (2005). Medically unexplained physical symptoms in the aftermath of disasters. *Epidemiologic Reviews*, 27(1), 92–106. DOI: 10.1093/epirev/mxi001 PMID: 15958430

Van Niel, M. S., & Payne, J. L. (2020). Perinatal depression: A review. *Cleveland Clinic Journal of Medicine*, 87(5), 273–277. DOI: 10.3949/ccjm.87a.19054 PMID: 32357982

Van Zutphen, A. R., Lin, S., Fletcher, B. A., & Hwang, S. A. (2012). A population-based case–control study of extreme summer temperature and birth defects. *Environmental Health Perspectives*, 120(10), 1443–1449. DOI: 10.1289/ehp.1104671 PMID: 23031822

Vandentorren, S., Bretin, P., Zeghnoun, A., Mandereau-Bruno, L., Croisier, A., Cochet, C., & Ledrans, M. (2006, December). August 2003 heat wave in France: Risk factors for death of elderly people living at home. *European Journal of Public Health*, 16(6), 583–591. Advance online publication. DOI: 10.1093/eurpub/ckl063 PMID: 17028103

Wahid, S. S., Raza, W. A., Mahmud, I., & Kohrt, B. A. (2023). Climate-related shocks and other stressors associated with depression and anxiety in Bangladesh: A nationally representative panel study. *The Lancet. Planetary Health*, 7(2), e137–e146. DOI: 10.1016/S2542-5196(22)00315-1 PMID: 36754469

Wisner, K. L., Sit, D. K. Y., McShea, M. C., Rizzo, D. M., Zoretich, R. A., Hughes, C. L., & Hanusa, B. H. (2013). Onset Timing, Thoughts of Self-harm, and Diagnoses in Postpartum Women with Screen-Positive Depression Findings. *JAMA Psychiatry*, 70(5), 490. DOI: 10.1001/jamapsychiatry.2013.87 PMID: 23487258

World Bank Group. (2018). Sustainable energy for all database. https://databank.worldbank.org/source/sustainable-energy-for-all/preview/on

World Health Organization. (1946). Preamble to the Constitution of the World Health Organization as adopted by the International Health Conference. *Official Records of the World Health Organization*, 2, 100. https://apps.who.int/gb/bd/PDF/bd47/EN/constitution-en.pdf

World Health Organization. (2001). World Health Organization. *Gender and women's mental health. Gender disparities and mental health: The Facts.*https://www.who.int/mental_health/media/en/242.pdf

World Health Organization. (2014). *Gender, climate change and health.*http://apps.who.int/iris/bitstream/10665/144781/1/ 9789241508186_eng.pdf

World Health Organization. (2017). *World Health Statistics 2017: Monitoring Health for the SDGs, Sustainable Development Goals.* https://www.who.int/gho/publications/world_health_statistics/2017/en/

Yadav, S. S., & Lal, R. (2018). Vulnerability of women to climate change in arid and semi-arid regions: The case of India and South Asia. *Journal of Arid Environments*, 149, 4–17. DOI: 10.1016/j.jaridenv.2017.08.001

Zhao, Q., Guo, Y., Ye, T., Gasparrini, A., Tong, S., Overcenco, A., Urban, A., Schneider, A., Entezari, A., Vicedo-Cabrera, A. M., Zanobetti, A., Analitis, A., Zeka, A., Tobias, A., Nunes, B., Alahmad, B., Armstrong, B., Forsberg, B., Pan, S. C., & Li, S. (2021). Global, regional, and national burden of mortality associated with non-optimal ambient temperatures from 2000 to 2019: A three-stage modelling study. *The Lancet. Planetary Health*, 5(7), e415–e425. DOI: 10.1016/S2542-5196(21)00081-4 PMID: 34245712

Chapter 2
Climate Change and Health Resilient Systems: Strengthening Public Health in a Changing World

Jeffrey Kurebwa
https://orcid.org/0000-0002-8371-8055
Bindura University of Science Education, Zimbabwe

Nyasha Yvonne Kurebwa
Women's University in Africa, Zimbabwe

ABSTRACT

The impacts of climate change pose serious threats to public health and healthcare systems globally. Extreme weather events, rising temperatures, shifting disease patterns, and other climate-driven changes are straining the ability of communities to protect human health and well-being. This book chapter explores the critical need to build health-resilient systems that can withstand and adapt to the challenges of a changing climate. Pathways through which climate change affects human health, including direct impacts from heatwaves, wildfires, and natural disasters, as well as indirect impacts on food and water security, the spread of infectious diseases, and population displacement. Practical guidance to policymakers, public health practitioners, healthcare administrators, and community leaders on how to build the adaptive capacity needed to protect human health in the era of climate change. Taking proactive steps to enhance health resilience, communities can reduce their vulnerability and ensure equitable access to quality healthcare, in the face of intensifying climate impacts.

DOI: 10.4018/979-8-3693-4183-4.ch002

INTRODUCTION

Climate change presents a significant challenge in the current era, leading to global temperature rises, altered rainfall patterns, and elevated sea levels (Prats, 2020). The repercussions of climate change include heat waves, rising sea levels, intense precipitation, floods, and droughts. While the average climate has remained relatively stable for millennia, it has escalated rapidly over the past five decades. Predictions indicate that the Earth's average temperature could surge by 1.8 to 4 °C by 2100 (Hansen et al. 2006). Climate, as defined by the American Meteorological Society (2023), encompasses the long-term weather conditions of a specific geographical area.

The term climate refers to the long-term weather patterns in a specific region, encompassing factors such as temperature, atmospheric pressure, wind velocity, humidity, and precipitation over at least 30 years. Climate change, on the other hand, denotes the alterations in the global climate over time, which can be influenced by internal earth processes, external forces like changes in sunlight intensity, and human activities. According to the United Nations Framework Convention on Climate Change (UNFCCC) (2023), climate change is defined as any modifications in climate that can be directly or indirectly linked to human activities, leading to changes in the composition of the global atmosphere beyond natural climate variability observed over similar time frames. The impacts of climate change are far-reaching, affecting essential aspects of life such as water, food security, and public health (Costell et al. 2009).

The social and environmental determinants of health are compromised by various factors, such as access to clean air, adequate food, safe drinking water, and secure shelter, as highlighted by the World Health Organization (WHO) (2018). Climate change is identified as a significant threat to sustainable global development, with its drivers and impacts having detrimental effects on human health. The negative consequences of climate change extend to the sustainable development of nations, resulting in reduced economic growth, heightened disparities in access to essential public services like education and healthcare, and even triggering conflicts and tensions among countries (WHO, 2018).

SDG 13 focuses on addressing climate change promptly, while SDG 3 aims to guarantee a healthy life and enhance well-being for everyone (Osborn et al. 2015). The connection between climate and health is significant, as actions taken to address one can impact the other. Climate change has a direct impact on human health, affecting both physical and mental well-being through alterations in weather patterns such as heat waves, droughts, floods, storms, and hurricanes. Additionally, climate change indirectly influences health by altering the quality of water, air, and food sources. Heat stress can result in various health issues like fatigue, digestive problems, skin irritations, heat stroke, and even fatalities. On the other hand, air pollution has been linked to increased mortality rates from respiratory infections, lung cancer, and heart diseases. According to WHO (2018), approximately 7 million individuals lose their lives annually due to exposure to outdoor and indoor air pollution stemming from the combustion of fossil fuels.

Defining the Scope of Health System Resilience

The term 'resilience' finds its roots in the Latin word 'resilire,' which translates to 'bounce back' (Holling, 1973). Initially, resilience was primarily associated with 19th-century shipbuilding and was first introduced academically in civil and mechanical engineering by Holling (1973) and Hollnagel et al. (2007). The essence of resilience, characterized by its ability to return to equilibrium following a disturbance, is known as resilience engineering (Folke, 2006). Over time, the concept of resilience expanded beyond its mechanical origins to encompass more intricate social systems. Resilience has evolved to encompass a wide array of disciplines, including ecology, economics, art, literature, and psychology. Scholars such as Alexander (2013), Bodin & Wiman (2004), and Fletcher & Sarkar (2013) have contributed to the multidisciplinary understanding of resilience in various complex systems.

Resilience is commonly understood as the capacity of a system to endure or assimilate the impacts of a disturbance while preserving the functionality and structure of the system. This notion has been utilized in examining Social-Ecological Systems (SES) both in theory and in practice (Adger et al. 2005). The characteristics of resilience in an SES have been delineated as follows: the system's ability to change and still maintain the same regulatory mechanisms on function and organization, the extent to which the system can self-organize, and the community's aptitude to cultivate and enhance its ability for learning and adaptation (Adger et al. 2005). Resilience was initially defined in ecology as the capability of a system to withstand change and disturbance while preserving the same relationships between populations or state variables, a concept that has since been adopted by various disciplines to describe the ability of individuals, populations, or systems to endure shocks while maintaining their original functions (Alexander, 2013). Discussions on livelihood approaches have emphasized the importance of social factors in determining health outcomes by fostering resilience to enhance healthcare accessibility within the broader societal context, laying the groundwork for research on resilience within the healthcare system (Obrist et al. 2007).

Resilience literature often highlights the importance of absorptive, adaptive, and transformative capacities in understanding and strengthening resilience levels (Béné et al. 2012, Lebel et al. 2006, Blanchet et al. 2017, Kagwanja et al. 2020). Absorptive capacity refers to the ability to absorb the impacts of events using predetermined coping responses (Cutter et al. 2008). Furthermore, a more dynamic interpretation of resilience suggests that social systems can respond to shocks in various ways, not only by absorbing the shocks and returning to their original state, but also by advancing through learning or adaptation, and even reaching a new state of equilibrium that may indicate transformation (Holling, 1986, Adger et al. 2005).

The concept of adaptive capacity revolves around a system's capability to adapt to changes, mitigate the impacts, and manage disruptions (Burton et al. 2002). On the other hand, transformative capacity involves the ability to establish a completely new system when the current one becomes unsustainable due to ecological, economic, or social factors (Walker et al. (2004). It is important to note that transformational changes are not solely technical but a blend of various elements such as technological advancements, institutional modifications, shifts in behavior, and cultural transformations (O'Brien, 2012). This combination challenges the existing norms and structures to bring about systemic changes. These capacities can result in different outcomes, including persistence, gradual adjustments, and transformative responses. By understanding and harnessing both adaptive and transformative capacities, systems can effectively navigate through changes and uncertainties, ultimately leading to sustainable and resilient outcomes (Alexander, 2013).

The varying results could be associated with the severity of the impact and the level of adjustment needed. As the system progresses from being able to absorb shocks, to being able to adapt, and ultimately to being able to transform, the costs and risks of transformation increase compared to maintaining the current state (Béné et al. 2012). The direct application and utilization of resilience in the context of health is a relatively recent development, with its initial use being to describe the significant structural changes that occurred in the health system of rural Ghana following a shock (Blanchet and James, 2013). The concept of resilience has enhanced the understanding of health systems as complex adaptive systems (Blanchet & James, 2013). Following the Ebola crisis in 2014, Kruk et al (2015) addressed the increasing demands of multilateral institutions by creating a framework for resilience that outlines five essential characteristics of a resilient health system. They also introduced a resilience index to assess resilience (Kruk et al. 2015, Kruk et al. 2017). In this context, health system resilience is described as the ability of health actors, institutions, and communities to prepare for and effectively respond to crises, maintain essential functions during emergencies, and adapt based on lessons learned from crises, encompassing all aspects of resilience.

Resilience is not only relevant in times of acute shocks such as wars, epidemics, or natural disasters, but also in cases of chronic stresses like leadership failures, staff turnovers, and drug shortages (Kruk et al. 2017). There is a growing emphasis on building everyday resilience, which refers to the ability of health systems to continue providing quality services despite ongoing challenges. This everyday resilience ultimately enables the systems to better respond to sudden shocks as well. In low- and middle-income countries, chronic stresses are often internally generated due to insufficient organizational capacities, making them the norm in these settings. (Kruk et al. 2017; Gilson et al. 2017a; Barasa et al. 2017). Resilience in daily life within a specific environment relies heavily on innovative styles of leadership that encourage problem-solving and embody mutual respect, allowing various stakeholders to impact learning outcomes through collaborative social connections. Gilson et al. (2017) emphasize that while leadership skills necessitate individual initiative, systemic resilience is achieved by extending this initiative to the organizational level and fostering organizational capabilities. The healthcare system must be able to address the healthcare requirements of the population in the face of climate change impacts. However, climate change poses a threat to the healthcare system's ability to safeguard and enhance public health. The disruption caused by rising sea levels, typhoons, hurricanes, and floods to delicate infrastructure, transportation networks, and the supply chain for essential goods and food is significant (Prats, 2020).

The health system is under immense pressure due to the effects mentioned. As a result, healthcare organizations may struggle to meet the demands brought about by climate change, potentially impacting the delivery of essential health services to the population. Conversely, healthcare organizations, with their round-the-clock operations and high energy consumption, are also contributing to climate change through the emission of greenhouse gases (Kruk et al. 2017). This presents a dual challenge for the health system, as it must address both the human health impacts of climate change and work towards reducing its carbon footprint. As a result, it is imperative to enhance the capacity of the health system to effectively respond to the health effects of climate change. This will be crucial in ensuring that the system can adequately address the challenges posed by climate change and continue to provide essential healthcare services to the population (Kruk et al. 2017).

The healthcare system is comprised of various entities and individuals tasked with formulating policies, securing funding, allocating resources, and delivering healthcare services to enhance and preserve the well-being of the population, while also meeting both clinical and non-clinical needs and safeguarding against excessive health-related expenses (Mosadeghirad, 2022). WHO (2007) identified governance and leadership, financing, health workforce, medical products, vaccines, technologies, information, and service delivery as the six fundamental pillars of the healthcare system that are indispensable for attaining its objectives. Furthermore, factors such as political, economic, social, technological, environmental, and legal elements play a significant role in influencing the overall health status of individuals within a community.

Hence, intricate and convoluted modern healthcare institutions operate within a demanding and ever-changing setting (Ghiasipour et al. 2017, Esfahani et al. 2018). Health system resilience refers to the health system's ability to forecast, prevent, prepare, absorb, adjust, and transform in the face of shocks and stresses while ensuring the continuous delivery of routine health services during crisis management (Ezzati et al. 2023). A health system that is resilient to climate change is defined as one that can anticipate, respond to, manage, recover from, and adapt to climate-related shocks and stresses, to bring about sustained enhancements in population health, despite an unpredictable climate (WHO, 2015). A climate-resilient health system presents an opportunity for sustainable human development.

Enhancing the resilience of the health system is crucial in mitigating the impacts of climate change on public health and improving overall well-being. It is imperative to fortify the six fundamental components of the health system to safeguard and enhance the health of individuals amidst a shifting and unpredictable climate. The health sector is particularly susceptible to the effects of climate change, yet there is a noticeable lack of initiatives in place, particularly in developing nations, to bolster the adaptability and robustness of health systems in the face of climate change challenges (Alexander, 2013). Fragile and unprepared health systems only serve to worsen the adverse consequences of climate change on public health. Enhancing the six fundamental components of the healthcare system, such as governance and leadership, financing, health workforce, facilities, equipment, and supplies, health information system, and health service delivery processes, is crucial to mitigating the adverse impacts of climate change on healthcare institutions, healthcare professionals, patients, and the community. Additionally, the external factors encompassing the healthcare system, such as political, economic, social, technological, environmental, and legal aspects, play a significant role in influencing the healthcare system's ability to address the challenges posed by climate change and should be taken into account in climate change and health initiatives (Mosadeghirad, 2023). In total, 87 actions were identified to enhance the health system's resilience to climate change, categorized into six key building blocks: governance, financing, workforce, equipment and medicine, information systems, and health services delivery processes. Strengthening these six building blocks of the health system is crucial to mitigating the impacts of climate change and ensuring the system remains resilient in the face of environmental challenges (Mosadeghirad, 2023). Effective governance, evidence-based policymaking, strategic planning, and collaboration with key stakeholders in water and sewage, nutrition, energy, and urban planning sectors are vital components in establishing a climate-resilient health system. Interventions focusing on governance and leadership play a critical role in proactively managing health system vulnerabilities to climate-related shocks and in formulating strategic health policies for the future.

It is crucial to prioritize health in the national climate change policy and strategy, leading many countries to develop their own national health and climate change policies as well as health adaptation plans to address current and future health risks. Strengthening inter-sectoral collaboration at both national and local levels is also essential, although the implementation of these plans and collaborations has proven to be challenging (WHO, 2022). In response to these challenges, the proposal of applying a 'crisis leadership model' for building a climate-resilient health system has been put forward. This model emphasizes the importance of planning and communication, adapting to change and updated risk assessment methods, engaging local communities and integrated early warning systems, developing a response culture, and reducing system-related greenhouse gas emissions (Tochkin et al. 2022).

Dynamic Attributes of Health System Resilience

Kruk et al. (2015) outlined that robust health systems exhibit common characteristics that appear to be relevant in various settings and responses to different types of crises.

Aware

A robust health system demonstrates an understanding of its capacity and vulnerabilities. This understanding involves the ability to recognize and interpret warning signs at the local level and promptly seek assistance when needed. This can be assessed by evaluating the knowledge and availability of health system resources, maintaining a proactive epidemiological surveillance system, and establishing effective communication channels to disseminate important information through different media platforms (Kruk et al. 2015).

Diverse

A robust healthcare system can effectively address a wide array of healthcare needs, encompassing the ability to tackle a broad spectrum of health issues stemming from a crisis and cater to the diverse needs of its populace (Kruk et al. 2015). This can be assessed by evaluating the range of health services offered in primary care, the sufficiency of public health spending, and the presence of financial safety nets (Kruk et al. 2017).

Self-regulating

A robust healthcare system is capable of self-regulation and ensuring continuous delivery of health services to its population. Self-regulation involves the capacity to manage and contain health risks while maintaining essential health services (Kruk et al. 2015). This can be assessed by examining the system's partnerships with local and international entities, including non-governmental and private healthcare providers, as well as its repository of service options for both affected and unaffected communities (Kruk et al. 2017).

Integrated

A robust healthcare system includes a structured, centralized coordination mechanism that is united and collaborates with non-health entities while involving the community in crisis response efforts. Integration refers to the ability to unite different stakeholders, concepts, and organizations to develop solutions and implement strategies (Kruk et al. 2015). This can be assessed by evaluating the presence of a national emergency coordination system, platforms for communication with community leaders, as well as different task forces and their respective functions and guidelines (Kruk et al. 2017).

Adaptive

A robust healthcare system is capable of adjusting to changing circumstances and handling emergencies with flexibility. Adaptability refers to the capacity to change in ways that enhance performance under extremely challenging conditions (Kruk et al. 2015). This can be assessed by examining the ability to reallocate resources during crises, the leadership skills of local healthcare personnel, the willingness to delegate decision-making authority, and financial preparedness during emergencies (Kruk et al. 2017).

The Impact of Climate Change on Human Health

Deitchman et al. (2021) stated that climate-related occurrences have a dual impact on human health. Firstly, they can directly lead to illness, injury, heightened mental health issues, and fatalities, such as in the case of tornadoes, hurricanes, or other extreme weather events. Secondly, these events can indirectly influence public health by disrupting living conditions, and access to healthcare and public services, causing damage to critical infrastructure, and impacting economic systems as a result of extreme weather events. Climate change has the potential to trigger population movements, resulting in a rise in population density in regions that were historically less populated due to extreme weather events prompting residents to relocate. Healthcare and social systems must adapt their plans to accommodate these shifts in population growth and migration. As highlighted by Crimmins, et al. (2016) and CDC (2021), climate change affects human health in various ways.

Temperature Fluctuations

The impact of global warming extends beyond altering arctic temperatures, as it also influences the jet streams that have a significant impact on weather patterns. In addition, extreme temperature events can disrupt the body's ability to regulate its internal temperature, leading to various health issues such as dehydration, heat exhaustion, edema, syncope, heatstroke, hyperthermia, and acute cerebrovascular accidents, which can contribute to thrombogenesis (Thompson, Hornigold, & Page, 2018). While direct deaths from heat are rare, most heat-related illnesses are a result of worsening existing medical conditions, including chronic pulmonary disease, cardiac conditions, kidney disorders, and psychiatric illness. Elevated temperatures have the potential to heighten the likelihood of mental health issues, including suicide (Thompson, Hornigold, & Page, 2018). Additionally, specific medications can pose health hazards by impacting thermoregulation, cognitive function, blood pressure, heart rate, perspiration mechanisms, and electrolyte equilibrium (The Lancet, 2015). Individuals with a compromised capacity to regulate body temperature, as well as homeless individuals who face heightened exposure risks, might be especially vulnerable to severe heat conditions. The impact of extreme heat in urban environments is a significant area of concern.

Extreme Cold

Frostbite and hypothermia can exacerbate pre-existing conditions like cardiovascular, respiratory, cerebrovascular, and diabetic diseases. Indirect consequences of extreme cold weather include exhaustion from snow shoveling, medical emergencies due to traffic accidents, transportation challenges hindering access to healthcare (e.g., blocked or icy roads), and incidents of fires or carbon monoxide poisoning from improper heating sources. Vulnerable populations, such as those with difficulty regulating body temperatures and individuals without homes, are especially at risk during extreme cold spells, similar to extreme heat situations (The Lancet, 2015).

Natural Disasters

According to the World Meteorological Organization, there has been a fivefold increase in weather-related disasters in the last 50 years. Each of these events has a significant impact on both public health and the healthcare system. Extreme precipitation, hurricanes, coastal storms, and tsunamis can lead to flooding and flash flooding, resulting in traumatic injuries and fatalities from drowning. Additionally, these events are linked to adverse outcomes such as preterm births, low birth weight, and carbon monoxide poisoning due to power outages and improper generator use. The aftermath of these disasters can also affect healthcare facilities, causing infrastructure damage and disruptions, as well as an influx of patients due to the spread of infectious diseases. Furthermore, as previously mentioned, acid rain can increase the risk of sinkholes, leading to severe structural damage and posing a threat to human life and safety (The Lancet, 2015).

Droughts

According to the U.S Drought Monitor (2024), climate change contributes to drought through:

(a) Warmer temperatures enhance evaporation, which reduces surface water and dries out soils and vegetation. This makes periods with low precipitation drier than they would be in cooler conditions.
(b) Climate change is also altering the timing of water availability.
(c) Warmer winter temperatures are causing less precipitation to fall as snow in the Northern Hemisphere, including in key regions like the Sierra Nevada of California.
(d) Decreased snowpack can be a problem, even if the total annual precipitation remains the same. This is because many water management systems rely on spring snowpack melt. Likewise, certain ecosystems also depend on snowmelt, which supplies cold water for species like salmon. Because snow acts as a reflective surface, decreasing snow area also increases surface temperatures, further exacerbating drought.
(e) Some climate models find that warming increases precipitation variability, meaning there will be more periods of both extreme precipitation and drought. This creates the need for expanded water storage during drought years and increases the risk of flooding and dam failure during periods of extreme precipitation.
(f) Climate change is making certain regions drier

The factors as mentioned earlier lead to a decline in water quality, availability, and accessibility, as well as respiratory issues caused by poor air quality, diminished sanitation and hygiene practices, higher rates of illnesses, and heightened mental health concerns.

Wildfires

Smoke inhalation, severe coughing, and wheezing can be triggered by these, while flames have the potential to cause severe burns and other traumatic injuries that may result in increased heart rate, chest pain, headaches, irritated sinuses, and a sore throat. For some individuals, wildfires and smoke can worsen conditions such as asthma, chronic obstructive pulmonary disease, and other respiratory illnesses. The effects of wildfires can also extend to mental health distress and illnesses related to exposure to hazardous materials due to damage to infrastructures, such as water system contamination (The Lancet, 2015).

Severe thunderstorms

Instances such as tornados, derechos (windstorms), and various forms of destructive winds have the potential to result in severe injuries and fatalities (such as cranial, chest, spinal, and peripheral nerve injuries). Following an injury, there are worries about fungal or necrotizing soft-tissue infections, respiratory infections-especially those triggered by unusual microbes found in people residing or working in tornado-prone areas, and carbon monoxide poisoning due to improper generator usage (for power outage management) (Kruk et al. 2017).

Air Quality Impacts

An illustration of this is the combination of air pollution and pollen levels. Elevated levels of air pollution and ground-level ozone have been linked to premature mortality and can worsen conditions related to asthma, chronic obstructive pulmonary disease, emphysema, and other respiratory issues. When extreme temperature changes coincide with air pollution, they can also impact how the human body responds to polluted air (Kruk et al. 2017).

Vector-borne Diseases

In the United States, the impact of climate change is anticipated to lead to a rise in disease-carrying vectors such as ticks, mosquitoes, and fleas, as well as an increase in the risk of zoonotic diseases. This could result in the spread of illnesses like Lyme disease, West Nile virus, locally acquired dengue (as evidenced in Florida and southern Texas), and chikungunya (which has recently been detected in southern Florida).

Water-related Illness

Diseases caused by pathogens found in water, such as bacteria, viruses, protozoa, and toxins from algae and cyanobacteria, have been on the rise due to shifting weather patterns. Exposure to contaminated water sources, whether through ingestion, inhalation, or direct contact, can have detrimental effects on human health. Symptoms of waterborne illnesses range from gastrointestinal issues like gastroenteritis and diarrhea to more severe conditions such as neurological disorders, respiratory problems, meningitis, organ failure, and skin infections (Kruk et al. 2017).

Food Safety

The systems and procedures concerning food safety, nutrition, and production are interconnected with the physical and biological surroundings. It is anticipated that cases of foodborne illnesses will increase due to the influence of climate change on environmental factors such as flooding, drought, and temperature changes. These changes can alter the seasonal and geographical distribution of bacteria, viruses, pests, parasites, fungi, and other harmful substances, resulting in potential food safety risks (CDC, 2021).

Mental Health

Gastrointestinal infections can vary in seriousness, from mild discomfort to potentially fatal outcomes. Severe cases of foodborne illnesses have the potential to result in miscarriages, stillbirths, premature births, or serious infections like meningitis and organ inflammation. The issue of food security is further exacerbated by the climate crisis, as shifts in weather patterns, rising natural disasters, and changes in delicate ecosystems all contribute to disruptions in food supplies, ultimately affecting nutritional resilience. The CDC (2021) highlights that individuals could face adverse mental health consequences due to climate change, irrespective of their medical backgrounds. Both immediate and prolonged repercussions may range from mild stress and signs of distress to severe psychiatric disorders such as anxiety, depression, and post-traumatic stress disorder. Predictable stress responses are often triggered by natural calamities, which can detrimentally impact individuals with pre-existing mental health issues. The cumulative effects of heat and cold stressors associated with certain occupations (e.g., farmers concerned about crop survival during droughts) can lead to significant mental health challenges, with symptoms varying in duration from short-term to chronic (Mullins & White, 2019).

Patients suffering from severe mental illnesses such as schizophrenia or dementia are more susceptible to hospitalization and mortality during heat waves due to the potential interference of their medications with temperature regulation or the contribution to hyperthermia. A study conducted by Mullins & White (2019) revealed a direct correlation between temperature and the occurrence of suicide in the United States. According to Crimmins et al. (2016) extreme precipitation and higher temperatures can lead to an increase in interpersonal violence and intergroup conflict. These conditions can result in heightened aggression, potentially leading to a rise in interpersonal violence and violent crime, which can have adverse effects on individual and societal mental health and well-being. Additionally, cumulative stressors can also elevate the incidence of child abuse, further emphasizing the impact of environmental factors on mental health and social well-being.

Healthcare Infrastructure Resiliency/Hardening

In response to the adverse effects of climate change on human health and mental well-being, various entities such as governmental agencies, healthcare systems, community organizations, and health experts are collaborating to develop strategies aimed at slowing down global warming, lessening its impacts, enhancing the resilience of health systems, and advancing research on climate-related health issues. According to Guenther & Balbus (2014), the rise in extreme weather events poses intricate challenges to existing infrastructure capabilities, redundancies, disaster preparedness, and response protocols. The evolving landscape of climate change introduces novel threats and necessitates a re-evaluation of building design standards (Guenther & Balbus, 2014). At different levels - regional, state, and local - the implementation of sustainable and resilient development practices, as well as preparedness measures, exhibit variations. Efforts are being made to adapt to the changing climate conditions and enhance the overall readiness to address climate-related challenges effectively.

The resource emphasizes that there is no universally accepted standard regarding the capacity of healthcare infrastructure to endure the effects of severe weather occurrences, a sentiment that remains relevant long after its initial publication. Healthcare leaders and emergency preparedness professionals bear the responsibility of guaranteeing the safety of their establishments for employees, patients, and guests. This duty encompasses ensuring that the physical structures of healthcare facilities are capable of withstanding extreme weather events, preparing for potential disruptions to critical services such as electricity, water, and Internet connectivity, as well as addressing challenges related to staffing and supply chain management.

Investing in backup systems and capabilities is crucial to maintaining the continuity of medical services in such circumstances (Crimmins et al. 2015).

SOLUTIONS AND RECOMMENDATIONS

The following solutions and recommendations are proffered:

(a) It is crucial to enhance the governance and leadership of the health system by creating a dedicated climate change and health department within the Ministry of Health. This department would be responsible for coordinating climate-resilient programs, fostering inter-sectoral collaboration, devising national health and climate change strategies, and establishing partnerships with key stakeholders.

(b) Sufficient resources and funding play a crucial role in ensuring that health systems can withstand the impacts of climate change. The detrimental effects of climate change on public health can result in a rise in healthcare expenses. Moreover, enhancing the adaptability and resilience of the healthcare system necessitates a sustainable financial plan. The annual costs for climate change adaptation in developing nations amount to approximately USD 70 billion and are anticipated to escalate to USD 140–300 billion by 2030 (Neufeldt et al. 2020). Hence, it is imperative to allocate financial resources towards fortifying the resilience of the health sector and backing climate change and health adaptation strategies. Innovative collaborative and inter-sectoral financing models should be explored to support the healthcare system. For example, the Government of Bangladesh established the Climate Change Trust Fund in 2010 to aid climate-resilient health initiatives (Rahaman, 2019). Such initiatives are essential in ensuring that health systems are equipped to address the challenges posed by climate change.

(c) The resilience of the health system and the delivery of high-quality, effective health services are heavily reliant on the health workforce. Health workers must have access to formal education, training, and mentorship programs focused on the impact of climate change on health. This will enable them to recognize, prevent, and address health risks associated with climate change. Furthermore, health workers must be equipped with the necessary knowledge and skills to provide essential health services during extreme climate conditions. Additionally, efforts should be made to enhance their capacity to effectively respond to the health challenges posed by climate change.

(d) Medical education at all levels, including undergraduate, graduate, and continuing medical education, should incorporate climate change and health. Philipsborn et al. (2019) suggested a framework for climate change and health education content for residents to enhance their professional capabilities. This framework covers the impact of climate change on health, the risks it poses to healthcare delivery, and the essential adjustments needed in clinical practice (Philipsborn et al. 2021). Furthermore, there is a need to enhance the ability to communicate with healthcare workers during emergencies.

(e) Healthcare organizations need to strengthen their institutional capacities in order to effectively address climate-related risks. It is crucial to have adequate facilities, equipment, and medications available to ensure the delivery of healthcare services during extreme weather events. Moreover, healthcare infrastructure and technologies should be designed to be resilient to climate change and sustainable in the long term. Strategies to achieve this goal may include relocating health facilities that are at risk (Schnitter et al. 2019) and implementing infrastructure adaptations. These adaptations could involve sustainable land use practices, the design of green and low-carbon buildings, the procurement of materials and products that are safe and sustainable, the use of energy-efficient and low-carbon technologies like telemedicine, the adoption of renewable energy sources, and the installation of emergency power generation systems (Aslan et al. 2019).

(f) By incorporating medical technologies and equipment with a lower environmental impact, healthcare facilities can enhance their climate resilience and contribute to a more sustainable healthcare system overall. Healthcare organizations need to prioritize these measures to effectively respond to the challenges posed by climate change and ensure the continuity of healthcare services in the face of climate-related risks.

(g) Healthcare organizations may face disruptions in their infrastructures and essential supplies like water, electricity, and gas during extreme weather conditions like floods or droughts. These organizations must construct their infrastructure in a manner that can endure such severe weather events.

(h) Health information systems are essential for gathering, examining, and disseminating data to aid managers in making decisions in the face of climate change. It is crucial to enhance the health information system to ensure that managers and health workers have access to pertinent information regarding the health system's vulnerability to climate risks, its ability to respond, and its adaptability and resilience to the impacts of climate change. This will enable evidence-based decision-making in the healthcare sector (Mosadeghrad, 2023).

(i) Vulnerability and adaptation assessments offer insights into the types and extent of possible health hazards linked to climate change, as well as pinpointing vulnerable groups and deficiencies in healthcare systems (Esfahan, 2018). This information enables managers to implement suitable adaptation strategies and resilience measures to mitigate or prevent the adverse health effects of climate change in the future.

FUTURE RESEARCH DIRECTION

Health policies and programs must take into account the impact of climate change, and political leadership plays a crucial role in addressing the health risks associated with climate change across various sectors. Collaboration between the health sector and other industries like agriculture and urban planning is necessary to tackle the underlying causes of climate change and its effects on health. Involving a wide range of stakeholders can enhance institutional capabilities to deliver improved health services (Sen et al. 2017). The health sector should prioritize managing environmental factors affecting health by implementing 'health in all policies' (HiAPs) and fostering a multi-sectoral collaborative approach. Policymakers need to prioritize public health considerations in all policy decisions, regardless of the domain, to ensure a comprehensive approach to addressing health challenges in the face of climate change.

CONCLUSION

Given the increasing worldwide phenomenon of climate change and its negative impact on health, it is essential to enhance the health system. Because health systems are intricate and multifaceted, a comprehensive approach is necessary to establish a resilient system that can withstand the effects of climate change. Implementing cohesive strategies that prioritize the six building blocks of the health system will lead to improved health results and ultimately save numerous lives on both local and global scales. Effective governance and leadership, along with increased public awareness, strategic allocation of resources, efforts to mitigate climate change, readiness for emergencies, reliable health services delivery, and investment in research are crucial for establishing a health system that is resilient to climate impacts. 2. Additionally, it is important to strengthen multi-sectoral cooperation by integrating health considerations into all policies and promoting collaboration among various departments such as environmental, economic, financial, energy, and education sectors.

REFERENCES

Adger, W. N., Hughes, T. P., Folke, C., Carpenter, S. R., & Rockström, J. (2005). Social-Ecological Resilience to Coastal Disasters. *Science*, 309(5737), 1036–1039. DOI: 10.1126/science.1112122 PMID: 16099974

Alexander, D. (2013). Resilience and Disaster Risk Reduction: An Etymological Journey. *Natural Hazards and Earth System Sciences*, •••, 13.

Aslan, M., & Yıldız, A. (2019). How blameless are hospitals in climate change? An example of a province in Turkey. *Environ Socio-economic Stud*, 7(4), 45–53. DOI: 10.2478/environ-2019-0023

Barasa, E. W., Cloete, K., & Gilson, L. (2017). From Bouncing Back, To Nurturing Emergence: Reframing The Concept Of Resilience In Health Systems Strengthening. *Health Policy and Planning*, 32(suppl_3), 91–94. DOI: 10.1093/heapol/czx118 PMID: 29149319

Béné, C., Wood, R. G., Newsham, A., & Davies, M. (2012). Resilience: New Utopia or New Tyranny? Reflection about the Potentials And Limits Of The Concept Of Resilience in Relation To Vulnerability Reduction Programmes. *IDS Working Paper*, 2012(405), 1–61. DOI: 10.1111/j.2040-0209.2012.00405.x

Berkes, F., & Jolly, D. (2001). Adapting to Climate Change: Social-Ecological Resilience in a Canadian Western Arctic Community. *Conservation Ecology*, 5(2), art18. DOI: 10.5751/ES-00342-050218

Blanchet, K., Nam, S. L., Ramalingam, B., & Pozo-Martin, F. (2017). Governance And Capacity To Manage Resilience Of Health Systems: Towards A New Conceptual Framework. *International Journal of Health Policy and Management*, 6(8), 431–435. DOI: 10.15171/ijhpm.2017.36 PMID: 28812842

Bodin, P., & Wiman, B. (2004). Resilience And Other Stability Concepts In Ecology: Notes On Their Origin, Validity, And Usefulness. *Ess Bulletin*, 2, 33–43.

Burton, I., Huq, S., Lim, B., Pilifosova, O., & Schipper, E. L. (2002). From Impacts Assessment To Adaptation Priorities: The Shaping Of Adaptation Policy. *Climate Policy*, 2(2-3), 145–159. DOI: 10.3763/cpol.2002.0217

Centre for Disease Control and Prevention (CDC). (2021). *Climate Effects on Health.*www.cdc.gov

Chapin, F. S., Peterson, G., Berkes, F., Callaghan, T. V., Anglestam, P., & Apps, M.. (2004). Resilience and vulnerability of northern regions to social and environmental change. *Ambio*, 33(6), 344–349. DOI: 10.1579/0044-7447-33.6.344 PMID: 15387072

Costello, A., Abbas, M., Allen, A., Ball, S., Bell, S., Bellamy, R., Friel, S., Groce, N., Johnson, A., Kett, M., Lee, M., Levy, C., Maslin, M., McCoy, D., McGuire, B., Montgomery, H., Napier, D., Pagel, C., Patel, J., & Patterson, C. (2009). Managing the health effects of climate change. *Lancet*, 373(9676), 1693–1733. DOI: 10.1016/S0140-6736(09)60935-1 PMID: 19447250

Crimmins, A. R., Balbus, J., Gamble, J. L., ... (2016). The Impacts of Climate Change on Human Health in the United States: A Scientific Assessment. U.S. Global Change Research Program. USA: Washington DC.

Cutter, S. L., Barnes, L., Berry, M., Burton, C., Evans, E., Tate, E., & Webb, J. (2008). A Place-Based Model For Understanding Community Resilience To Natural Disasters. *Global Environmental Change*, 18(4), 598–606. DOI: 10.1016/j.gloenvcha.2008.07.013

Deitchman, S. D., Kirsch, T. D., Auerbach, P. S., & Hill, A. C. (2021). Climate Resilience: It is Time for a National Approach. Health Security. *Health Security*, 19(6), 10–24. DOI: 10.1089/hs.2021.0108 PMID: 34698530

Esfahani, P., Mosadeghrad, A. M., & Akbarisari, A. (2018). The success of strategic planning in health care organizations of Iran. *International Journal of Health Care Quality Assurance*, 31(6), 563–574. DOI: 10.1108/IJHCQA-08-2017-0145 PMID: 29954262

Ezzati, F., Mosadeghrad, A. M., & Jaafaripooyan, E. (2023). Resiliency of the Iranian healthcare facilities against the Covid-19 pandemic: Challenges and solutions. *BMC Health Services Research*, 23(1), 1–16. DOI: 10.1186/s12913-023-09180-6 PMID: 36859284

Folke, C. (2006). Resilience: The Emergence of A Perspective For Social-Ecological Systems Analyses. *Global Environmental Change*, 16(3), 253–267. DOI: 10.1016/j.gloenvcha.2006.04.002

Forbes, B., Stammler, F., Kumula, T., Meschtyb, N., Pajunen, A., & Kaarlejärvi, E. (2009). High resilience in the Yamal-Nenets social-ecological system, West Siberian Arctic, Russia. *Proceedings of the National Academy of Sciences of the United States of America*, 106(52), 22041–22048. DOI: 10.1073/pnas.0908286106 PMID: 20007776

Ganesh, C., Schmeltz, M., & Smith, J. (2020). Introduction climate change and the legal, ethical, and Health issues facing Healthcare and Public Health Systems. *The Journal of Law, Medicine & Ethics*, 48(4), 636–642. DOI: 10.1177/1073110520979370 PMID: 33404343

Ghiasipour, M., Mosadeghrad, A. M., Arab, M., & Jaafaripooyan, E. (2017). Leadership challenges in health care organizations: The case of Iranian hospitals. *Medical Journal of the Islamic Republic of Iran*, 31(1), 560–567. DOI: 10.14196/mjiri.31.96 PMID: 29951397

Gilson, L., Barasa, E., Nxumalo, N., Cleary, S., Goudge, J., Molyneux, S., Tsofa, B., & Lehmann, U. (2017a). Everyday Resilience In District Health Systems: Emerging Insights From The Front Lines In Kenya And South Africa. *BMJ Global Health*, 2(2), 2. DOI: 10.1136/bmjgh-2016-000224 PMID: 29081995

Guenther, R., & Balbus, J. (2014). *Primary Protection: Enhancing Health Care Resilience for a Changing Climate*. U.S. Department of Health and Human Services.

Gunderson, L. H., & Holling, C. S. (Eds.). (2002). *Panarchy: understanding transformations in human and natural systems*. Island Press.

Hansen, J., Sato, M., Ruedy, R., Lo, K., Lea, D. W., & Medina-Elizade, M. (2006). Global temperature change. *Proceedings of the National Academy of Sciences of the United States of America*, 103(39), 14288–14293. DOI: 10.1073/pnas.0606291103 PMID: 17001018

Holling, C. S. (1973). Resilience and Stability Of Ecological Systems. *Annual Review of Ecology and Systematics*, 4(1), 1–23. DOI: 10.1146/annurev.es.04.110173.000245

Hollnagel, E., Woods, D. D., & Leveson, N. (2007). *Resilience Engineering: Concepts and Precepts*. Ashgate Publishing, Ltd.

Kruk, M. E., Ling, E. J., Bitton, A., Cammett, M., Cavanaugh, K., Chopra, M., El-Jardali, F., Macauley, R. J., Muraguri, M. K., Konuma, S., Marten, R., Martineau, F., Myers, M., Rasanathan, K., Ruelas, E., Soucat, A., Sugihantono, A., & Warnken, H. (2017). Building Resilient Health Systems: A Proposal For A Resilience Index. *BMJ (Clinical Research Ed.)*, •••, J2323. DOI: 10.1136/bmj.j2323 PMID: 28536191

Kruk, M. E., Myers, M., Varpilah, S. T., & Dahn, B. T. (2015). What Is A Resilient Health System? Lessons From Ebola. *Lancet*, 385(9980), 1910–1912. DOI: 10.1016/S0140-6736(15)60755-3 PMID: 25987159

Lebel, L., Anderies, J. M., Campbell, B., Folke, C., Hatfield-Dodds, S., Hughes, T. P., & Wilson, J. (2006). Governance And The Capacity To Manage Resilience In Regional Social-Ecological Systems. *Ecology and Society*, 11(1), 11. DOI: 10.5751/ES-01606-110119

Mosadeghrad, A. M. (2022). A practical model for health policy making and analysis. *Payesh*, 21(1), 7–24. DOI: 10.52547/payesh.21.1.7

Mosadeghrad, A.M., Isfahani, P., Eslambolchi, L., Zahmatkesh, M., Afshari, M. (2023). Strategies to strengthen a climate-resilient health system: a scoping review. *Global Health*, 28;19(1):62.

Mosadeghrad, A. M., & Shahsavani, A. (2023). Strengthening the resilience of Iran's health system against climate change. *Payesh*, 22(4), 479–482. DOI: 10.61186/payesh.22.4.479

Mullins, J. T., & White, C. (2019). Temperature and Mental Health: Evidence from the Spectrum of Mental Health Outcomes. *Journal of Health Economics*, 68, 102240. DOI: 10.1016/j.jhealeco.2019.102240 PMID: 31590065

Neufeldt, H., Christiansen, L., & Dale, T. W. (2020). *Adaptation Gap Report 2020*. United Nations Environment Programme.

O'brien, K. (2012). Global Environmental Change Ii: From Adaptation to Deliberate Transformation. *Progress in Human Geography*, 36(5), 667–676. DOI: 10.1177/0309132511425767

Obrist, B., Iteba, N., Lengeler, C., Makemba, A., Mshana, C., Nathan, R., Alba, S., Dillip, A., Hetzel, M. W., Mayumana, I., Schulze, A., & Mshinda, H. (2007). Access To Health Care In Contexts Of Livelihood Insecurity: A Framework For Analysis And Action. *PLoS Medicine*, 4(10), 4. DOI: 10.1371/journal.pmed.0040308 PMID: 17958467

Osborn, D., Cutter, A., & Ullah, F. (2015). Universal Sustainable Development Goals. *Underst Transformational Chall Developed Ctries*, 2(1), 1–25.

Philipsborn, R. P., Shefield, P., White, A., Osta, A., Anderson, M. S., & Bernstein, A. (2021). Climate change and the practice of medicine: Essentials for resident education. *Academic Medicine*, 96(3), 355–367. DOI: 10.1097/ACM.0000000000003719 PMID: 32910006

Prats, E. V. (2020). WHO guidance for climate-resilient and environmentally sustainable health care facilities. www.who.int

Rahaman, M. A., Rahman, M. M., & Rahman, S. H. (2019). Pathways of climate-resilient health systems in Bangladesh. Confronting Climate Change in Bangladesh. *Adaptation and Resilience*, 3, 119–143.

Schnitter, R., Verret, M., Berry, P., Chung, T., Fook, T., Hales, S., Lal, A., & Edwards, S. (2019). An assessment of climate change and health vulnerability and adaptation in Dominica. *International Journal of Environmental Research and Public Health*, 16(1), 70. DOI: 10.3390/ijerph16010070 PMID: 30597870

Sen, B., Dhimal, M., Latheef, A. T., & Ghosh, U. (2017). Climate change: Health effects and response in South Asia. *BMJ (Clinical Research Ed.)*, 5, 359. DOI: 10.1136/bmj.j5117 PMID: 29146578

The Lancet. (2015). *Health Professionals: Be Prepared for Heatwaves.*www .thelancet.org

Tochkin, J., Richmond, J., & Hertelendy, A. (2022). Healthcare system leadership and climate change: Five lessons for improving health systems resiliency. *BMJ Leader*, 4(7), 21–32. PMID: 37013877

Walker, B., Holling, C. S., Carpenter, S. R., & Kinzig, A. (2004). Resilience, Adaptability, and Transformability In Social–Ecological Systems. *Ecology and Society*, 9(6), 15–28. DOI: 10.5751/ES-00650-090205

WHO. (2007). *Everybody's business–strengthening Health Systems to Improve Health Outcomes: WHO's Framework for Action.*www.who.int

WHO. (2015). *Operational framework for building climate resilient health systems.* www.who.int

WHO. (2022). *Health and climate change: country profile 2022: Czechia.*www .who.int

World Health Organization (WHO). (2018). *COP24 special report: health and climate change*. www.who.int

ADDITIONAL READINGS

Ahmed, B., Kelman, I., Fehr, H. K., & Saha, M. (2016). Community resilience to cyclone disasters in coastal Bangladesh. *Sustainability (Basel)*, 8(805), 1–29. DOI: 10.3390/su8080805

Bi, P., Shi, X.-M., & Liu, Q.-Y. (2020). Climate Change and Population Health Research in China: Knowledge Gaps and Further Directions. *Advances in Climate Change Research*, 11(3), 273–278. DOI: 10.1016/j.accre.2020.07.001

Cai, H., Lam, N. S., Zou, L., Qiang, Y., & Li, K. (2016). Assessing community resilience to coastal hazards in the Lower Mississippi River Basin. *Water (Basel)*, 8(2), 1–18. DOI: 10.3390/w8020046

Hess, J. J., McDowell, J. Z., & Luber, G. (2012). Integrating climate change adaptation into public health practice: Using adaptive management to increase adaptive capacity and build resilience. *Environmental Health Perspectives*, 120(2), 171–179. DOI: 10.1289/ehp.1103515 PMID: 21997387

Kais, S. M., & Islam, M. S. (2016). Community capitals as community resilience to climate change: Conceptual connections. *International Journal of Environmental Research and Public Health*, 13(12), 1–16. DOI: 10.3390/ijerph13121211 PMID: 27929448

Onwutuebe, C. J. (2019). Patriarchy and Women's Vulnerability to Adverse Climate Change in Nigeria. *Climate Change Original Research*, 9(1), 1–7.

Springmann, M., Mason-D'Croz, D., Robinson, S., Garnett, T., Godfray, H. C., Gollin, D., Rayner, M., Ballon, P., & Scarborough, P. (2016). Global and regional health effects of future food production under climate change: A modeling study. *Lancet*, 387(10031), 1937–1946. DOI: 10.1016/S0140-6736(15)01156-3 PMID: 26947322

KEY TERMS AND DEFINITIONS

Adaptation-: adjustment in natural or human systems in response to actual or expected climatic stimuli or their effects, which moderates harm or exploits beneficial opportunities."

Adaptive capacity-: is the ability of a system to adjust to climate change, to moderate potential damages, to take advantage of opportunities, or to cope with the consequences.

Climate-: the typical or average weather for an area.

Climate change-: a change in the state of the climate that can be identified by changes in the mean and/or the variability of its properties and that persists for an extended period, typically decades or longer.

Climate resilient health system-: is capable of anticipating, responding to, coping with, recovering from, and adapting to climate-related shocks and stress, to bring sustained improvements in population health, despite an unstable climate.

Health system-: comprises all the organizations, institutions, and resources that are devoted to produce actions principally aimed at improving, maintaining or restoring health.

Resilience- a system's ability to absorb change and disturbance and still maintain the same relationships between populations or state variables,' which has influenced other disciplines to embrace the concept of resilience as the capacity of individuals: populations, or systems to absorb a shock whilst retaining functions of its original state.

Vulnerability-: is the degree to which individuals and systems are susceptible to or unable to cope with adverse effects of climate change including climate variability and extreme

Chapter 3
Mental Health Risks Due to Climate Change:
Eco-Anxiety Amidst the Challenges of Climate Change

Anchal Bhaskaran
https://orcid.org/0009-0000-9724-2660
PES University, India

Srishti Muralidharan
PES University, India

ABSTRACT

Eco-anxiety encompasses concerns not only for one's immediate survival but also for the wellbeing of future generations. Media, societal interpretations, and personal experiences are identified as some of the amplifiers of eco-anxiety, contributing to a broader understanding of its factors. Emphasizing both the negative and positive effects of eco-anxiety indicates to the urgency of addressing individual distress within the broader context of climate action. Examining the impact of eco-anxiety, or climate distress, on mental health is essential due to its prevalent occurrence worldwide and the considerable adverse effects it imposes on individuals' functioning. Eco anxiety is rising daily with the increase in the population's concern for climate change proving to leave drastic and dramatic environmental changes.

DOI: 10.4018/979-8-3693-4183-4.ch003

INTRODUCTION: IMPACT OF CLIMATE CHANGE AND ITS INFLUENCE ON ECO-ANXIETY

Eco-anxiety originated with Glenn Albrecht. According to the definition given by the author, it is “a fear of environmental catastrophe”. A prominent report published by Eco America and the American Psychological Association (APA) in 2017 describes eco-anxiety as “a fear of environmental cataclysm that arises out of the observation of the seemingly irreversible effects of climate change as well as the accompanying concern for one's future and the future generations” (Clayton et al., 2017, p.68; Iberdrola, 2023). Solastalgia is a form of emotional or existential distress caused by environmental change (Wikipedia contributors, 2024a).

Eco-anxiety, which is short for the term ecological anxiety and sometimes known as climate anxiety and is also recognized as climate distress, refers to an ongoing apprehension about environmental catastrophes resulting from the acknowledgment of the inevitable consequences of global effects of climate change. Eco-anxiety is a challenging emotional response to the effects of climate change and other environmental concerns. Eco-anxiety is not considered to be a medical diagnosis yet but due to the rising trends in fear of climate change, there is a chance of eco-anxiety being a mental health concern globally. With climate change, disasters are more likely to occur more frequently and can lead to disorders associated with post-traumatic stress, adjustment disorders, depression, etc. (Lawrance et al., 2022). Padhy et al (2015) says, “Increased exposure to heat is likely to become more common with global temperatures. Various suggestions have been made regarding the relationship that exists between temperature rise and aggressive behavior”. Climate change affects eco-anxiety in an individual through severe weather events and occurences that last for quite a while, natural disasters like heat waves, cyclones, hurricanes, increases in pollution, etc. (Palinkas & Wong, 2020).

It is commonly seen that how people engage and interact with each other is affected by the difference in their climate in many instances. For example, in regions experiencing prolonged droughts, climate-induced migration can reshape social dynamics. Research in parts of Africa, where droughts impact agriculture, highlights that migration becomes a coping strategy, influencing social interactions as displaced individuals integrate into new communities (Gray & Mueller, 2012). This migration-driven interaction can lead to both cultural exchange and challenges in social integration. In other cases, stress from a changing environment, eco-migration, and/or eco-anxiety may affect a community's mental well-being through a loss of social identity and cohesion, hostility, violence, and interpersonal violence (Clayton et al., 2017).

In terms of the positive consequences in contrast to the large data of negative consequences concerning eco-anxiety, people suffering from serious anxiety may find it difficult to maintain a healthy lifestyle, both psychologically and physically. Still, the simple reality that the masses are experiencing eco-anxiety can pose a clue to the very clear indication that individuals are becoming more conscious as well as understand and accept the seriousness of the hazards that climate change pose to our future (Pikhala, 2020).

According to Clayton, Manning, Krygsman, and Speiser (2017), some emotional responses are natural, and even unpleasant feelings are essential to living a full life. In severe circumstances, they can impair our capacity to think sensibly, organize our activities, and consider alternate options. A severe weather occurrence can be traumatic, resulting in crippling feelings. More subtle and indirect consequences of climate change might cause stress in people's lives to varied degrees. Although everyone can handle a certain amount of stress, the cumulative effects of compound stress can push a person from mentally healthy to mental illness. Hearing about other people's terrible experiences, as well as anxieties about their possible vulnerability, may have a detrimental impact on people (Clayton et al., 2017).

It is commonly recognized and known that uncertainty is a major contributor to anxiety in such a way that, the uncomfortable feeling of not knowing something that is of importance to you or not knowing something that could potentially be dangerous to you, fills the person up with dread and panic which switches off their rational thinking and logical reasoning, making them go into a state of survival mode in most of extreme scenarios. Individuals who pursue to attain and control their lives to the degree and extent of perfection are more at a higher risk of the dangers of such phenomena known as eco-anxiety as the uncertainty of environmental changes such as global warming due to climate change leads to uncertainty that could skew the hold individuals have over their daily routine and functioning because of the uncontrollability of the changing environment that is considered a threat to their survival and not normal due to societal and individual norms. Eco-anxiety has ties with other emotions as well, and researchers have been trying to understand the link between feelings, emotions, and climate change that a term exists that describes it and is known as “ecological emotions” (Vakoch and Mickey, 2022).

Climate change can increase eco-anxiety in several ways, reflecting both direct and indirect impacts on individuals and communities. Frequent news coverage and scientific reports on the accelerating pace of climate change heighten awareness and concern about the future of the planet, contributing to feelings of anxiety and helplessness. Personal or community experiences of extreme weather events, such as hurricanes, floods, and wildfires, lead to immediate and long-term psychological distress, exacerbating eco-anxiety.

The perception that climate change poses a direct threat to personal safety, health, and livelihood intensifies feelings of anxiety. Concerns about food security, water scarcity, and displacement play significant roles in this. The unpredictable nature of climate change impacts and the uncertainty about the future also lead to existential anxiety, including worries about the long-term viability of ecosystems and the well-being of future generations. Witnessing the decline of biodiversity and the degradation of natural landscapes evokes feelings of grief, loss, and helplessness, contributing further to eco-anxiety.

Understanding the global interconnectedness of climate change impacts amplifies anxiety as individuals realize the widespread and far-reaching consequences of environmental degradation. Awareness of the social and economic disruptions caused by climate change, such as increased inequality, resource conflicts, and migration, heightens anxiety about societal stability and justice. Perceived inadequacies in governmental and organizational responses to climate change foster feelings of frustration, helplessness, and anxiety, especially when the scale of the problem appears to far exceed the actions being taken.

Ethical concerns about the impact of human activities on the planet and the responsibility to protect the environment lead to moral distress and heightened eco-anxiety, particularly among those with a strong sense of environmental stewardship. Social and cultural contexts also shape how individuals perceive and respond to climate change, with communities that have strong environmental values potentially experiencing higher levels of eco-anxiety if they feel their collective efforts are insufficient. Addressing eco-anxiety requires comprehensive strategies, including promoting mental health and resilience, enhancing public understanding of climate change, fostering community support, and advocating for effective climate action.

Studies on eco-anxiety and climate change utilize a combination of quantitative, qualitative, and mixed-methods approaches to gain a comprehensive understanding of the phenomenon. Quantitative methodologies, such as large-scale surveys and psychometric assessments, are essential for assessing the prevalence and intensity of eco-anxiety across diverse populations. These tools enable researchers to collect demographic data and analyze correlations, revealing differences in eco-anxiety levels among various groups. Longitudinal studies are particularly valuable as they track changes in eco-anxiety over time, helping to identify trends and long-term effects influenced by individual experiences and external factors like climate events and media coverage.

Qualitative methodologies, including in-depth interviews and focus group discussions, provide rich, detailed data about personal experiences, coping strategies, and emotional responses to eco-anxiety. These approaches allow participants to express their thoughts and feelings freely, offering deep insights into the phenomenon that numerical data alone cannot capture. Content analysis of media and public discourse further complements this by revealing how eco-anxiety is represented and discussed in the public sphere, highlighting the narratives that shape public perceptions and emotional responses to climate change.

Mixed-methods approaches, which integrate quantitative and qualitative data, offer a holistic understanding of eco-anxiety by combining numerical trends with personal experiences. Designs like the convergent parallel, explanatory sequential, and exploratory sequential approaches ensure a comprehensive analysis. Additionally, experimental and intervention studies, such as randomized controlled trials and pilot studies, evaluate the effectiveness of specific treatments in reducing eco-anxiety, providing evidence-based strategies for addressing the issue. Cross-cultural and comparative studies expand this understanding globally, identifying cultural factors that influence eco-anxiety and highlighting practical implications for localized interventions. These diverse methodologies collectively enable researchers to explore the multifaceted nature of eco-anxiety, its causes, manifestations, and potential solutions.

For a thorough grasp and understanding of what eco-anxiety is let us look at a scenario that displays what it is like. Jessica, a university student, has been feeling increasingly worried, stressed, and overwhelmed by the state of the environment. She often reads news articles and studies about climate change and its devastating effects on ecosystems and communities worldwide. Recently, she came across a report highlighting the rapid decline of polar ice caps and the subsequent rise in sea levels, which could pose as a threat to coastal cities and ecosystems. As Jessica learns more about these environmental challenges, she becomes deeply concerned about the future of the planet. She worries about the impact on future generations and feels a sense of powerlessness in the face of such vast and complex issues. Every time she sees images of wildfires, floods, or species extinction, her anxiety intensifies, and she finds it difficult to focus on her studies or daily activities. She experiences feelings of remorse and guilt for not doing or accomplishing enough to mitigate environmental harm and wonders if individual actions can make a meaningful difference. Despite her efforts to reduce her carbon footprint and advocate for environmental causes, she feels overwhelmed by the scale of the problem and uncertain about the future. This is what makes climate anxiety very difficult for many people. Living in the middle of climatic disasters for the rest of one's life can be difficult, leading to feelings of powerlessness and uncertainty about one's future despite measures taken. Living amidst climate crises leads to uncertainty about their future and the feeling of helplessness regardless of what steps they may take to change.

ECO-ANXIETY AND MENTAL HEALTH

People of all ages, from children to young adults have been experiencing eco-anxiety more and more over the past few years because of the active participation of climate change and environmental activists in showing the world the massive consequences that global warming in particular has brought along on the onset of climate change among other environmental phenomena and changes. Together with the media's representation of these changes which are to a large extent exorbitant, play into the people's fear of what the future could potentially hold for themselves and their future generations and offspring. Physical health is also something that will affected by eco-anxiety as psychological and physiological issues go hand in hand. Physical health will be affected by phenomena such as heat waves, which in some instances raise the aggression levels of an individual.

During the COVID-19 pandemic, the majority of individuals experienced at least one of the following: anxiety, depression, chronic stress, sadness, anger, loneliness, etc. It left a lasting imprint on a person that resulted in them sustaining trauma. Before climate change or global warming was considered an international threat, extreme weather events were seen as a rare occurrence that happened every once in a while in a few years, and the subsequent survivors could recover and rebuild without much difficulty later on. With the rapid succession of climate change, communities are experiencing recurring disaster events from which it is difficult to recover (Ecoanxiety Is a Growing Mental Health Crisis, 2022). As such climate change and environmental degradation are some of the most serious issues facing the world today (Lutz et al., 2023).

Other Ecological Emotions

Emotionally, eco-anxiety manifests as feelings of sadness, anger, frustration, or numbness, as individuals struggle to understand as well as come to terms with the enormity of environmental challenges and their stake in the planet's future. In a paper released by Stanley et al. (2021), they delved into the emotions experienced according to emotional distress which was eco-anger and eco-depression that showed different levels of responses to climate change. Generally, depression is a deactivating emotion while anger is activating which means that depression induces the inability to motivate any form of action and anger induces the ability to approach. Concerning this, general anxiety is an activating emotion-provoking avoidance of anything that is considered to be of danger or threat. Eco-grief is also another term that is related to ecological emotions and can be presumed that much of eco-anxiety is commonly associated with eco-grief as well (Pikhala, 202). Eco-grief encompasses feelings of sadness, mourning, and distress stemming from the recognition and awareness of ecological losses due to climate change.

Individuals experiencing eco-grief mourn the loss of iconic landscapes, biodiversity, and natural environments that are directly impacted by climate-related events such as hurricanes, wildfires, or coral bleaching.

The psychological impact of eco-anxiety can be profoundly different and varies for everyone, affecting them across distinctive age groups, cultures, countries, and socio-economic backgrounds. For example, a systematic scoping review of the literature published by Coffey et al. (2021) included various studies done on eco-anxiety and concluded regarding their findings which were, indigenous groups and those that care more and are connected to the environment were most impacted by eco-anxiety. A growing body of research indicates that children and young people are more likely to experience emotional and psychological impacts due to climate change.

Cognitive symptoms such as rumination, intrusive thoughts, and negative thinking may dominate the mental landscape of those experiencing eco-anxiety, leading to difficulties in concentration, decision-making, and problem-solving.

Eco-anxiety is often accompanied by physiological responses such as elevated heart rate, muscle tension/strain, and changes in sleep patterns, reflecting the body's heightened state of arousal in the face of perceived threats. Chronic stress associated with eco-anxiety can compromise immune function, disrupt hormonal balance, and increase the risk of stress-related health conditions such as cardiovascular disease, depression, and anxiety disorders.

Eco-anxiety can also have significant interpersonal and social implications, straining relationships, undermining social stability, and diminishing trust in health and environmental institutions that are placed to help protect the environment and the climate. Individuals may experience conflicts and disagreements among friends, family members, or colleagues who hold opposing views on environmental issues, leading to feelings of isolation, alienation, or social withdrawal. Positively, eco-anxiety may fuel activism, advocacy, and collective action as individuals seek to channel their concerns into meaningful efforts to address environmental challenges and promote sustainability.

Managing eco-anxiety

Although eco-anxiety is not regarded as a mental disorder it is possible for it to soon be labeled as such due to the rise in symptoms of eco-anxiety being of concern. Severe forms of eco-anxiety can clash with clinical mental disorders and may require professional treatment (Wang et al., 2023). Media coverage of climate-related disruption can induce eco-anxiety and as such current messages displayed and viewed by the audience are not pleasant and quite gloomy with a sense of doom etched into every image shown. Strategic message framing can used as a coping tool to combat eco-anxiety and needs to be implemented on news coverage channels in a way that appeals to the audience to take action in a positive manner rather than to feel as if nothing can be done. It builds a feeling and sense of self-determination rather than self-pity and destruction.

Instead of trying to undermine or minimize an individual's feelings or concerns, it is more important to not depreciate these feelings and embrace these emotions in a way that increases and makes them more meaningful and valued. An intervention method that incorporates this is mindfulness as it involves learning to live with, tolerate, and accept your feelings so that you can move forward positively. As a result, you are not trying to change your feelings, but rather, you are learning to manage the distress that climate change can result in (Heid, 2022). Often, anxiety is associated with feelings of uncertainty or a lack of control. Engaging in the fight against climate change is an excellent way to alleviate these feelings. Getting involved would include movements like volunteering in local communities and by also becoming media active to spread the message as well as working with non-profit organizations. These ventures could help remove the feeling of helplessness. Engaging and interacting with those who share the same feelings as you who are intensely upset about climate change reduces feelings of loneliness.

In a scoping review published by Bauden and Jachens (2021), one of the main themes explored concerning eco-anxiety was "fostering clients" inner resilience, and the cognitive intervention involved "Shifting from catastrophizing towards a less black-and-white picture". This type of intervention focuses on helping an individual to broaden their narrowed view of thinking about climate-related issues and to embrace the ideation that not just distressful events happen but good affairs can also coexist alongside.

Resilience is a very crucial aspect of the ability to respond to the torrent of climate change information. Related terms are psychological and emotional resilience (Coppola, 2021).

Psychological and emotional resilience is the ability to cope mentally and emotionally with a crisis and to withstand any stress that arises and to recover smoothly (Wikipedia contributors, 2024b). Mindfulness is a very essential factor in cultivating psychological and emotional resilience. Selfefficacy could be included as one of the aspects of an individual that can be improved to combat eco-anxiety. The concept of self-efficacy refers to a person's belief that they will be able to succeed in a given situation or accomplish a given task (Clayton et. al., 2017).

Many individuals succumb to eco-anxiety because of a lack of self-efficacy. This can be improved through forming strategies such as implementing achievable goals, and regular public discussions with peers of alike interests, especially on the interpersonal level, which increases self-efficacy and confidence in oneself. Knowledge-based interventions can be very useful in terms of managing eco-anxiety. It promotes engagement and informs the person that not all is lost and with the confidence of understanding what is occurring with the environment and climate change, the individual's self-efficacy improves drastically and their feelings of helplessness decrease. But overconsumption of information is also not helpful because it might lead to the person feeling overwhelmed with the burden of "saving the world" from doom.

As their coping capacities are still developing, young people are more susceptible and vulnerable to climate change's negative effects. In an interview-based study conducted by Coppola (2021), the findings suggested that the main reason participants were motivated to be active in environmental and climate change movements was that they felt a sense of moral duty or responsibility. Coppola (2021) emphasized that "Action through involvement with environmental organizations appears to serve as an antidote for eco-anxiety among participants through the cultivation of a sense of community, and the cultivation of efficacy (self and community)".

EFFECTS OF ECO-ANXIETY ON BEHAVIOR AND COGNITION

Eco-anxiety has various effects on behavior and cognition which influences multiple aspects of an individual's life. It can also have significant implications on an individual and society as a whole.

Behavioral Responses

Eco-anxiety can lead individuals to engage themselves in multiple and various behavioral responses. These can be both negative and positive. Some of the positive behavioral responses may include, adopting sustainable lifestyle practices, and being involved in environmental activism, and some of the negative aspects of behavioral responses include experiencing feelings of helplessness and avoidance (Pikhala, 2020). It is possible for individuals to experience practical anxiety as a response to eco-anxiety, which can result in the gathering of new information and reevaluating behavioral options (Pikhala, 2020).

On the contrary, some individuals may cope with eco-anxiety by avoiding or denying information about environmental problems. They may disengage from discussions about climate change or actively avoid news and media coverage related to environmental issues to reduce their feelings of distress. Eco-anxiety can also affect interpersonal relationships. Individuals may experience conflict or tension with friends, family members, or colleagues who hold differing views on environmental issues. This can lead to strained relationships and social isolation. Eco-anxiety can change the way individuals purchase items. It can influence individuals' consumption patterns, hence, leading to changes in purchasing behavior. Those experiencing eco-anxiety may prioritize sustainable and ethically sourced products, opting for environmentally friendly alternatives even if they are more expensive. They may also reduce consumption overall, practicing minimalism to lessen their environmental footprint.

Cognitive Responses

Eco-anxiety can impact cognitive processes, leading to increased levels of stress, worry, and fear related to environmental concerns. The ecological crisis, including the climate crisis, can cause feelings of uncertainty, unpredictability, and uncontrollability, which are classic ingredients in anxiety and can affect cognitive functioning. Eco-anxiety may manifest as existential anxiety, prompting individuals to grapple with deep questions related to meaning, guilt, and mortality in the face of environmental challenges (Pikhala, 2020). Healthcare professionals and other professionals need to be knowledgeable about the various forms of eco-anxiety to respond adequately to the mental health impacts of the ecological and climate crises. Understanding the role of personality traits, anxiety sensitivity, and other factors in eco-anxiety is crucial for designing effective healthcare practices and interventions for individuals experiencing ecoanxiety.

Individuals experiencing eco-anxiety may have persistent and intrusive thoughts about environmental issues, leading to rumination and preoccupation with negative outcomes. Ecoanxiety can also impair concentration and focus as individuals may find it challenging to divert their attention from environmental concerns to other tasks. A new variation of eco-anxiety is observed where while some individuals may be spurred into action by eco-anxiety some few individuals become overwhelmed and anxious and this anxiety is so intense that they "become paralyzed and cannot act" (Usher et al., 2019). Such a phenomenon is known as eco-paralysis and was coined by Glenn Albrecht. People with eco-anxiety may engage in catastrophic thinking, imagining worst-case scenarios connected to climate change and its consequences, which can contribute to heightened anxiety levels. In most of the scenarios observed, eco-anxiety influences decision-making processes which may lead the individual to make decisions and choices based on fear and distress about the future rather than to think rationally and logically about the risks and benefits. The cognitive load of eco-anxiety can affect memory functions, leading to difficulties in recalling information or events due to heightened stress and worry about environmental issues. Eco-anxiety is associated with symptoms of depression, anxiety, stress, insomnia, cognitiveemotional impairment, and reluctance to have children, primarily among younger generations, women, and poorer low-income countries (Boluda-Verdu et al., 2022). These cognitive responses can contribute to the overall distress and leave its influence and impact on mental well-being experienced by individuals grappling with eco-anxiety in the light of climate change. In summary, eco-anxiety can have profound effects on both behavior and cognition, influencing how individuals respond to environmental challenges and navigate their emotional and psychological well-being in the face of ecological crises.

CHALLENGES OF ADAPTATION AND CLIMATE CHANGE MITIGATION MEASURES OR SOLUTIONS

As climate change gets worse, it is very important to take action to adapt to and lessen its effects. But it is not easy – there are a lot of different challenges that make it hard to respond effectively. These challenges make people and communities more worried about the environment, which we call eco-anxiety. One of the foremost challenges in implementing adaptation and climate change mitigation measures is the scarcity of resources and inadequate infrastructure, particularly in vulnerable regions. Developing countries, in particular, often lack the financial means and technological capabilities to undertake large-scale initiatives to reduce carbon emissions or fortify infrastructure against climate-related hazards. As a result, marginalized communities are disproportionately affected, further widening existing socio-economic disadvantages and heightening eco-anxiety. In situations where people lose their homes, businesses, and valuables or potentially witness the loss of life, the resulting impact on their mental health is most likely and expected to be traumatic (Arcanjo, 2019).

Some new technologies meant to help assist in addressing climate change, like capturing carbon emissions or manipulating the environment, are not proven to work well yet as they carry inherent risks and ethical considerations. People are cautious about utilizing them since they may have risks and raise ethical questions. Also, we are not yet certain how climate change will affect different places, so it is hard to plan how to adapt. This uncertainty makes people concerned about what may happen in the future. Additionally, scientific uncertainties regarding the precise impacts of climate change on different regions and ecosystems pose challenges for formulating targeted adaptation strategies, leading to anxiety about the unknown future.

Post-disaster counseling for victims of environmental disasters has the potential to limit the intensity of eco-anxiety even though it is an expensive program to consider (Arcanjo, 2019). Being evacuated before the event of a disaster occurring because of climate change can greatly reduce most of the symptoms of eco-anxiety in a great deal. However, most people are unwilling to leave or evacuate their homes until a disaster has struck because of fear of losing their valuables. Changes associated with climate change are most likely to increase migration along with conflict between nations. People will be forced to leave their destroyed and desolate homes due to many climate change-induced factors such as melting glaciers, rise in sea level, droughts, etc. All this will make it almost unbearable to remain or stay any longer. There is, however, the possibility of migration causing great stress, especially when it involves crossing borders, the journey itself is uncertain and the process of adjusting to a new home is challenging, especially if the existing residents are not welcoming.

Climate change activism can be viewed as a form of adaptation, as opposed to just coping with change (Arcanjo, 2019). Increasing the number of trips taken on bicycles in towns, reducing the number of flights, and taking more trips by train or shared car are all ways to promote public transportation and sustainable mobility. This can be taken as a mitigation measure to reduce the feeling of helplessness and loss of control over how climate change will impact our lives. It is imperative that international mitigation actions be taken quickly in order to reduce energy demand and transition to green energy.

While policy and systematic change are the only means of addressing the overall issue, practitioners of mental health are also required to take concrete actions (Rose, 2023). In order to adapt psychologically, it is necessary to develop coping skills and engage in behavioral and psychological activities.

Inefficient waste management practices, including landfilling and incineration, contribute to methane emissions and air pollution. Mitigation measures in the waste management sector include promoting recycling and composting to remove organic waste from landfills and reduce food waste through improved storage and distribution practices, and implementing biogas capture systems to capture methane emissions from organic waste. Implementing waste management strategies that promote and prioritize recycling, composting, and methane capture, can lead to environmental benefits, such as reduced greenhouse gas emissions and improved air quality. Knowing that their actions are making a positive difference in mitigating climate change and reducing pollution can provide individuals with a sense of purpose and hope, counteracting feelings of despair and anxiety about the future. Engaging and participating in a community-based waste management program, such as neighborhood clean-up events or composting cooperatives, can foster a sense of belonging and social connection among individuals. Engaging with like-minded individuals who share similar environmental concerns can provide valuable social support networks, allowing individuals to share experiences, exchange ideas, and collectively work towards common goals, thereby reducing feelings of isolation. Policymakers and governments play a very important role in mitigation and adaptation towards the changes brought by climate change. Implementing carbon pricing mechanisms, such as carbon taxes or cap-and-trade systems, can incentivize emissions reductions and promote investment in low-carbon technologies. Additionally, setting ambitious emissions reduction targets and providing financial incentives and regulatory assistance for renewable energy deployment can accelerate the transition to a low-carbon economy.

Raising awareness about the environmental impacts of waste management practices and the significance of adopting sustainable alternatives can help individuals make educated choices, and informed decisions and take meaningful action to address eco-anxiety. Providing education and resources on waste reduction strategies, such as recycling guidelines and composting tutorials, can empower individuals with the knowledge and skills needed to make positive changes in their daily lives, reducing feelings of uncertainty and anxiety about the future of the planet.

Burning fossil fuels has put our planet on a dangerous path towards climate change that could have extremely negative consequences. As a result, the Paris Agreement is a legally enforceable international treaty which brings together nations across the globe to acknowledge and address climate change collectively (O'Brien & Elders, 2022). This agreement represents a landmark in the climate change process, as it brings together all nations for the first time for the purpose of combating and adapting to the effects of climate change (United Nations, n.d.). Around the world, we are witnessing health impacts ranging from reduced water quality and availability, and respiratory disorders resulting from air pollution, to direct effects of extreme heat, depletion of the ozone layer, and fires (O'Brien & Elders, 2022).

CASE STUDIES HIGHLIGHTING MENTAL HEALTH IMPACTS OF ECO-ANXIETY

Anxiety about climate change is experienced by a wide range of individuals. The symptoms of this condition are similar to those of accepted anxiety disorders, including subjective distress, rumination, sleep disturbance, somatic disturbance, negative affect, and hopelessness (O'Brien & Elders, 2022). Climate change has emerged as a global crisis, affecting not only ecosystems and biodiversity but also human health and well-being. In recent years, the mental health impacts of eco-anxiety have come to the forefront, prompting researchers and mental health professionals to delve deeper into comprehending its complexities and implications.

Four case studies will be presented to explore and illuminate the diverse manifestations of ecoanxiety and its profound effects on mental health. As a preview, the first case study talks about how rising temperatures and extreme changes in the climate affect the farmers of India's mental state and drive them to the brink of suicide due to their primary source of livelihood being compromised by things that are out of their control and by external factors apart from climate change like dept, loan sharks, low crop yield, no water for irrigation, etc.

Case Study 1: Nearly 60,000 Indian Farmers' Suicides are Connected to Climate Change

Rising temperatures and severe weather events have been associated with indirect impacts on suicide rates. Particularly in regions where agriculture is the primary source of livelihood, the correlation between climate-related stressors and suicide is pronounced. India has experienced a notable surge in suicide cases, underscoring the urgency of addressing the mental health implications of climate change. Climate change may have resulted in the suicides of nearly 60,000 Indian farmers and farm laborers over the past three decades, according to a study done by Carleton (2017), which examines that, Rising temperatures are already having an impact on vulnerable societies.

This case study looks into a research paper that explores the profound impact of climate variability on suicide rates in India. The study sheds light on the relationship between changing climatic conditions, agricultural outcomes, and human well-being, particularly in a country where agriculture is a cornerstone of the economy.

The study utilizes data from India's National Crime Records Bureau spanning from 1967 to 2013, coupled with agricultural crop yield information and high-resolution climate data. By analyzing the relationship between suicide rates, temperature, and rainfall across different states and union territories in India, the research aims to identify the mechanisms by which climatic fluctuations influence suicide rates. The methodology employed in Carleton's (2017) research allows for a clear understanding of the complex interplay between climate, agriculture, and mental health outcomes. The findings of the study reveal a very significant effect of climate variability on suicide rates in India. Specifically, the research highlights the role of temperature fluctuations during the agricultural growing season in exacerbating suicide rates. The study underscores the economic mechanism through which climate impacts are manifested, emphasizing the link between crop yields, income stability, and mental health outcomes. Furthermore, the research indicates an absence or lack of adaptive behavior over the years, suggesting that there is an urgent need to take proactive steps to solve and address the escalating challenges posed by climate change. Farmers tend to borrow money due to poor harvests which are because of variations in the climate that the farmers were not able to predict and calculate. However, climate fluctuations can cause multiple failed years, resulting in a cycle of debt. Their inability to sustain a livelihood, combined with growing debt, leads many farmers to take their own lives. For example, The Hindustan Times interviewed the wife of a farmer who took his own life and stated that her late spouse owed the bank 80,000 rupees, and unable to pay such a high debt, consumed a toxic potion outside the bank branch, and died on the post.

The Hindustan Times (Safi, 2017) identified 1309 suicides in three years in a traditionally prosperous farming region in the Punjab. While it is difficult to establish temperature as a direct cause, many are pointing to it as a major contributing factor. As a result of global warming, it is estimated that 59,300 suicides have occurred in the agricultural sector over the past 30 years.

In conclusion, this case study emphasizes the critical link between climate variability and suicide rates in India and lays stresses on the need for proactive measures to mitigate the adverse effects of climate change on mental health outcomes. Behind each statistic lies a story of human suffering, desperation, and loss, reflecting the profound impact of climate change eco-anxiety on the lives and livelihoods of vulnerable communities. The suicides of Indian farmers although directly linked to rising temperatures are indirectly affected by eco-anxiety. Climate change and environmental degradation instill feelings of fear, helplessness, and hopelessness in communities that entirely depend on farming and the weather for their livelihood. The psychological burden of eco-anxiety extends beyond immediate concerns about food security and economic stability, permeating every aspect of daily life and shaping individual and collective responses to environmental challenges.

Case Study 2: Exploring Eco-Anxiety Through Psychodrama

The case study was taken by Raile P (2023) due to their work on the relation between psychodrama and eco-anxiety. Psychodrama is a structured form of therapy in which a person dramatizes a personal problem or conflict, usually in front of a group of other therapy participants. The other participants usually take part in the drama, though each performance focuses on a single person's concerns (Villines, 2022). Psychodrama aims to foster empathy, help a person better understand themselves and how they view themselves, support the person, and help a person "resolve psychological issues from trauma and relationship conflicts" (Villines, 2022). Psychodrama can be taken as a form of a larger treatment plan or only catered solely for the individual seeking the treatment. Usually, psychodrama occurs in three steps according to Villines's (2022) article. These are warm-up, action, and sharing. During warm-up, the therapist works with the client to reflect on the goals for the session and to start playing and forming the drama. Action is the second step where the client/protagonist acts out the situation. Sharing, which is the last step and stage is where they share feedback on the drama if it is a group program or just sharing of feedback between the client and the therapist. They will normally talk about how the drama emotionally resonated and affected them.

Taken from Greenwood & Loewenthal (2005) from the book Psychoanalytic Psychotherapy volume 19, Raile (2023) describes and provides an elaborative description of a case study where the subject had severe eco-anxiety and tried psychodrama as a form of coping. In this case study we will briefly go through the subject's experience with eco-anxiety and psychodrama.

Sven O, a 26-year-old sociology doctoral student and university lecturer, sought help for his severe eco-anxiety. Participating in a psychodrama group, Sven expressed how the ongoing climate crisis and media reports on environmental concerns intensified his fears. In a psychodramatic session, led by a group leader which is the therapist, Sven was encouraged to explore his feelings and fears surrounding the climate crisis.

Sven described how fear would sneak up on him, particularly when reading news or browsing social media. He expressed feeling paralyzed and blocked by fear, hindering his ability to concentrate on his work. Through psychodrama, Sven aimed to confront and overcome his fear, or at least reduce its paralyzing effects.

The leader guided Sven to embody his fear and engage in role reversals to better understand his emotions. In one scene, Sven portrayed bad news while another participant embodied him, simulating the experience of fear creeping up on him. This exercise helped Sven externalize his fear and gain some understanding of its impact on his daily life.

As the session progressed, Sven explored deeper connections between his past experiences and his current eco-anxiety. He recalled memories of his sister providing him with strength and courage during challenging times. By revisiting these memories in a psychodramatic setting, Sven uncovered the roots of his fear and its manifestations in his present life. Through creative roleplaying and scene constructions, Sven confronted his fear head-on. In one powerful scene, Sven personified fear, allowing another participant to embody it physically. This intense interaction helped Sven confront the overwhelming nature of his fear and its suffocating effects on his wellbeing. The leader facilitated discussions on how fear was blocking Sven's creativity and suggested exploring past life events to uncover underlying triggers. By externalizing his fear and engaging in role reversals, Sven gained new perspectives on his emotions and learned to confront them constructively.

In conclusion, this case study exemplifies and presents the transformative power of psychodrama in helping individuals like Sven overcome eco-anxiety. Through creative expression, role-playing, and guided interventions, individuals can confront their fears, gain insights into their emotions, and develop coping strategies to manage eco-anxiety effectively. Psychodrama offers a unique therapeutic approach to address environmental concerns and promote mental well-being in the face of the climate crisis.

Although there is no professional text that deals with eco-anxiety from a psychodramatic perspective, many practitioners' websites do make this connection. A case study of eco-anxiety will be used as an example in order to give psychotherapists who have not yet received any training in psychodrama an insight into its possibilities for treatment. This case study can provide new ideas for practitioners who treat patients suffering from severe forms of eco-anxiety (Raile, 2023).

Case Study 3- Eco-Anxiety and Heat-Health Vulnerability in Informal Settlements: A Case Study from Dar es Salaam, Tanzania

Taken from Pasquini et al. (2020) in their work on the heat-health vulnerability of informal settlement residents in Dar es Salaam, Tanzania, this case study examines the intersection of ecoanxiety and heat-health vulnerability among informal settlement residents in Dar es Salaam, Tanzania (Pasquini et al., 2020). By exploring the psychological impacts of climate change on vulnerable populations, the study aims to raise awareness of the mental health implications of environmental challenges and propose holistic strategies for addressing eco-anxiety in the context of heat-related health risks. This case study investigates the relationship between eco-anxiety and heat-health vulnerability among informal settlement dwellers in Dar es Salaam, where rising temperatures exacerbate existing health risks.

Using a mixed-methods approach, this study includes qualitative interviews with the settlement residents, mental health professionals, and climate experts to explore the psychological impacts of heat-health vulnerability. Data collection includes narratives of lived experiences, observations of community responses to extreme heat events, and analysis of coping mechanisms in the face of environmental stressors. In Dar es Salaam, the findings of the study indicate that informal settlement residents are vulnerable to eco-anxiety and heat-related illness.

Upon being interviewed on, the residents expressed feelings of helplessness, fear, and uncertainty about the future, due to the immediate health risks posed by extreme heat. Eco-anxiety is found to amplify existing vulnerabilities and hinder the ability to adapt to different climate-related challenges. The findings highlight the need for a holistic approach to addressing eco-anxiety and heat-health vulnerability in informal settlements. Strategies that integrate mental health support, community resilience-building, and climate adaptation measures are essential to mitigate the psychological impacts of environmental stressors and enhance overall well-being.

Based on the study findings, it is recommended that policymakers and health practitioners in Dar es Salaam prioritize mental health support services for informal settlement residents facing ecoanxiety and heat-related health risks. Interventions should focus on building community resilience, fostering social support networks, and promoting adaptive coping strategies to address both psychological and physical health challenges. The study reveals a relationship between ecoanxiety and heat-health vulnerability among informal settlement residents in Dar es Salaam.

Residents express heightened levels of anxiety, uncertainty about the future, and feelings of powerlessness in the face of climate-related threats.

In conclusion, this case study underscores the interconnected nature of eco-anxiety and heat-health vulnerability in informal settlements, emphasizing the importance of addressing mental health concerns in climate adaptation efforts. By recognizing and responding to the psychological impacts of environmental stressors, it is possible to promote resilience and well-being among vulnerable populations in the face of climate change.

Case Study 4- The Silent Storm: Exploring Eco-Anxiety in the Aftermath of Urban Flooding in the United Kingdom

Taken from Fewtrell & Kay (2008) on their work on the health impacts of flooding in the UK using an urban case study. Researchers in the Bradford area of the United Kingdom conducted a study in order to quantify the health effects of flooding, using Disability Adjusted Life Years (DALYs) to compare different flood events and their health impacts.

In recent years, the intersection of climate change and mental health has gained an immense amount of attention. One particular aspect of this relationship is eco-anxiety, a form of psychological distress arising from environmental concerns. This case study examines into the concept of eco-anxiety in the context of urban flooding, using a study on the health impacts of flooding in the UK as a foundation. As a result of flooding, physical health effects were relatively minor in the case study areas; however, mental health problems, particularly psychological discomfort or distress, dominated the discussion. Urban flooding, increased by climate change, not only poses physical risks but also instills emotional responses in individuals. Eco-anxiety, characterized by feelings of fear, helplessness, and despair related to environmental issues, can intensify in the aftermath of flooding events. The uncertainty and disruption caused by flooding can amplify existing eco-anxiety or induce it in individuals previously unaffected.

Two urban neighborhoods experienced flooding, affecting a small number of properties and residents. While the physical health impacts were minimal, the mental health repercussions were profound. Residents reported heightened levels of eco-anxiety following the flooding, manifesting as increased stress, worry about future events, and a sense of powerlessness in the face of climate change. The case study was conducted at two locations in Keighley (Utley and Devonshire Park).

The eco-anxiety experienced by individuals in the community had drastic effects on their wellbeing. The lingering effects of eco-anxiety, such as sleep disturbances, mood swings, and difficulty concentrating, persisted long after the floodwaters receded.

In response to the prevalence of eco-anxiety in the community post-flooding, mental health support services and resilience-building initiatives were implemented. In the intent to combat eco-anxiety and foster a sense of empowerment in the face of environmental challenges, counseling sessions, group therapy, and community workshops were conducted.

The case study delved into the sensitive relationship between urban flooding, eco-anxiety, and mental health. By recognizing and addressing the psychological impacts of environmental events, communities can better prepare for in reference to the challenges posed by climate change. Through a holistic approach that integrates mental health support, community engagement, and environmental education, individuals can navigate eco-anxiety and build resilience in the face of a changing climate.

FUTURE RESEARCH DIRECTIONS

Future research directions for eco-anxiety in the context of climate change could encompass several key areas. Investigating the prevalence of eco-anxiety across different demographics, including age, gender, socioeconomic status, and cultural backgrounds, is crucial. Understanding how eco-anxiety manifests differently in various regions and communities globally can provide insights into tailored interventions.

Studying the psychological mechanisms that underlie eco-anxiety, including cognitive, emotional, and behavioral components, is another vital area. Examining the long-term mental health effects of eco-anxiety and its interactions with other forms of anxiety and depression can help in developing comprehensive mental health strategies. Developing and evaluating interventions and coping strategies to help individuals manage eco-anxiety is essential. This includes assessing the effectiveness of various therapeutic approaches, such as cognitive-behavioral therapy, mindfulness, and nature-based therapies.

The role of education and awareness is also significant in this research. Investigating the impact of environmental education and awareness campaigns on eco-anxiety levels, and exploring how different types of information and framing influence eco-anxiety and public engagement with climate change, can guide more effective communication strategies. Examining how social support networks and community engagement can mitigate eco-anxiety is crucial. Understanding the role of collective action and activism in addressing eco-anxiety and fostering a sense of agency and empowerment can help build resilient communities.

Analyzing how media representations of climate change influence eco-anxiety and investigating the role of social media in spreading awareness and its impact on individuals' eco-anxiety levels is another important research direction. Integrating perspectives from psychology, environmental science, sociology, and public health can provide a holistic understanding of eco-anxiety. Exploring the intersections of eco-anxiety with other societal issues, such as economic inequality and social justice, can reveal broader implications and solutions.

Assessing how governmental and organizational policies can address eco-anxiety by promoting mental health and environmental sustainability is necessary. Investigating the role of policy in creating supportive environments for individuals experiencing eco-anxiety can lead to more effective governance. Studying factors that contribute to resilience and positive adaptation in the face of climate change and eco-anxiety is also essential. Identifying ways to foster hope, optimism, and constructive engagement with environmental issues can empower individuals and communities.

Finally, conducting cross-cultural studies to understand how eco-anxiety is experienced and addressed in different cultural contexts can provide valuable insights. Exploring indigenous and traditional ecological knowledge systems as sources of resilience and coping strategies can offer alternative perspectives and solutions. These research directions can provide a comprehensive understanding of eco-anxiety and inform strategies to support individuals and communities in navigating the psychological challenges posed by climate change.

CONCLUSION

Climate change can manifest as a natural phenomena or disaster in some circumstances. Natural catastrophes result from predictable natural processes, yet human disasters can occur. Human disasters are the consequence of human error or inactivity, such as the Deepwater Horizon oil leak (Lindsey & Saut, 2020). The oil spill is a human disaster since it was caused by human activity in the environment and might have been avoided. So, climate change is a human calamity since it was caused by human action in the environment and may be avoided. Natural catastrophes are related with faster recovery and less mental health problems since they occur naturally. Human disasters are connected with more complex grief and healing processes because humans know that the disaster might have been prevented if action had been done. As people lose their sense of belonging in their surroundings and community, violence, drug and alcohol misuse, and suicide rates increase. This is particularly obvious among minority populations, such as the Inuit. Individually, climate change exacerbates current mental health concerns while also causing new ones, such as PTSD and increased anxiety (Lindsey & Saut, 2020).

Anxiety caused by climate change is becoming increasingly recognized as one of the consequences of climate change on mental health. There is no doubt that climate change is not only an environmental issue but also a psychological one (Clayton, 2020). Despite the fact that psychologists have been involved in climate change research for many decades, the majority of their research has focused on risk perception, methods to communicate about climate change, attitudes toward climate change, and interventions aimed at promoting mitigation through more sustainable behavior. The phenomenon is not only observed in populations that experience the most direct effects of climate change but also can be triggered by merely contemplating such impacts (Tam et al., 2023) . There has been very little recognition of the mental health implications of climate change, despite the fact that the last decade has brought increased attention to the topic, and based on the rising data, it is time to carefully consider how climate change might harm mental health. Climate change should be on the minds of psychologists and those in the mental health field. This might imply that clinical training should include knowledge and teaching about the issue. It may imply a focus on creating best practices for assisting those who are suffering what may be described as climate anxiety. Climate change is also a societal issue. The worry that defines certain people's responses to climate change is shaped in part by how society addresses or does not handle the issue. Individual mental health interventions are unlikely to be entirely effective without social, if not global, attention to the issue. Climate anxiety symptoms are not always sensations to be overcome, but rather lessons to be learned, however only if they can be felt safely, via meaningful action and constructive social, psychological, and ecological transformation (Dodds, 2021).

Finally, climate change is a real problem. If people do not find effective strategies to cope with emotional responses to climate change, increased levels of worry may endanger society's functioning. International adaptation, mitigation, and resilience planning must emphasize the need to provide resources to help people cope with this new reality.

REFERENCES

Arcanjo, M. (2019). Eco-anxiety: mental health impacts of environmental disasters and climate change. *A Climate Institute Publication. Washington.*

Baudon, P., & Jachens, L. (2021). A scoping review of interventions for the treatment of eco-anxiety. *International Journal of Environmental Research and Public Health*, 18(18), 9636. DOI: 10.3390/ijerph18189636 PMID: 34574564

Boluda-Verdu, I., Senent-Valero, M., Casas-Escolano, M., Matijasevich, A., & Pastor-Valero, M. (2022). Fear for the future: Eco-anxiety and health implications, a systematic review. *Journal of Environmental Psychology*, 84, 101904. DOI: 10.1016/j.jenvp.2022.101904

Carleton, T. A. (2017). Crop-damaging temperatures increase suicide rates in India. *Proceedings of the National Academy of Sciences of the United States of America*, 114(33), 8746–8751. DOI: 10.1073/pnas.1701354114 PMID: 28760983

Clayton, S. (2020). Climate anxiety: Psychological responses to climate change. *Journal of Anxiety Disorders*, 74, 102263. DOI: 10.1016/j.janxdis.2020.102263 PMID: 32623280

Clayton, S., Manning, C. M., Krygsman, K., & Speiser, M. (2017). *Mental Health and Our Changing Climate: Impacts, Implications, and Guidance*. American Psychological Association, and ecoAmerica.

Coffey, Y., Bhullar, N., Durkin, J., Islam, M. S., & Usher, K. (2021). Understanding eco-anxiety: A systematic scoping review of current literature and identified knowledge gaps. *The Journal of Climate Change and Health*, 3, 100047. DOI: 10.1016/j.joclim.2021.100047

Coppola, I. G. (2021). Eco-Anxiety in "the Climate Generation": Is Action an Antidote?.

Dodds, J. (2021). The psychology of climate anxiety. *BJPsych Bulletin*, 45(4), 222–226. DOI: 10.1192/bjb.2021.18 PMID: 34006345

Fewtrell, L., & Kay, D. (2008). An attempt to quantify the health impacts of flooding in the UK using an urban case study. *Public Health*, 122(5), 446–451. DOI: 10.1016/j.puhe.2007.09.010 PMID: 18222511

Gray, C., & Mueller, V. (2012). Drought and population mobility in rural Ethiopia. *World Development*, 40(1), 134–145. DOI: 10.1016/j.worlddev.2011.05.023 PMID: 22523447

Greenwood, D., & Loewenthal, D. (2005). The use of 'case study'in psychotherapeutic research and education. *Psychoanalytic Psychotherapy*, 19(1), 35–47. DOI: 10.1080/02668730512331341564

Heid, M. (2022,April 22). *How to cope with Eco-Anxiety: 5 expert tips*. Everyday-Health.com. https://www.everydayhealth.com/eco-anxiety/how-to-cope-with-eco-anxiety/

Iberdrola. (2021, April 22). *What Is Eco-Anxiety*. Iberdrola. https://www.iberdrola.com/social-commitment/what-is-ecoanxiety

Lawrance, E. L., Thompson, R., Newberry Le Vay, J., Page, L., & Jennings, N. (2022). The impact of climate change on mental health and emotional wellbeing: A narrative review of current evidence, and its implications. *International Review of Psychiatry (Abingdon, England)*, 34(5), 443–498. DOI: 10.1080/09540261.2022.2128725 PMID: 36165756

Lindsey, L., & Saut, S. (2020). The Interconnection Between Climate Change and Mental Health. *Climate Change and Human Responses*, 115.

Lutz, P. K., Zelenski, J. M., & Newman, D. B. (2023). Eco-anxiety in daily life: Relationships with well-being and pro-environmental behavior. *Current Research in Ecological and Social Psychology*, 4, 100110. DOI: 10.1016/j.cresp.2023.100110

O'Brien, A. J., & Elders, A. (2022). Climate anxiety. When it's good to be worried. *Journal of Psychiatric and Mental Health Nursing*, 29(3), 387–389. DOI: 10.1111/jpm.12817 PMID: 34967950

Padhy, S. K., Sarkar, S., Panigrahi, M., & Paul, S. (2015). Mental health effects of climate change. *Indian Journal of Occupational and Environmental Medicine*, 19(1), 3. DOI: 10.4103/0019-5278.156997 PMID: 26023264

Palinkas, L. A., & Wong, M. (2020). Global climate change and mental health. *Current Opinion in Psychology*, 32, 12–16. DOI: 10.1016/j.copsyc.2019.06.023 PMID: 31349129

Pasquini, L., van Aardenne, L., Godsmark, C. N., Lee, J., & Jack, C. (2020). Emerging climate change-related public health challenges in Africa: A case study of the heat-health vulnerability of informal settlement residents in Dar es Salaam, Tanzania. *The Science of the Total Environment*, 747, 141355. DOI: 10.1016/j.scitotenv.2020.141355 PMID: 32777515

Pihkala, P. (2020). Eco-anxiety and environmental education. *Sustainability (Basel)*, 12(23), 10149. DOI: 10.3390/su122310149

Pihkala, P. (2020). Anxiety and the ecological crisis: An analysis of eco-anxiety and climate anxiety. *Sustainability (Basel)*, 12(19), 7836. DOI: 10.3390/su12197836

Raile, P. (2023). Psychodrama and eco-anxiety—A case study. *Environment and Social Psychology, 9*(1).

Rose, W. M. (2023). Climate Anxiety: The Emotional Impact of Climate Change.

Safi, M. (2017). Suicides of nearly 60,000 Indian farmers linked to climate change, study claims. *The Guardian, 31*.

Stanley, S. K., Hogg, T. L., Leviston, Z., & Walker, I. (2021). From anger to action: Differential impacts of eco-anxiety, eco-depression, and eco-anger on climate action and wellbeing. *The Journal of Climate Change and Health*, 1, 100003. DOI: 10.1016/j.joclim.2021.100003

Tam, K. P., Chan, H. W., & Clayton, S. (2023). Climate change anxiety in China, India, Japan, and the United States. *Journal of Environmental Psychology*, 87, 101991. DOI: 10.1016/j.jenvp.2023.101991

Usher, K., Durkin, J., & Bhullar, N. (2019). Eco-anxiety: How thinking about climate change-related environmental decline is affecting our mental health. *International Journal of Mental Health Nursing*, 28(6), 1233–1234. DOI: 10.1111/inm.12673 PMID: 31724833

ADDITIONAL READINGS

United Nations. (n.d.). *The Paris Agreement|United Nations*. https://www.un.org/en/climatechange/paris-agreement

Vakoch, D. A., & Mickey, S. (Eds.). (2022). *Eco-Anxiety and Pandemic Distress: Psychological Perspectives on Resilience and Interconnectedness*. Oxford University Press.

Villines, Z. (2022, April 29). *What to know about psychodrama therapy*. https://www.medicalnewstoday.com/articles/psychodrama-therapy#what-it-is

Wang, H., Safer, D. L., Cosentino, M., Cooper, R., Van Susteren, L., Coren, E., Nosek, G., Lertzman, R., & Sutton, S. (2023). Coping with eco-anxiety: An interdisciplinary perspective for collective learning and strategic communication. *The Journal of Climate Change and Health*, 9, 100211. DOI: 10.1016/j.joclim.2023.100211

Wikipedia contributors. (2024a, February 5). Solastalgia. Wikipedia. https://en.wikipedia.org/wiki/Solastalgia

Wikipedia contributors. (2024b, March 18). *Psychological resilience*. Wikipedia. https://en.wikipedia.org/wiki/Psychological_resilience

World Economic Forum. (2022, November 1). *Eco-anxiety is a growing mental health crisis*. https://www.weforum.org/agenda/2022/11/children-mental-health-eco-anxiety/

World Economic Forum. (2024, January 19). *How to turn eco-anxiety into bold action on nature*. https://www.weforum.org/agenda/2024/01/how-to-turn-eco-anxiety-into-bold-action-on-nature/

Znachko, C., Anzellini, A., Parker, K., & Hicks, C. (2022). Climate Change and Human Responses.

KEY TERMS AND DEFINITIONS

Action: The process of doing something to achieve a goal or effect; in the context of environmental issues, it often refers to initiatives and measures taken to mitigate or adapt to climate change and ecological problems.

Climate Change: Long-term alterations in temperature, precipitation, wind patterns, and other aspects of the Earth's climate system, primarily caused by human activities such as burning fossil fuels, deforestation, and industrial processes.

Eco-anxiety: A form of anxiety that arises from concerns about ecological disasters and the long-term sustainability of the environment, often driven by awareness of climate change and environmental degradation.

Ecological: Relating to the relationships between living organisms, including humans, and their physical environment; concerned with the interdependence of organisms within their habitats.

Environment: The natural world, including the land, water, air, plants, and animals; also refers to the conditions and influences affecting the development and survival of organisms.

Global Warming: The ongoing increase in Earth's average surface temperature due to rising levels of greenhouse gases, such as carbon dioxide and methane, resulting from human activities like burning fossil fuels and deforestation.

Media: Various channels of communication that provide news, entertainment, education, data, or promotional messages to the public; includes television, radio, newspapers, the internet, and social media.

Representation: The depiction or portrayal of someone or something in a particular way; in media and communication, it refers to how ideas, people, events, and issues are presented to audiences.

Stress: A physical, mental, or emotional response to external pressures or demands, which can be acute or chronic and affect overall well-being and health.

Chapter 4
The Impact of Climate Change on Persons With Disabilities

Jeffrey Kurebwa
https://orcid.org/0000-0002-8371-8055
Bindura University of Science Education, Zimbabwe

Nyasha Yvonne Kurebwa
Women's University in Africa, Zimbabwe

ABSTRACT

The purpose of the chapter is to provide an understanding of the impact of climate change on persons with disabilities. The concepts of climate change, climate adaptation, ableism, and models of persons with disabilities are discussed in greater detail. The chapter shows that persons with disabilities are seriously affected by climate change and provides recommendations that can be taken by various actors to reduce the impact of climate change on persons with disabilities. These include ensure meaningful participation of persons with disabilities and their representative organizations across the climate change policymaking and planning cycle at all levels, keep investing in strengthening inclusive approaches in high-impact sectors that offer greater potential for building climate-resilient and inclusive societies and promote a programmatic approach to climate resilience that is holistic and equitable, integrating social, economic, health, and climate considerations and generating far-reaching benefits for all vulnerable groups.

DOI: 10.4018/979-8-3693-4183-4.ch004

INTRODUCTION

According to international human rights law, governments are mandated to uphold the rights of Persons with Disabilities (PWD) while addressing climate change. The preamble of the Paris Agreement emphasizes the importance of respecting, promoting, and considering human rights obligations, including those of individuals with disabilities, in climate action efforts. Various entities such as the United Nations Human Rights Council, the Committee on the Rights of Persons with Disabilities, and the Office of the United Nations High Commissioner for Human Rights have acknowledged the heightened vulnerability of people with disabilities to the impacts of climate change. These organizations have urged States to implement disability-inclusive human rights-based strategies in combating climate change.

In particular, State parties to the United Nations Convention on the Rights of Persons with Disabilities (UNCRPD, 2006) are obliged to:

(a) Assess and take into account the differential impacts of climate change on the rights of persons with disabilities through an intersectional approach;
(b) Develop and implement evidence-based climate mitigation and adaptation policies to prevent and minimize the adverse impacts of climate change on persons with disabilities;
(c) Provide access to information on climate change to persons with disabilities, strengthen their capacity to participate in climate decision-making that concerns them, and ensure that they have access to judicial or administrative proceedings and effective remedies when they suffer harm from climate action and inaction; and
(d) Support international efforts to combat climate change and enhance the climate resilience of persons with disabilities in developing countries.

Climate change affects the entire world, but its adverse effects are particularly severe for impoverished nations and disadvantaged communities that depend heavily on natural resources. These groups struggle to withstand, rebound from, and adjust to extreme climate events. Among these marginalized communities and nations, individuals with disabilities are especially at risk from the impacts of climate change, which worsen their existing vulnerabilities. People with disabilities account for approximately 15% of the global population (WHO, 2011).

The majority of people with disabilities, about 80%, reside in countries with lower- and middle-income economies, which further exacerbates the challenges they already face (UNICEF, 2022). This results in lower levels of education, poorer health, higher unemployment, and increased poverty rates for individuals with disabilities, limiting their opportunities to fully participate in society due to stigma, exclusion, and inaccessible environments. These societal inequalities hinder the ability of people with disabilities to develop assets, resources, and social capital, making it difficult for them to build resilience against the negative impacts of climate change. In developing countries that are highly susceptible to climate change, these challenges are intensified, leaving millions of individuals with disabilities disproportionately at risk and without the opportunity to enhance their resilience and contribute to climate change response efforts.

The CRPD (2006) aims to safeguard and advance the rights of individuals with disabilities. Rather than establishing novel human rights, the CRPD serves as a declaration that the rights of individuals with disabilities have been infringed upon, emphasizing the necessity for further measures (UNICEF, 2022). It delineates the entitlements of individuals with disabilities and outlines the obligations that States Parties must uphold to safeguard and uphold those rights (United Nations, 2015). The United Nations ratified the treaty in 2006 and made it available for countries to sign a year later.

The seventh article of the Convention focuses on the rights of children with disabilities and emphasizes the importance of prioritizing the best interests of the child in all decisions that impact them. It also highlights the need for States Parties to ensure that their laws treat children with disabilities equally to other children (United Nations, 2015). Even though individuals with disabilities make up 15% of the global population, there has been a lack of adequate representation of their rights in climate action, particularly at the multilateral level.

The 2015 Paris Agreement preamble acknowledges that persons with disabilities are among the groups most severely impacted by climate change. However, they have been significantly marginalized from the decision-making processes and results under the UN Framework Convention on Climate Change (UNFCCC), as well as from the climate change policies and strategies of individual countries. This exclusion of persons with disabilities from global efforts to address the climate crisis has serious repercussions. Inaccessible disaster preparedness plans, systemic discrimination, and widespread poverty have resulted in people with disabilities being left behind in relief and response efforts. This exclusion has had a significant impact on their ability to cope with the effects of climate change and has further exacerbated their vulnerability.

Climate change has a significant impact on the lives, well-being, and livelihoods of individuals with disabilities worldwide, ranging from sudden disasters like typhoons and wildfires to gradual changes like droughts, rising temperatures, and sea level rise. These effects are particularly pronounced for members of the disability community who face multiple forms of discrimination, including women, children, Indigenous peoples, older individuals, displaced populations, and other marginalized groups. Additionally, those living in poverty and underrepresented groups of persons with disabilities, such as those with intellectual disabilities, psychosocial disabilities, and deaf-blindness, are especially vulnerable to the consequences of climate change.

Defining Climate Change

Climate change refers to a shift in the condition of the climate, which can be recognized by alterations in its average and/or the range of its characteristics, and that endures for a prolonged period, usually spanning decades or more (Field et al. 2012). The United Nations Framework Convention on Climate Change (UNFCCC) (1994) makes a clear distinction between climate change caused by human activities that modify the atmospheric makeup, and climate variability caused by natural factors (Field et al. 2012). 2. According to this differentiation, climate change is defined as a modification in climate that is directly or indirectly attributed to human actions that change the composition of the global atmosphere, and that is in addition to the natural climate variability observed over similar time frames (UNFCC, 1994).

Individuals with disabilities face a disproportionate impact during climate disasters and emergencies, making them more vulnerable to the effects of climate change. They often find themselves isolated from support systems as their families may need to relocate in search of employment opportunities, leaving them behind due to their disabilities. 2. Moreover, the disruption of essential services like healthcare further hinders people with disabilities from exercising their rights. Studies indicate that they encounter additional challenges in receiving humanitarian aid during times of crisis, exacerbating their already precarious situation.

Persons with disabilities are frequently disregarded in climate research, as highlighted by Moran et al. (2017). While studies have explored the effects of climate change on vulnerable populations, there remains a gap in addressing the specific requirements of individuals within these groups, such as those with disabilities, as noted by Lewis & Ballard (2012). Despite existing research on the rights of people with disabilities and climate change, the intersection of these two areas has not received sufficient attention, according to Peek & Stough (2010) and UNICEF (2022).

Defining Disability

Individuals with disabilities have faced widespread discrimination worldwide in multiple aspects of societal life, encompassing social interactions, work environments, economic and political arenas, cultural realms, and the fundamental rights of human beings (Bufdir, 2021). According to Russel (2019), the capitalist perspective on disability portrays individuals with disabilities as a social issue due to their bodies not aligning with the conventional concept of a worker's physique. Consequently, they have been marginalized from the workforce and have not had equal opportunities to earn a living and enjoy the benefits associated with employment (Russel, 2019).

Disability is conceptualized as a construct that is socially constructed within the context of labor relations in capitalist societies, according to Russel (2019), who argues that the accumulation of wealth by the capitalist class is directly linked to the oppression of individuals with disabilities. The CRPD (2006) defines persons with disabilities as individuals who experience long-term physical, mental, intellectual, or sensory impairments that, when combined with societal barriers, may impede their full and equal participation in society (Article 1, Convention on the Rights of Persons with Disabilities, 2006). Disability is commonly categorized into four main types: physical disabilities, sensory disabilities, intellectual disabilities, and mental health disabilities. Each category presents unique challenges and barriers that individuals with disabilities may face in their daily lives.

Globally, there are approximately 1 billion individuals with disabilities, but this figure may be higher due to social stigma, resulting in some people being overlooked, and a lack of comprehensive data (Priddy, 2019). In Norway, around 17% of the population is living with a disability (Skarstad, 2019). Disabilities often require adjustments, assistance, and support systems to ensure equal involvement in society.

Disability and Climate Change

In the realm of climate change and its consequences, it is widely acknowledged by researchers that individuals with disabilities are among the most susceptible populations (Kett et al. 2021). This vulnerability stems from the nature of their impairments and their marginalized status in society, as well as their tendency to be impoverished and lack the means to adapt to climate change (United Nations, 2018). There is extensive literature on the impact of climate change on vulnerable groups, who are anticipated to bear a disproportionate burden. However, there is a dearth of research exploring the vulnerability of individuals with disabilities within these groups (Lewis and Ballard, 2012). Studies have been carried out to investigate the potential impact of climate change on various vulnerable populations, including children, women, the elderly, and individuals living in both rural and urban areas. Researchers have explored how People with Disabilities may be disproportionately affected by changing environmental conditions (Abedin et al. 2013, Pichdara et al. 2022).

The Concept of Climate Adaptation

Climate adaptation encompasses the disparities within society and the unequal impacts of climate change and adaptation on individuals (IPCC, 2022). Climate justice and fair adaptation strategies seek to address these inequalities and injustices, promoting a more inclusive approach to climate adaptation that avoids perpetuating discrimination (Juhola et al. 2022). The evolving climate necessitates adjustments to new environmental conditions. Adaptation efforts must consider various societal elements such as infrastructure, building regulations, safety standards for housing, agricultural practices, and irrigation systems. Effective adaptation is crucial for preparing for the complex challenges posed by climate change.

The concept of adapting to climate change, as outlined by the IPCC (2022), involves making adjustments to ecological, social, and economic systems in response to current or anticipated climatic conditions and related events. In essence, climate adaptation aims to minimize the negative impacts of climate change while capitalizing on the opportunities it presents, as highlighted by Jones & Boyd (2011). This necessitates structural changes across various societal levels, encompassing efforts to reduce greenhouse gas emissions and address factors contributing to vulnerability and risk. It is important to acknowledge the barriers and constraints that may impede climate adaptation, particularly with the inclusion of individuals with disabilities (Leichenko & O'Brien, 2019). In today's world, the challenges associated with adaptation go beyond the physical aspects of climate change. Simply suggesting that people should find shelter in shaded or cooler areas during a heat wave is not enough, as a combination of economic, social, cultural, and political influences creates different outcomes for various groups. The discussion critically analyzes how these factors affect the ability of individuals and communities to adapt (Jones & Boyd, 2011).

Malloy & Ashcraft (2020) delve into the exploration of just adaptation, a specific area within climate adaptation that centers on the involvement of socially vulnerable groups in climate adaptation policy. The concept of just adaptation draws from principles of social justice theory. In the context of just adaptation, social justice theory pertains to the integration of principles that tackle disparities to ensure fairness in the distribution of the adverse effects and advantages linked to climate impacts and adaptation endeavors (Malloy & Ashcraft, 2020). The theoretical framework is constructed on the necessity to take into account the rights, necessities, and susceptibilities of marginalized and disadvantaged groups in societies. These groups frequently bear an unequal burden of climate change impacts, owing to restricted access to resources and opportunities (Schlosberg, 2013).

Furthermore, the theory seeks to address the underlying structural inequalities that worsen the impact of climate change on marginalized and disadvantaged communities (Schlosberg, 2013). Malloy & Ashcraft (2020) contend that climate adaptation is often carried out through traditional methods that prioritize the interests of the elite and those who are already privileged. This suggests that the ongoing marginalization of vulnerable groups is likely to persist under current climate adaptation policies. However, this should not come as a surprise, as the continued marginalization of certain groups, such as people with disabilities, in various aspects of society raises questions about why these patterns persist despite being acknowledged as problematic.

The persistence of these patterns can be attributed to a variety of factors, including the lack of inclusivity in mainstream policy development for individuals with disabilities. When adaptation efforts fall short, it is often due to inadequate planning or the implementation of measures based on incomplete or unreliable information, as noted by Schipper (2020). Maladaptation, or the failure to adapt effectively, can lead to heightened vulnerability and increased risk in extreme cases. Preventing maladaptation requires a thorough understanding of the factors and circumstances that shape vulnerability and risk. To conduct a comprehensive examination of the incorporation of individuals with disabilities into adaptation strategies, it is crucial to address the specific challenges and limitations that may hinder effective climate adaptation efforts.

Limits to Climate Adaptation

Jones & Boyd (2011) present a framework for understanding the various constraints that can hinder adaptation efforts. They identify three main categories of limitations: natural, social, and human, each with its subcategories. The natural limitations encompass ecological and physical factors such as ecosystem resilience, biodiversity, and geographical constraints, all of which can impact the ability of human systems to adapt to changing conditions. Furthermore, the authors highlight the increasing severity of threats posed by ecological degradation and loss of biodiversity to human systems and livelihoods. On the other hand, the human and informational limitations revolve around economic, technological, and knowledge-based barriers to adaptation. These can include factors such as a lack of reliable weather forecasts, limited financial resources for adaptation measures, and restricted access to essential technologies like renewable energy. Jones and Boyd's framework provides a comprehensive overview of the diverse challenges that can impede successful adaptation efforts.

The social limits, which constitute the third category of limits, encompass normative, cognitive, and institutional factors that may hinder adaptation measures. These limits are embedded within social structures and may not be immediately apparent as obstacles until efforts to adapt are underway. Normative barriers pertain to the influence of social norms on climate adaptation actions, including how a society collectively perceives and responds to the threat of climatic events based on shared values and understanding (Jones & Boyd, 2011).

One illustration of this phenomenon is how individuals who opt out of consuming meat for environmental and climate-related reasons may find greater support and validation for their decision within a community that shares similar values and beliefs, as opposed to a community that dismisses the environmental and climate impacts of meat consumption. Persons with disabilities may have unique concerns about the impact of climate change compared to those without disabilities, stemming from their encounters with severe weather events. Institutional barriers encompass the formal and informal systems that either facilitate or hinder individuals' capacity to adjust to these challenges. Meanwhile, cognitive barriers to adaptation pertain to the psychological reactions and responses to climate-related risks and vulnerability (Jones & Boyd, 2011).

Models on Disability

Throughout time, there have been several different approaches to explaining and comprehending disability. These range from interpreting disability as a consequence of divine punishment in the moral/religious model to understanding disability as a product of societal barriers created by exclusive and limited structures in the social model (Retief & Letšosa, 2018). Different models may be applicable in specific situations, taking into account social and cultural factors. However, some models, like the moral/religious and medical models, are based on discriminatory and segregationally beliefs and viewpoints. These diverse models will be explored further below.

The Charitable Model

The charitable model perceives individuals with disabilities as incapable of acquiring or accessing sufficient resources independently, thus necessitating charitable assistance. While this model views disability as a misfortune (Oliver, 1990), embracing it can result in charitable contributions being directed towards enhancing the circumstances of those with disabilities. However, this perspective on disability diminishes the autonomy of individuals with disabilities, making it unacceptable to many. The criticism surrounding the erosion of personal autonomy has significantly impacted the charitable framework in developed countries, where self-reliance and individual agency are highly valued. Nevertheless, the charitable approach has been disseminated from the developed countries to the developing countries, notably facilitated by missionaries during the colonial period (Grech, 2015).

The Medical Model

The medical model of disability categorizes individuals with disabilities based on their medical diagnosis. In this framework, disability is viewed as a condition that requires treatment for the individual to be included in society (Thomas, 2004). Consequently, individuals with disabilities are often perceived as unequal and are deprived of the same rights as others. Nonetheless, disability encompasses more than just medical considerations and treatment. While medical care is crucial for receiving necessary interventions, it should not dictate how individuals with disabilities lead their lives. As highlighted by Brisenden (1986), the medical aspect is significant for appropriate care, but it should not limit the autonomy and choices of individuals with disabilities. In contrast to the charitable model, which does not recognize individuals with disabilities as independent contributors to society, the medical model suggests that the disabilities of people can be treated (Osborne, 2015).

According to this approach, disability is seen as a physical or mental abnormality that hinders a person's ability to function (Mitchell &Snyder, 2000). By addressing these abnormalities, especially through medical interventions, the disability can be improved, reduced, or managed to enable a higher level of functioning. The medical model has resulted in the creation of specialized services, such as rehabilitation, aimed at improving, reducing, and managing physical or mental abnormalities. The development of specialized services based on a medical model understanding of disability has led to significant progress but has also brought about significant drawbacks (CDC, 2015). The belief that disability can be remedied is rooted in, and perpetuates, the idea that disability is a problem. According to Brisenden (1986), the medical model has resulted in a scenario where healthcare professionals and others are constrained in their actions by a definition that they have constructed. They are unable to deviate from the framework of their perception of disabled individuals.

The Social Model of Disability

Disability activists and scholars challenge the medical model of disability by advocating for the social model of disability, which emphasizes the impact of the environment on a person's disability (Oliver, 1983). According to this model, the focus is on eliminating barriers that hinder the participation of individuals with disabilities, such as inaccessible buildings for people in wheelchairs. By removing these barriers, the goal is to promote independence and equality for people with disabilities, allowing them the same opportunities for participation as others (Shakespeare, 2006; United Nations, 2018; Goering, 2015).

Viewing disability through a social lens allows for the framing of disability as a matter of social equity (Oliver, 1983). This perspective highlights how disability can be seen as a form of unfair distribution, leading to economic marginalization of disabled individuals who may face challenges in securing employment or receiving lower wages compared to their non-disabled counterparts. Moreover, the social interpretation of disability sheds light on the concept of procedural injustice, where disabled individuals are often left out of decision-making processes due to various barriers such as physical inaccessibility or attitudinal obstacles within organizations Shakespear & Watson, 2001). This exclusion can further perpetuate inequalities and hinder the full participation of disabled individuals in society. Lastly, the aspect of recognition in social justice, as outlined by Fraser (2000), emphasizes the importance of acknowledging the rights and dignity of disabled individuals. By recognizing the unique experiences and contributions of disabled individuals, society can work towards creating a more inclusive and equitable environment for all.

People with disabilities often face misrecognition due to institutionalized cultural norms that tend to exclude or ignore disability, resulting in their inability to fully engage in society (Fraser, 2000). The social model of disability, unlike the charitable and medical models, was explicitly formulated as a deliberate response to this exclusion. While the charitable and medical models viewed disability as a personal issue, the social model aimed to shift the focus onto societal barriers and structures that hinder full participation. The concept of the social model revolutionized the understanding of disability by distinguishing impairment from disability. This distinction aimed to separate the individual's characteristics from the societal barriers that contribute to disability (Kafer, 2013). While impairments were seen as inherent traits of individuals, disability was recognized as a product of a society that fails to be inclusive. The social model originated from the advocacy efforts of the Union of the Physically Impaired Against Segregation (UPIAS) in the UK and was further developed by disability studies pioneer Mike Oliver in 1983.

The social model has been disseminated worldwide from its origins, offering valuable contributions to the disability rights movement by changing the perspective on disability, especially in regions where charity and medical models were predominant (Thomas, 1999). Despite its advantages, the diffusion of the social model may not be suitable for communities and environments with unique challenges different from those that led to its development. Grech (2009) pointed out that the social model's applicability across cultures is questionable, as it reflects the viewpoints of specific groups from a particular time and place, mainly white, educated, middle-class, and Western disabled individuals.

The Critical Realist Model

The critical realist model of disability, which combines the medical and social models, posits that individuals are disabled by both society and their bodies (Williams, 1999). This perspective acknowledges the complex interplay between personal characteristics and external factors in shaping disability experiences. While emphasizing the role of societal structures in creating disabling conditions, the critical realist model also recognizes the significance of nonhuman natural environments in contributing to disability (King & Gregg, 2022). This nuanced approach challenges the notion that disability is solely a result of social interactions and relationships. By advocating for an interactional approach that considers both individual and structural factors, the critical realist model offers a comprehensive framework for understanding disability. This model highlights the importance of examining medical conditions, and impairments, as well as environmental, cultural, and policy influences on disability experiences. Scholars like Thomas (1999) have further expanded on this model by incorporating the impact of social oppression into the definition of disability. This broader perspective underscores the systemic barriers and discrimination faced by individuals with disabilities, adding another layer to the understanding of disability as a complex and multifaceted phenomenon (King & Gregg, 2022).

The integration of both disability models aligns with the World Health Organization's International Classification of Functioning, Disability, and Health framework. This approach enables us to conceptualize physical disability as the convergence of an individual's health and the environmental factors that restrict their full participation in society (WHO, 2011). The critical realist model aims to recognize the personal experiences of people with disabilities without pathologizing or making assumptions about their bodies (Williams, 1999).

The Human Rights-Based Model

The human rights-based approach to disability expands upon the social model by incorporating a rights perspective (Disability Advocacy Resource Unit, 2019). It places the onus on the state to uphold the rights of individuals with disabilities and eliminate obstacles that hinder their equal participation in society. Moreover, this model underscores the importance of empowering individuals with disabilities to assert their rights and have a say in matters that impact them directly (Disability Advocacy Resource Unit, 2019). The human rights-based model shares commonalities with the social model in its portrayal of disability and discrimination against individuals with disabilities as systemic issues ingrained in society, rather than medical concerns. The assertion made by the human rights-based model suggests that the UN Declaration of Human Rights falls short of adequately safeguarding the rights of individuals with disabilities. For human rights to effectively protect the rights of persons with disabilities, it is imperative to address societal barriers, prevent violations, and promote equal participation within society (Priddy, 2019).

The Identity Model

The focus of this model is on highlighting the unique and shared identity of individuals with disabilities, positioning them as part of a minority group similar to gender, race, LGBTQ+, and others (Brewer et al. 2012). Within this identity framework, disability is viewed as an experience shaped by interactions within a society that often fails to fully integrate disabled individuals. This perspective encompasses any situation where people with disabilities are marginalized within the larger societal context, leading to a sense of being outside the norm. While it allows for individual variation, the underlying premise is that they are consistently treated differently due to their disability (Brewer et. al. 2012).

The International Classification of Functioning (ICF) Framework

The International Classification of Functioning, Disability, and Health (ICF) was established by the World Health Organization (WHO) as part of its classification system in 2001. In addition to serving as a classification tool, the WHO also presented the ICF as a framework in 2002, which includes a definition of disability. This definition is based on a set of constituent constructs that are interconnected through a model. According to the ICF, disability is described as "an umbrella term for impairments, activity limitations, and participation restrictions" (WHO, 2002, p. 2). Each of the constituent constructs within the ICF is carefully defined and elaborated upon to provide a comprehensive understanding of disability and health. These detailed definitions and descriptions not only establish a structure for the ICF but also facilitate the classification of each construct within the system. By categorizing impairments, activity limitations, and participation restrictions, the ICF aims to provide a holistic approach to understanding and addressing disability and health issues (WHO, 2001).

The primary focus of the ICF that holds the most significance in this conversation is the ICF's endeavor to blend the medical and social models by presenting the ICF as a biopsychosocial model (WHO, 2002, p. 9). As a result, the ICF has evolved from two models that originated in the global North: the medical model due to the global North's growing medicalization and the social model as a specific reaction to the medical model by individuals with disabilities.

The Traditional Model

The conventional concept of disability stands apart from earlier interpretations of disability. Referred to as 'the myth of disability in developing nations' by Ingstad (1999, p. 757), the traditional model posits that in all "traditional cultures" (of the Global South), disability is viewed as a spiritual occurrence with severe repercussions for individuals with disabilities. Roush & Sharby (2011, p.1769) illustrate a 'moral model' that encompasses both positive and negative attitudes towards disabilities, although negative perceptions may be more widespread. Central to Roush & Sharby's (2011) depiction is a recognition of nuance, particularly about how spiritual interpretations may align or diverge across various societies or belief systems.

Roush & Sharby (2011) introduced a set of concepts by outlining a moral framework, but this set is characterized by its variety. The conventional framework is based on a notion of uniformity. According to the conventional framework, any diversity in the understanding of disability within traditional societies is considered inconsequential. As Ingstad (1999) contends, while it is acknowledged that there are instances of spiritual beliefs leading to the mistreatment of individuals with disabilities in the global North, these instances are not used to form a broad representation of behavior in Europe or the USA.

Ableism

There is an increasing recognition of the fundamental issues surrounding individuals with disabilities, which is referred to as ableism. Ableism encompasses a set of beliefs that contribute to the experience of disability within society. It serves as a contrast to able-bodiedness, which can be understood as the antithesis of disability (Tarvainen, 2019; Jammaers, 2021). Ableism emerges as a response to the perception of a 'normal' individual as someone who functions at an average or typical level (Grue, 2014, p. 141). The language and terminology used to describe individuals with disabilities and their relationship to climate change, such as 'common' and 'uncommon' bodies, are subjects of critical examination in studies on ableism (Tarvainen, 2019, p. 292).

Grue (2014) argues that criticism of ableism often portrays individuals with disabilities as victims, overlooking the need for in-depth analysis and meaningful discussions on specific challenges they face. Drawing parallels to the struggles of men of color in the 1950s in the USA, Grue (2014) highlights the various obstacles they encountered in areas such as employment, education, family life, and other aspects of society. While being a man of color may have contributed to their hardships, it was not the sole reason behind their difficulties (Grue, 2014). Similarly, the discourse on ableism suggests that the challenges faced by individuals with disabilities are not solely due to their disability, but are often rooted in societal interactions and attitudes.

The Impact of Climate Change on Persons with Disabilities

Individuals with physical disabilities face heightened risks and challenges in the face of climate change compared to the general population. These vulnerabilities stem from a range of physical limitations, including impairments in hearing, cognition, and symptom exacerbation, as well as difficulties in regulating body temperature (Stella, Mari, Gabrieli, Barbante, & Bertuzzo, 2020). Additionally, limitations in mobility and the ability to repair or replace adaptive equipment further compound the vulnerabilities of individuals with disabilities during emergency preparedness, evacuation, and recovery efforts (Aryankhesal, Pakjouei, & Kamali, 2018). Extreme weather events can disrupt access to essential healthcare services for individuals with disabilities, while their assistive devices may be lost or damaged, further exacerbating their vulnerability (Jodoin, Lofts, & Ananthamoorthy, 2020; UNHCHR, 2020). These factors underscore the need for targeted interventions and support systems to address the unique challenges faced by individuals with physical disabilities in the context of climate change.

Individuals with physical disabilities or chronic conditions often rely on consistent electrical power sources to sustain their health and safety, such as respirators or dialysis machines. The absence of reliable electricity can pose life-threatening risks to these individuals, while other disruptions, like power outages affecting power chairs or CPAP machines, can jeopardize their independence and well-being if prolonged. Climate-related disasters exacerbate these challenges further, as physical barriers in the environment and systemic oversights create additional obstacles for individuals with disabilities. Emergency planners frequently overlook the specific needs of this population, resulting in evacuation centers that are inaccessible or unsafe during crises (Aryankhesal, Pakjouei, & Kamali, 2018).

People with physical disabilities face significant challenges during disasters, as they are often excluded from evacuation centers, rendering essential services such as food and medical assistance inaccessible (King & Gregg, 2022). This exclusion can have compounding effects, particularly for women with disabilities who are already at a higher risk of experiencing gender-based violence. In times of crisis, health systems may struggle to meet the needs of individuals with disabilities, as services like medication refills and adaptive equipment replacements are often inflexible and not designed to accommodate evacuation or displacement scenarios. This lack of flexibility can further exacerbate vulnerabilities and increase the risks faced by people with disabilities during emergencies (King & Gregg, 2022). Taking a critical realist perspective, it becomes evident that the health vulnerabilities of individuals with disabilities are heightened during disasters due to poor planning and exclusionary policies. This increased exposure to environmental risks underscores the urgent need for inclusive emergency preparedness and response strategies that prioritize the needs of all individuals, including those with disabilities (Zhang, Bi, & Hiller, 2007).

Individuals living with disabilities face the brunt of the consequences of climate change due to their heightened exposure to social determinants of health influenced by climate-related factors, such as water accessibility (Wolbring, 2011). Despite the significance of social determinants of health in public health discussions, the impact on people with disabilities is often overlooked, leading to their exclusion from critical studies in sociology and broader health discourse. The absence of individuals with disabilities in conversations surrounding social determinants of health can be attributed to societal norms that prioritize able-bodied individuals, perpetuating ableism and overlooking the unique vulnerabilities faced by those with disabilities in the face of environmental challenges (Okoro et al. 2018). This exclusion highlights the need for a more inclusive approach to public health that considers the diverse needs and experiences of all individuals, including those with disabilities (US Centre for Disease Control, CDC, 2020).

Inclusive discourse plays a critical role in effectively preventing and planning future policies. The combined impact of increased vulnerability and lack of visibility raises concerns about inequality and climate injustice. Climate change not only poses challenges for individuals with existing disabilities, but also contributes to the spread of both chronic and infectious diseases (Zhang, Bi, & Hiller, 2007). Furthermore, the thawing of permafrost could lead to the resurgence of diseases that have been absent from our planet for thousands of years (Stella, Mari, Gabrieli, Barbante, & Bertuzzo, 2020). These diseases have the potential to disable individuals who were previously unaffected and exacerbate the vulnerability of those already living with disabilities.

Individuals with disabilities are more inclined to report lower overall health and reduced access to adequate healthcare, as indicated by the US Centre for Disease Control (CDC, 2020). Consequently, infectious diseases will have a disproportionate impact on this demographic. Research studies focusing on these effects would address the need for investigations that can enhance our comprehension of the various factors contributing to vulnerability, such as genetic, physiological, social, and behavioral aspects, and their causal mechanisms concerning the specific health impacts of climate change, as highlighted by the US Global Change Research Program (USGCRP, 2018, p. 263). It is important to recognize that not only are individuals with disabilities at an elevated risk of being more sensitive to climate change, but climate change also has the potential to further expand the scope of disability.

SOLUTIONS AND RECOMMENDATIONS

From the above discussion, several solutions and recommendations are given as to how persons with disabilities can adapt to climate change.

(a) A perspective on climate governance through the lens of disability rights emphasizes the significance of acknowledging and safeguarding the rights, both substantive and procedural, of individuals with disabilities throughout the process of formulating, executing, overseeing, and assessing climate-related policies and initiatives. This approach also mandates that governments implement climate adaptation strategies that safeguard individuals with disabilities from harm during climate-related crises and bolster their ability to withstand the effects of climate change. These responsibilities stem from the overarching obligations outlined in the UNCRPD to uphold, protect, and fulfill the rights of individuals with disabilities, particularly in the face of challenges posed by climate change.

(b) It is crucial to guarantee that individuals with disabilities and their advocacy groups are actively involved in all stages of climate change policy development and planning at various levels. This helps create a conducive and supportive atmosphere that empowers individuals with disabilities to engage and take proactive steps in addressing climate change issues.

(c) It is crucial to continue investing in enhancing inclusive strategies within high-impact sectors that have the potential to foster the development of climate-resilient and inclusive societies. By focusing on these sectors, we can make significant strides towards creating a more sustainable and equitable future for all.

(d) Encourage the adoption of a comprehensive and fair programmatic strategy for climate resilience, which encompasses social, economic, health, and climate factors, and delivers extensive advantages to all marginalized communities.

(e) Guarantee that climate funding for the general population also includes and promotes the active and significant involvement of disability organizations. This should involve forming partnerships with a variety of disability organizations, sharing knowledge, conducting intersectional analyses of climate vulnerabilities and capacities during needs assessments, and disaggregating data.

(f) Allocate specific funding for climate action initiatives that are inclusive of individuals with disabilities, ensuring that organizations representing persons with disabilities are actively involved in all stages of the project - from design and planning to implementation and monitoring.

(g) Advocate for initiatives that enhance collaboration among different levels of government and promote the advancement of decentralization efforts, with a specific focus on distributing climate finance to support those who are most at risk.
(h) Participate in advocacy efforts based on evidence at the national level, leveraging improved internal collaboration between disability organizations and pursuing collaborative advocacy campaigns. Emphasize the creation of a supportive policy environment that recognizes the diverse effects of climate change, allocates specific resources to address these impacts, and ensures fair and accessible treatment of individuals with disabilities by service providers.

FUTURE RESEARCH DIRECTIONS

The global community is currently grappling with significant challenges that highlight the urgency to dismantle obsolete institutional policies, systematic structures, and ideologies, while also emphasizing the importance of reconstructing evidence-based approaches for capacity building. Climate change stands out as a complex issue, often referred to as a wicked problem due to the uncertainty surrounding its nature and the knowledge required to address it. This underscores the necessity to dispel misconceptions and outdated narratives. A comprehensive approach involving disability studies, climate change studies, and policy studies is essential to address these pressing global issues. Disability studies offer a humanistic perspective on individuals with disabilities, while climate change studies provide scientific and technological insights, and policy studies facilitate community-level implementation and resource allocation.

Addressing the critical shortage in disability, climate change, and policy studies, along with limited resources and the lack of participants from underrepresented communities, is crucial to unlocking the full potential of individuals with disabilities and their communities. Scholars and researchers play a key role in bridging this participation gap, potentially leading to advocacy for disability-inclusive solutions. By engaging individuals with disabilities in these discussions, researchers can catalyze a ripple effect of change agents.

CONCLUSION

An emphasis on disability rights within climate governance highlights the necessity of upholding the rights of individuals with disabilities throughout all climate policy development and implementation stages. This includes safeguarding their substantive and procedural rights and ensuring their inclusion in monitoring and evaluating climate programs. States are urged to implement climate adaptation strategies that specifically address the needs of individuals with disabilities during climate-related emergencies. These policies should aim to shield them from harm and bolster their ability to withstand the effects of climate change. The foundation for these obligations lies in the overarching principles outlined in the UN Convention on the Rights of Persons with Disabilities (UNCRPD). States are called upon to not only recognize but also protect and fulfill the rights of individuals with disabilities, particularly in the face of challenges posed by climate change. While the impacts of climate change are felt on a global scale, solutions must be developed at the local level. These solutions should be inclusive, taking into account the perspectives of those most affected by climate change, and should prioritize a bottom-up approach to address the diverse needs of communities.

The impacts of climate change on humans are extensive, yet they are not evenly distributed across society, with persons with disabilities bearing a disproportionate burden. Unfortunately, decision-makers and planners often overlook persons with disabilities, rendering them "invisible" in discussions. Regrettably, current climate action plans do not adequately consider how countries will mitigate the unequal effects of climate change on individuals with disabilities. This failure to address the issue exacerbates existing disparities and exposes people with disabilities to significant, and potentially life-threatening, risks in both the short and long term. It is crucial for climate action plans to take into account the disproportionate impact of climate change on individuals with disabilities. The effects of climate change are not uniform across all populations. There is a growing awareness that people with disabilities are particularly vulnerable to the compounded effects of climate change, including poverty, discrimination, stigma, and intersecting vulnerability factors, which place them at a heightened risk.

REFERENCES

Abedin, M. A., Habiba, U., & Shaw, R. (2013). Gender and climate change: Impacts and coping mechanisms of women and special vulnerable groups. *Climate change adaptation actions in Bangladesh,* pp.165-184.

Albrecht, G. L., Ravaud, J. F., & Stiker, H. J. (2001). L'émergence des disability studies: état des lieux et perspectives. The emergence of disability studies: state of places and of perspectives. *Sciences Sociales et Sante*, 19(4), 43–73. DOI: 10.3406/sosan.2001.1535

Aryankhesal, A., Pakjouei, S., & Kamali, M. (2018). Safety Needs of People with Disabilities During Earthquakes. *Disaster Medicine and Public Health Preparedness*, 12(5), 615–621. DOI: 10.1017/dmp.2017.121 PMID: 28965529

Brewer, E., Brueggemann, B., Hetrick, N., & Yergeau, M. (2012). Introduction, background, and history. In Brueggemann, B. J. (Ed.), *Arts and Humanities* (pp. 1–62). SAGE Publications. DOI: 10.4135/9781452218335.n1

Brisenden, S. (1986). Independent Living and the Medical Model of Disability. *Disability, Handicap & Society*, 1(2), 173–178. DOI: 10.1080/02674648666780171

Centers for Disease Control and Prevention. (2020). Disability and Health Related Conditions. www.cdc.gov

Convention on the Rights of Persons with Disabilities. (2006). Convention on the Rights of Persons with Disabilities. www.ohchr.org

Field, B. C. (2012). *Summary for Policymakers in Managing the Risks of Extreme Events and Disasters to Advance Climate Change Adaptation: A Special Report of Working Groups I and II of the Intergovernmental Panel on Climate Change*. Cambridge University Press. DOI: 10.1017/CBO9781139177245

Fraser, N. (2000). Rethinking Recognition. *New Left Review*, 3, 107–120.

Grech, S. (2009). Disability, Poverty, and Development: Critical reflections on the majority world debate. *Disability & Society*, 24(6), 771–784. DOI: 10.1080/09687590903160266

Grech, S. (2015). Decolonizing Eurocentric disability studies: Why colonialism matters in the disability and global South debate. *Social Identities*, 21(1), 6–21. DOI: 10.1080/13504630.2014.995347

Hemingway, L., & Priestley, M. (2006). Natural Hazards, Human Vulnerability and Disabling Societies: A Disaster for Disabled People? *Review of Disability Studies*, 2(3).

Ingstad, B. (1999). The myth of disability in developing nations. *Lancet*, 354(9180), 757–758. DOI: 10.1016/S0140-6736(99)06049-3 PMID: 10475204

IPCC. (2022). Summary for Policymakers in H.-O. Pörtner, D.C. Roberts, M. Tignor, E.S. Poloczanska, K. Mintenbeck, A. Alegría, M. Craig, S. Langsdorf, S. Löschke, V. Möller, A. Okem, B. Rama (Eds.). *Climate Change 2022: Impacts, Adaptation, and Vulnerability.Contribution of Working Group II to the Sixth Assessment Report of the Intergovernmental Panel on Climate Change.* Cambridge University Press, Cambridge, UK and New York, NY, USA, pp. 3-33.

Jammaers, E. (2021). Embodied reflections of an able-bodied disability scholar. *Gender, Work and Organization*, 28(5), 1885–1892. DOI: 10.1111/gwao.12714

Jodoin, S., Lofts, K. A., & Ananthamoorthy, N. (2020). A Disability Rights Approach to Climate Governance. *SSRN*, 47(1). Advance online publication. DOI: 10.2139/ssrn.3610193

Jones, L., & Boyd, E. (2011). Exploring social barriers to adaptation: Insights from Western Nepal. *Global Environmental Change*, 21(4), 1262–1274. DOI: 10.1016/j.gloenvcha.2011.06.002

Juhola, S., Heikkinen, M., Pietilä, T., Groundstroem, F., & Käyhkö, J. (2022). Connecting climate justice and adaptation planning: An adaptation justice index. *Environmental Science & Policy*, 136, 609–619. DOI: 10.1016/j.envsci.2022.07.024

Kafer, A. (2013). *Feminist, Queer, Crip*. Indiana University Press.

Kett, M., Sriskanthan, G., & Cole, E. (2021). *Disability and Climate Justice: A Research Project.* Open Society Foundations.

King, M.M., & Gregg, M.A. (2022). Disability and Climate Change: a critical realist model of climate justice. *Sociology Compass* 1 6(1): e12954.

Leichenko, R., & O'Brien, K. (2019). *Climate and Society, Transforming the Future*. Polity Press.

Lewis, D., & Ballard, K. (2012). *Disability and Climate Change: Understanding vulnerability and building resilience in a changing world.* CBM.

Malloy, J. T., & Ashcraft, C. M. (2020). A framework for implementing socially just climate adaptation. *Climatic Change*, 160(1), 1–14. DOI: 10.1007/s10584-020-02705-6

Metcalf, H. (2009). Pay gaps across the equality strands: a review. Equality and Human Rights Commission Research Report Series. www.equalityhumanrights.com

Moran, M., Hillman, W., Mann, J., Kinnear, S., Taylor, D., & Gaskin, C. J. (2017). Of People with Disability: A Systematic Review. *Weather, Climate, and Society*, 9(4), 801–814. DOI: 10.1175/WCAS-D-16-0126.1

Office of the United Nations High Commissioner for Human Rights (UNHCHR). (2020). Analytical study on the promotion and protection of the rights of persons with disabilities in the context of climate change. www.undocs.org

Okoro, C. A., Hollis, N. D., Cyrus, A. C., & Griffin-Blake, S. (2018). Prevalence of Disabilities and Health Care Access by Disability Status and Type Among Adults — United States, 2016. *MMWR. Morbidity and Mortality Weekly Report*, 67(32), 882–887. DOI: 10.15585/mmwr.mm6732a3 PMID: 30114005

Oliver, M. (2013). The social model of disability: Thirty years on. *Disability & Society*, 28(7), 1024–1026. DOI: 10.1080/09687599.2013.818773

Ortoleva, S., & Lewis, H. (2012). *Forgotten Sisters - A Report on Violence Against Women with Disabilities: An Overview of its Nature, Scope, Causes and Consequences (No. 104–2012).* Northeastern University School of Law Research Paper.

Peek, L., & Stough, L. M. (2010). Children with Disabilities in the Context of Disaster: A Social Vulnerability Perspective. *Child Development*, 81(4), 1260–1270. DOI: 10.1111/j.1467-8624.2010.01466.x PMID: 20636694

Pichdara, L., Monin, N., Marong, C., Sivmuy, D., & Saren, K. (2022). Challenges of targeting poor and vulnerable groups to reduce climate change vulnerability: The case of a Water and Sanitation project in Kampong Svay District, Cambodia. Financial Crises, Poverty and Environmental Sustainability: Challenges in the Context of the SDGs and Covid-19 Recovery, pp.155-169.

Priddy, A. (2019). Disability in Armed Conflict. The Geneva Academy. www .geneva-academy.ch

Ray, S. J., & Sibara, J. (Eds.). (2017). *Disability Studies and the Environmental Humanities: Toward an Eco-Crip Theory*. University of Nebraska Press. DOI: 10.2307/j.ctt1p6jht5

Roush, S. E., & Sharby, N. (2011). Disability reconsidered: The paradox of physical therapy. *Physical Therapy*, 91(12), 1715–1727. DOI: 10.2522/ptj.20100389 PMID: 22003161

Russell, M. (2019). *Capitalism and disability: Selected writings by Marta Russell.* Haymarket Books.

Schipper, E. L. F. (2020). Maladaptation: When adaptation to climate change goes very wrong. *One Earth*, 3(4), 409–414. DOI: 10.1016/j.oneear.2020.09.014

Schlosberg, D. (2013). Theorizing environmental justice: The expanding sphere of a discourse. *Environmental Politics*, 22(1), 37–55. DOI: 10.1080/09644016.2013.755387

Shakespeare, T. (2013). *Disability Rights and Wrongs Revisited* (second). New York: Routledge.

Shakespeare, T. (2013). *Disability Rights and Wrongs Revisited* (second). New York: Routledge.

Stella, E., Mari, L., Gabrieli, J., Barbante, C., & Bertuzzo, E. (2020). Permafrost dynamics and the risk of anthrax transmission: A modeling study. *Scientific Reports*, 10(1), 16460. DOI: 10.1038/s41598-020-72440-6 PMID: 33028874

Tarvainen, M. (2019). Ableism and the life stories of people with disabilities. *Scandinavian Journal of Disability Research*, 21(1), 291–299. DOI: 10.16993/sjdr.632

Thomas, C. (1999). *Female Forms: Experiencing and understanding disability. Buckingham.* Open University Press.

Thomas, C. (2004). Developing the social-relational in the social model of disability: a theoretical agenda. In Barnes, C., & Mercer, G. (Eds.), *Implementing the Social Model of Disability: Theory and Research* (pp. 32–47). Disability Press.

Ton, K. T., & Adamson, C. (2020). Disaster justice for people with disabilities. *Disaster Prevention and Management*, 30(2), 125–136. DOI: 10.1108/DPM-08-2020-0248

UN General Assembly. (2007). *Convention on the Rights of Persons with Disabilities: resolution / adopted by the General Assembly, A/RES/61/106.* www.refworld.org

UNICEF. (2022). *UNICEF Fact Sheet Children with Disabilities.* www.unicef.org

Union of the Physically Impaired Against Segregation (UPIAS). (1976). *Fundamental Principles of Disability.* UPIAS.

United Nations. (2015). *General Comment on equal recognition before the law.* www.ohchr.org

United Nations. (2018). *Persons with disabilities: Breaking down barriers.* www.un.org

U.S. Global Change Research Program (USGCRP). (2016). *The Impacts of Climate Change on Human Health in the United States: A Scientific Assessment.*

WHO. (2002). *Towards a common language for functioning, disability and health.* www.who.org

Williams, S. J. (1999). Is anybody there? Critical realism, chronic illness and the disability debate. *Sociology of Health & Illness*, 21(6), 797–819. DOI: 10.1111/1467-9566.00184

Wolbring, G. (2009). A Culture of Neglect: Climate Discourse and Disabled People. *M/C Journal,* 12(4).

Wolbring, G. (2011)... *People with Disabilities and Social Determinants of Health Discourses.*, 102(4), 317–319.

World Health Organization (WHO). (2001). *International classification of functioning, disability and health.* www.who.org

Zhang, Y., Bi, P., & Hiller, J. E. (2007). Climate change and disability-adjusted life years. *Journal of Environmental Health*, 70(3), 32–36. PMID: 17941401

ADDITIONAL READING

Gaskin, C. J., Taylor, D., Kinnear, S., Mann, J., Hillman, W., & Moran, M. (2017). Factors associated with the climate change vulnerability and the adaptive capacity of people with disability: A systematic review. *Weather, Climate, and Society*, 9(4), 801–814. DOI: 10.1175/WCAS-D-16-0126.1

Kellock, A. (2020). Children's well-being in the primary school: A capability approach and community psychology perspective. *Childhood*, 27(2), 220–237. DOI: 10.1177/0907568220902516

Keogh, M. (2020). *Climate Change: This Century's Defining Issue*. CBM.

Russell, M. (2019). *Capitalism and disability: Selected writings by Marta Russell.* Haymarket Books.

Shakespeare, T. (2006). The social model of disability. *The Disability Studies Reader*, 2, 197–204.

Ton, K. T., Gaillard, J. C., Adamson, C. E., Akgungor, C., & Ho, H. T. (2019). Expanding the capabilities of people with disabilities in disaster risk reduction. *International Journal of Disaster Risk Reduction*, 34, 11–17. DOI: 10.1016/j.ijdrr.2018.11.002

Verchick, R. R. M. (2012). Disaster Justice: The Geography of Human Capability. *Duke Environmental Law & policy. Forum*, 23(23).

KEY TERMS AND DEFINITIONS

Ableism-: the systemic discrimination against persons with disabilities.

Adaptive capacity-: the ability of communities, institutions, or people to adjust to potential hazards, to take advantage of opportunities, or to respond to consequences.

Climate adaptation-: adjusting ecological, social, or economic systems to the experienced or expected climatic conditions and following climatic events

Climate change-: a change in the state of the climate that can be identified by changes in the mean and/or the variability of its properties and that persists for an extended period, typically decades or longer.

Just adaptation-: a field within climate adaptation that focuses on the role of socially vulnerable groups in climate adaptation policy.

People with disabilities-: those who have long-term physical, mental, intellectual, or sensory impairments which in interaction with various barriers may hinder their full and effective participation in society on an equal basis with others.

Vulnerability-: the likeliness of someone being negatively or adversely affected by something.

Chapter 5
Natural Disaster:
Prevalence of Post-Traumatic Stress and Psychological Outcome of Survivors in Joshimath, Uttarakhand, India

Sudhanshu Negi
Graphic Era University, India

Ravindra Singh
https://orcid.org/0000-0003-3062-1812
Magadh University, Bodh Gaya, India

Ajay Kumar Singh
https://orcid.org/0000-0003-0429-0925
Graphic Era University, India

ABSTRACT

This chapter examines the implication of natural disaster on mental health of children and adolescents in Joshimath of Uttarakhand, India. Mental health is caused to increase death anxiety and psychological disorder among the children due to the existence of natural disasters. Natural disasters cause increased problems of post-traumatic stress disorder (PTSD) among the children and adolescents. For the mentioned investigation, this chapter includes that sample of two hundred eighty children and adolescents in the age group of 12 – 18 years. The sample consists of 116 boys and 164 girls with mean and standard (SD) of 14.52 years and 1.70, respectively. The sample of 280 students are taken from the different schools that were affected from the natural disaster. The results highlight that policy makers should develop intervention and prevention tools to overcome the stress of natural

DOI: 10.4018/979-8-3693-4183-4.ch005

disasters on mental health of children & adolescents.

1. INTRODUCTION

Climate change has a significant direct and indirect impact on mental health and well-being of people (Shivanna, 2022). It is a slow process in which the values of most climatic factors increase from their mean in a long-period (Kumar et al., 2017). Changes in climatic factors in different weather seasons are responsible for increasing many health issues among the students in academic institutions (Ashraf & Singh, 2022). Also, climate change has an adverse impact on multiple activities like agricultural production, food security, livelihood security, income, health security, and ecosystem services that are associated with livelihood security of people (Bazga, 2015; Kumar & Singh, 2023; Singh et al., 2023a,b). While unexpected increase in intensity of storms, rising sea levels and coastal flooding, droughts and wildfires, heatwaves and heat-related illnesses & changing rainfall patterns are the crucial events of natural disasters (Agbola & Ojeleye, 2007; Makwana, 2019; Singh et al., 2024).

Land sinking that is also known as land subsidence or land settlement. It also refers to the gradual or rapid sinking of the Earth's surface (Neria et al., 2008). Land sinking due to human activities has several adverse consequences on the livelihood security of people. Moreover, land sinking is responsible for increased damages of infrastructural development, and increased flood risk, saltwater intrusion, loss of ecosystems, and mental health issues among the people (Wade, 2022). It is a downward movement in the earth's surface due to change in natural processes or soil erosion (Ashraf & Singh, 2022; Davydzenka et al., 2024). In the last two to three decades frequency of natural calamities has become more severe and it is reducing human capital (Malhi et al., 2021). It also creates mental health hazards in the natural disaster affected area across the globe. Natural calamities are also hampering multiple activities and creating enormous negative impact on health and nutritional security, and mental health.

Climate change is a threat to the well-being and mental health of rural and urban communities (Cianconi et al., 2020). It is also caused to increase mental sickness and health hazards (Alam et al., 2011). Natural calamities or disasters include heavy storms, floods, landsliding, wildfires, and industrial accidents (Faridi & Wadood, 2010; Alam et al., 2011). The mentioned activities have profound psychological and emotional impacts on mental health of local communities (Uddin et al., 2021). In the last 20 years, natural disasters have become responsible for millions of deaths across countries (Agyapong et al., 2022). Natural disaster, therefore, is a serious public health concern due to several reasons. It contributes to loss of life and property, social disturbance, health services, economic shutdown, and disrupts psycho-social well-being (Yorulmaz & Karadeniz, 2021).

A cross-sectional study systematically comparing the multiple traumatic events and cumulative mental health in FMM on the residents of Fort McMurray between April 24 to June 2 2021 (Agyapong et al., 2022). The goal to assess if due to the increased number of traumatic events (coronavirus pandemic, flooding 2020, wildfire which occurred in 2016) and a significant impact was found on the intensity and prevalence of disorders like Major Depressive Disorder (MDD), PTSD, low levels of Resilience and Generalized anxiety Disorder (Agyapong et al., 2022). The hypothesis, thus, formulated suggested that residents who endured the traumatic implications of the 2020 flooding or 2016 wildfire have experienced worse mental conditions than the population just undergoing the Covid19 pandemic. The result of this study could be successful for concluding that those respondents who were affected by 2020 flooding and 2016 wildfires in Fort McMurray along with the Covid19 pandemic had worse mental health in comparison to population with just trauma under the Covid19 pandemic (Agyapong et al., 2022). Further, there exists a correlation between traumatic disasters experienced and mental health burdens (Agyapong et al., 2022). This study successfully added research and clinical evidence relating to disaster and trauma. It also increased the attention of the government and policy makers for providing support to the vulnerable communities to cope with the ongoing pandemic.

Another study with a sample of 655 adults who have experienced a disaster event investigated the mediating role of disaster resilience between disaster exposure and post-traumatic stress and depressive symptoms in the USA using the structural equation model (First et al., 2024). The association between disaster exposure as a predictor, and depression and post-traumatic stress disorder as dependent variables was found to be mediated by catastrophe resilience, according to the findings of First et al. (2024). These findings can guide future disaster mental health interventions and practice models, and they have significant implications for our knowledge of the mechanisms by which catastrophe resilience improves post-disaster mental health. Another research attempted to understand the relationship between disasters and their effect on mental health through a systematic review and meta -analysis adhering to the PECO framework across Asia, America and Europe. The study was successful in establishing a connection between disaster and poorer mental health outcomes (Keya et al., 2023).

Disruption of essential services and relocation lead to an increase in fatalities and psychological morbidities. The most emphasized calamity being flooding. The meta -analysis revealed the highest prevalence rate of mental health disorders in "medium human development countries". Following catastrophic events, a high prevalence rate of mental disorders was also found in the "very high human development" and "high human development" nations (Keya et al., 2023). Strategies to avoid and mitigate mental problems during natural disasters can be developed using this study. A suitable mitigation strategy, improved access to healthcare services and increased community resilience, can aid to better the circumstances of the disaster's vulnerable population (Wang et al., 2022). Most scholars have identified following health issues due to unexpected change in climatic factors and natural disasters:

1.1. Post-Traumatic Stress Disorder (PTSD)

Natural disasters often expose individuals to traumatic experiences. These traumatic events can trigger symptoms of PTSD, including intrusive memories, flashbacks, nightmares, hyperarousal, and avoidance of reminders associated with the natural disaster. The severity and duration of PTSD symptoms can vary among individuals. However, natural disasters have a significant and negative impact on daily functioning activities of people and their overall well-being (Neria et al., 2008).

1.2. Anxiety and Depression

Different types of anxiety disorders can emerge due to exposure to natural disasters. The uncertainty and unpredictability in natural disasters and climate change can lead to increased excessive worry and panic attacks among the people. Further, it also increases safety concerns and re-experiencing of traumatic memories (Grupe & Nitschke, 2013). Additionally, natural disasters are also accountable to decrease development, and increase progression of severe depression. The loss of assets, livelihood opportunities, and the disruption of day-to-day routines can lead to increased feelings of grief, hopelessness, and despair among the people. The emotional and psychological toll of the disaster, coupled with the challenges of recovery and rebuilding, can contribute to the onset of depressive symptoms (Shultz et al., 2013).

1.3. Stress and Psychological Distress

Natural disasters also create significant stressors that can overwhelm individuals' ability to cope with the impact of it. The immediate and long-term challenges of dealing with evacuation, loss of homes, financial strain, displacement, and the uncertainty of recovery can contribute to high levels of stress and psychological distress. This distress can manifest as irritability, difficulty concentrating, sleep disturbances, and physical health problems (Keya et al., 2023). Bastami et al. (2024) also observed that flood victims also face post-traumatic stress disorder due to natural disasters.

1.4. Death Anxiety

Natural disasters can also significantly develop or trigger death anxiety. Individuals may experience a heightened awareness of mortality and an increased fear of their own or their loved ones' death (Hoelterhoff & Chung, 2013). The uncertainty and vulnerability exposed by the disaster can intensify existential concerns and thoughts about mortality.

2. REVIEW OF LITERATURE

Victims of a natural disaster feel negative self-evaluation, anxiety and distress. Natural disaster, therefore, has a direct impact on livelihood security of people (Kar & Bastia, 2006; Singh et al., 2022). Natural disasters also destroy the physical assets like houses, buildings, and agricultural land which also increase post-traumatic stress disorder (PTSD) of survivors. Natural disaster affected families have five times more PTSD as compared to non-natural disaster affected families. All activities related to natural disasters are causing serious mental health problems among the people. PTDS can remain prevalent amongst survivors even after three years of incidence (Pandey et al., 2023). The WHO has suggested that post-disaster psychiatric intervention can definitely help and prevent the chances of PTSD 1995 (Yehuda et al., 2015). Suicidal ideation and drug abuse may also occur in natural disaster-prone areas (Fergusson et al., 2014).

PTSD has the significant association between being buried after an earthquake, loss of loved ones, and witnessing close ones being buried, injured or died (Cénat & Derivois, 2014). Collectively people can be accounted as autonomous factors for PTSD (Kuo et al., 2007). Survivors of a natural disaster may also feel more negative trauma due to family and social responsibilities. It can also produce instance financial pressure and psychological distress, which further can contribute to higher PTSD among the people (Zhou et al., 2013). The frequency of PTSD cases among the people is relatively higher in the flood affected areas. Therefore, it also creates a negative attitude, loss of interest and disturbed lifestyle, and increases mental disorders. While gender has emerged as a key factor which may contribute to the prevalence of PTSD in the survivor of a natural disaster. More women were found under the risk of PTSD after the natural disasters. In case of earthquakes, females have reported higher prevalence of PTSD then men's (Dai et al., 2016). The mentioned finding shows assimilation and accordance with the previous studies (Acierno et al., 2007). Furthermore, it is noticed that resilience plays a mediating role between tragedy experience as a predictor and depression, and PTSD as dependent variables (First, 2024).

Natural disasters also increase the extensive vulnerability for poor people and marginally poor communities to sustain their livelihood security (Greg et al., 2011; Samhsa, 2017; Singh et al., 2022). Past evidence implies that rapid industrialization and urbanization are making natural disasters more frequent (Prasad & Francescutti, 2017; Kumar & Singh, 2023). Human activities are the sole cause for the rising quantity of greenhouse gases in the atmosphere and it is increasing overburden on the environmental system (USGCRP, 2017). The high occurrence in most climatic factors is reported by many scientific research communities and international organizations (Malhi et al., 2021). The scientific research community is universally admitting that climatic factors are changing due to rising quantities of CO_2 and GHGs emissions in the environment (Malhi et al., 2021). Subsequently, climate change and natural disasters have several direct implications on human health that are related to mental well-being.

2.1. Impact of Climate Change on PTSD Cases

Increased intensity of storms, rising sea levels and coastal flooding, droughts and wildfires, heatwaves and heat-related illnesses & changing rainfall patterns are showing the existence of climate change and natural disaster (Mall et al., 2006; Agbola, & Ojeleye, 2007; Alam et al., 2011; Banholzer et al., 2014; de Almeida and Mostafavi 2016). Land sinking, also known as land subsidence or land settlement that is refers to the gradual or rapid sinking of the Earth's surface. Human activities can significantly contribute to land sinking, often through the extraction of groundwater from underground aquifers and extraction of minerals from earth. Climate change and environmental degradation have a significant impact on mental health. It, therefore, is obvious that most people face emotional and mental health problems (e.g. distress, depression, anxiety, loneliness, fear, and anguish) due to natural disasters. These negative psychological effects may appear for shorter duration, but significantly elevate the level of stress and anxiety. In the long run the presence of these symptoms due to natural disasters can develop post-traumatic stress disorder (PTSD) and depression (Goldmann & Galea, 2014). It also reported that the occurrence of PTSD among individuals leads to increased psychological issues. It also found that persons above 35 years are adversely affected due to floods. They are more vulnerable due to low education level and low income that further caused a high prevalence of PTSD. Passive coping strategies along with social coping strategies were significantly correlated with the average PTSD.

To deal with PTSD social support from NGOs that can provide training programs at various levels to build resilience and improve mental health and healthcare facilities. Therefore, an effective mass level screening of PTSD is essential for promoting positive coping strategies, and psychological well-being for survivors. Furthermore, there is a positive correlation between passive and social coping strategies, with PTSD (Charuvastra & Cloitre, 2008). Findings of Thompson et al. (2018) also suggested that there is a significant correlation between social support and the course of development of PTSD. Passive and social coping strategies were much lower in individuals with low level education (Thompson et al., 2018). The cases of PDST after a natural disaster are much higher among affected people's comparison to the nearby population.

These impacts are being felt by people all over the world, and they are likely to become more severe in the future. Mental health issues are critical due to increasing climate change and environmental degradation (Benevolenza & DeRinge, 2018). Humanity faces several health issues due to increasing environmental degradation and natural disaster. Accordingly, these may be responsible for increasing psychological disorders in terms of anxiety, depression and mental stress. Natural disasters are highly unfavourable events. Although, natural disasters are the past of earth's natural processes that are frequently linked with casualties, trauma, and loss of property (Chaudhary & Awais, 2021). However, these events can pose a serious threat to our psychological well-being in several ways. These also cause both immediate and long-term psychological distress. Collectively these significantly increase the burden of mental health disorders on both the affected individuals and the community (Saeed & Gargano, 2022).

Cianconi et al. (2020) looked at how the general public, especially vulnerable communities and at-risk groups, were affected by global climate change. To reduce extreme fluctuations in temperatures, heat waves, floods, droughts, tornadoes, hurricanes, and wildfires should be the prime concern for global makers to sustain the livelihood security of common people (Seneviratne et al., 2012). In terms of distress symptoms, suicide rates, and clinical illnesses like depression, anxiety, sleep difficulties, and PTSD are also investigated by the existing researchers and scientists (Bell et al., 2018). Even though some situations may unfold gradually, the majority start off quickly and end in catastrophes, which frequently result in PTSD (Neria et al., 2008; Makwana, 2019). Furthermore, those who are more susceptible to meteorological and atmospheric phenomena may be more affected by climate change. The gradual changes in climatic factors are also caused to increase sea levels and terrain changes (Agbola, & Ojeleye, 2007). People use different coping mechanisms to reduce the adverse impact of natural disasters on their livelihood style (Kumar et al., 2017; Rahmani et al., 2022; Singh et al., 2022).

In addition to above, past studies have investigated how young people have experienced natural disasters, symptoms of internalizing and externalizing difficulties, including PTSS, frequently co-occur? The international community will take effective action to combat climate change and natural disaster that could be strengthened by gaining a thorough understanding of the myriad hazards. Global warming is always dangerous for people's well-being and health, particularly for the health of migrant's people (Agbola, & Ojeleye, 2007; McMichael et al., 2012). However, given the fact that change in climatic factors is already present and temperature increases that is unavoidable. While it is essential to create and discover adaptable methods that can be positive to increase human health in the presence of natural disasters. The negative effects of climate-related migration and health can be greatly reduced by abating GHGs emissions and implementing extensive socio-ecological adaptation mechanisms worldwide.

Understanding the relationship between natural disasters and mental health is essential for developing effective strategies to support and assist individuals and communities from the above-mentioned problems. Most communities in lower Himalayan regions are highly dependent on climate and natural resources. While they are facing the direct impact of climate change and natural disaster on their mental health and well-being. Traumatic events have a profound impact on individuals, and children. Also, they face many struggles to cope with the emotional distress due to natural disasters. The Himalayan regions of Uttarakhand, India is the most climate sensitive and natural disaster-prone area. International boundary of China is also nearest to the Himalaya regions. Politically it is a very important geographical area of India in multiple perspectives. Therefore, it is necessary to provide social and economic security to the vulnerable group of the society to prevent them from natural disaster and climate change in this region. It will be useful to boost confidence and enhance the resilience capacity of people to cope with climate change and natural disaster.

3. OBJECTIVE AND HYPOTHESIS OF THE STUDY

This chapter, therefore, has following hypothesis:

- H_1: There would be a significant and positive correlation between natural disaster and mental well-being of children & adolescents.
- $H_{2:}$ There would a significant and positive correlation between natural disaster and death anxiety of children & adolescents.
- $H_{3:}$ There would be a significant and positive correlation between death anxiety and depression of children & adolescents.
- $H_{4:}$ There would be a significant and positive correlation between death anxiety and stress of children & adolescents.

In order to check the validity of above-mentioned hypothesis, the present chapter has following objectives:

- To determine the correlation between natural disaster (PTSD) and depression, anxiety, & stress.
- To determine the correlation between natural disaster (PTDS) and death anxiety.
- To determine the correlation between death anxiety and depression & stress.

4. RESEARCH METHODOLOGY

4.1. Study Area

The research is conducted in Joshimath, a small town of Uttarakhand state of India. Joshimath is a small town located in Chamoli district of Uttarakhand. It is located at an altitude of 1,875 meters. The livelihood security of people of Joshimath primarily depend on tourism, seasonal agriculture activities and ecosystem services. It is situated in a geographically vulnerable area of Himalayas. The population of Joshimath is completely exposed to natural disasters and climate change every year. The region is prone to various environmental hazards, such as land slide, land sinking, unlikely flash floods, and avalanches. Most people often face heavy rainfall, melting glaciers, and mild to heavy seismic activity. The sinking of land has several significant effects on the local population, leading to various challenges, and increasing disruptions in their daily livelihood security of common people, children and women.

Earlier studies argued that most people face different problems due to climate change and natural disasters in Joshimath. Natural disasters show a negative impact on physical safety of children, and put their livelihood and well-being in risk due to incidence of injury. Moreover, natural disasters also have an adverse impact on emotion in terms of intense fear, anxiety, trauma and multiple range of mental health problems of the people. It also creates disturbance in the education of children and difficulties in adjusting to the new relocated environment. Displacement and social disruption also increase as natural disasters increase. Further, it also causes increased loss of social connections, separation from friends, and feelings of isolation. Several health concerns increase due to inadequate sanitation, disrupted healthcare services, and unsanitary living conditions after natural disasters.

4.2. Data and Resource: Participants

The chapter is based on primary data that include the information of 280 children who are selected from the study site. The sample of 280 respondents includes 116 boys and 64 girls with mean age of 14.52 years, standard deviation (SD) of 1.7, and range of 12-18 years. The students are studying in various schools (private & government), and these are also affected by the natural disaster (i.e., land sinking). These schools are selected purposely as per the prescribed objectives of this research.

Demographic and Family History: Semi-structured Proforma is used for recording details of the participants. The information comprises age, education, residence, gender, total number of family members, and monthly income of the participants. Further, it used the following tools to assess the perceptions of individuals on natural disaster and its impact on desired variables.

Post-Traumatic Stress: It used the Hindi version of the children's impact of event scale -13 (CRIES-13) and it is a design to assess the impact of traumatic events on children and adolescents. The technique is developed by the Children and War Foundation (Perrin et al., 2005).

Depression, Anxiety, and Stress: The DASS-21 is widely used in research and clinical settings (Lovibond et al., 1995). It consists of three subscales, each focusing on a specific domain: depression, anxiety and stress. Depression is observed as a situation of hopelessness, and lack of interest or pleasure in activities, feelings of sadness and the presence of depressive symptoms in the respondents. Anxiety is defined as excessive worry, apprehension, physiological arousal feelings of nervousness, restlessness, and tension. While stress is evaluated as tension, irritability, and difficulty coping with life's demands, difficulty relaxing, and a sense of being under pressure. For each subscale, respondents rate the extent to which they have experienced on a Likert scale ranging from 0 (Did not apply to me at all) to 3 is applied. Scores for each subscale are summarized to provide a measure of the severity of symptoms. Higher scores indicate a higher level of emotional distress in the corresponding domain (depression, anxiety, or stress). It is important to note that the DASS-21 is not intended as a diagnostic tool but rather as a measure of symptom severity.

Death Anxiety Questionnaire: It used the Hindi version of the Death Anxiety Questionnaire (DAQ). DAQ is a 15–item measure that assesses worries about the process of dying and death. Developed by Conte et al. (1982). The total DAQ of each item is rated on a three-point response scale. The respondents are requested to mark their answers using a three-point scale (0= never to 2= very much).

Procedure of Data Collection: We first take formal permission from the principal of the school to conduct a personal interview of the children. Accordingly, we make an approach to the children for filling the prescribed questionnaires. The sample of 280 children are above 12 years and having enrolment in above 6th standard classes. We have provided detailed information about this study to all the participants. The entire exercise is conducted in a classroom/area ensuring participant's privacy and confidentiality. The complete assessment (CRIES-13, DASS-21, & DAQ-15) took about 15 – 20 minutes. The study is also approved by the ethical committee of Graphic Era (Deemed to be) University Dehradun, India.

5. DATA ANALYSIS AND RESULTS

The correlation coefficient analysis is used to examine the association between depression, anxiety, and stress. Similarly, correlation between the children's impact of event scale (CRIES-13) & death anxiety, and depression & anxiety are also estimated using correlation coefficient analysis. Also, the linear regression equation is used to identify the relationship between a dependent variable (i.e., CRIES) and one or more independent variables (i.e., depression, anxiety, and stress). It is assumed that changes in the independent variables (i.e., depression, anxiety, and stress) are directly and proportionally related to changes in the dependent variable (CRIES-13).

Table 1 shows the descriptive statistics for children's revised impact of event scale scores. The total scores on the Children's Revised Impact of Event Scale (CRIES) are reported in the range between 0 to 65. Higher scores indicate the higher levels of PTSD symptoms among the children. The average score on the CRIES for the sample of children is found 30.62. This suggests that, on average, the children experienced a moderate level of PTSD symptoms (SD=11.04).

Table 1. Children's revised impact of event scale scores

Variables	M	SD	N
CRIES	30.63	11.04	280

Source: Estimated by authors. *CIES = Children's Impact of Event Scale.*

Table 2 provides the descriptive statistics for death anxiety scores of children. The total scores on the Death Anxiety Questionnaire (DAQ) of children are reported in the range of 0 – 30. Therefore, the average score on the DAQ for the sample of individuals appeared 11.60. The results imply that on average, the individuals experienced a moderate level of death anxiety. The standard deviation is found at 5.90146 and it shows that the scores in the sample varied around the mean. With a higher standard deviation, it implies that there is a significant diversity in the scores within the sample. Some individuals are reporting lower levels of death anxiety and others reporting higher levels.

Table 2. Death anxiety scores of respondents

Variables	M	SD	N
Death Anxiety	11.60	5.90	280

Source: Estimated by authors.

Table 3 provides the descriptive statistics for depression, anxiety, and stress scales of the DASS. The mean score on the depression is found 14.59. This suggests that, on average, the individuals in the sample experienced a moderate level of depression symptoms (SD- 8.10). The mean score on the Anxiety is noted 14.11 (SD- 7.90) and it reveals that there is a variability in the severity of anxiety symptoms across samples. The mean score of stress is 15.38 (SD- 8.13) and implies that, on average, the individuals in the sample experienced a moderate level of stress symptoms.

Table 3. Depression, Anxiety, and Stress Scales of the DASS

Variables	Depression	Anxiety	Stress
Depression	14.59	8.10	280
Anxiety	14.11	7.90	280
Stress	15.38	8.13	280

Source: Estimated by authors. *DASS = Depression Anxiety Stress Scales.*

Table 4 presents the descriptive statistics for the male and female respondents across samples. The table includes the number of participants (N), the mean, standard deviation, for each variable. For the variable "CRIES," the mean score is reported as 26.79 (SD = 10.7) for boys, and 33.3 (SD = 10.4) for girls. The estimate infer that girls have a higher prevalence of PTSD score as compared to boys. In the case of the "Death Anxiety" variable, boys have a mean score of 11.0 (SD = 6.2), and girls have a mean score of 12.0 (SD = 5.6). Girls, on average, have a slightly higher mean score. It indicates that girls have a higher prevalence of Death Anxiety score as compared to boys. For the "Depression" variable, boys have a mean score of 14.29 (SD = 8.49), while girls have a slightly higher mean score of 14.8 (SD = 7.8). Regarding the "Anxiety" variable, boys have a mean score of 14.36 (SD = 8.22), and girls have a slightly lower mean score of 13.93 (SD = 7.77). The "Stress" variable, boys have a mean score of 14.10 (SD = 8.72), while girls have a higher mean score of 16.29 (SD = 7.59). Girls have a higher mean score that is suggesting that female respondents have higher levels of stress on average as compared to male respondents.

Table 4. Diversity in male and female respondents

Variables	Boys			Girls		
	N	**M**	**SD**	**N**	**M**	**SD**
CRIES	116	26.79	10.70	164	33.33	10.49
Death Anxiety	116	11.05	6.21	164	12.00	5.56
Depression	116	14.29	8.49	164	14.80	7.82
Anxiety	116	14.36	8.22	164	13.93	7.77
Stress	116	14.10	8.72	164	16.29	7.59

Source: Estimated by authors.

The correlation coefficient among the CRIES, depression, anxiety, and stress is given in Table 5. The results indicate that there is a significant and positive correlation coefficient of natural disasters (PTSD) with depression (r = 0.37), anxiety (r = 0.33) & stress (r =0.48) of children and adolescents. Also, there is reported a significant positive correlation coefficient of natural disasters (PTSD) with death anxiety (r = 0.36). Death anxiety and depression also have a positive and significant correlation (r =0.45). While there seems to be a significant positive correlation between death anxiety and stress (r = 0.43). These findings suggest that natural disasters have a profound impact on the health and well-being of children and adolescent mental health.

Table 5. Correlations between CRIES, depression, anxiety, and stress

Variables	CRIES	Depression	Anxiety	Stress
CRIES	1			
Depression	0.37**	1		
Anxiety	0.33**	0.68**	1	
Stress	0.48**	0.71**	0.63**	1

Source: Estimated by authors. *CRIES = Children's Revised Impact of Event Rating Scale.*

There is reported a positive association of CRIES with depression, anxiety and stress. The total children's revised impact of event rating scale (CRIES) is positively associated with depression. The estimates are also suggesting that CRIES is caused to increase depression, anxiety & stress. The results also indicate a higher level of impact of an event, the level of depression tends to be increased. It means that children who experience a greater impact of an event are more likely to exhibit a symptom of depression. The magnitude of the correlation coefficient (0.37) suggests a moderate relationship between CRIES and depression. While not extremely strong, it still indicates a significant association between the two variables.

The correlation coefficient (*r*= 0.33) between CRIES and anxiety, suggests a weak to moderate positive relationship between them. The correlation coefficient between the Total Children's Revised Impact of Event Rating Scale (CRIES) and stress is *r*= 0.48.). The positive correlation coefficient of 0.48 indicates a moderately positive relationship between CRIES and stress.

There is found a positive correlation coefficient (*r*=0.68) between depression and anxiety, and it indicates that depression increases due to increase in anxiety. The correlation between depression and stress is found *r*= 0.71. This means that stress and depressive symptoms exist together. The correlation coefficient between anxiety and stress of *r*=0.63, indicates a moderately positive relationship between these. This correlation suggests that there is a tendency to be increased in anxiety due to increase in stress.

Table 6. Correlations between CRIES & death anxiety

	CRIES	**Death Anxiety**
CRIES	1	
Death Anxiety	0.36**	1

Source: Estimated by authors. Note: N=280, ** P<0.01, CRIES = Children's Revised Impact of Event Rating Scale

Table 6 indicates the correlations between CRIES & death anxiety. The correlation coefficient of *r*=0.36 indicates a positive relationship between the Children's Revised Impact of Event Rating Scale (CRIES) and death anxiety. The positive correlation suggests that as scores on the CRIES increase, then there tends to be a slight increase in death anxiety scores. Table 7 indicates the correlations between death anxiety and depression. The correlation coefficient of *r*= 0.45 indicates a positive correlation between death anxiety and depression. The positive correlation suggests that as levels of death anxiety increase, so does the likelihood of experiencing symptoms of depression. The correlation coefficient between death anxiety and stress is found *r*= 0.43 at 0.01. It suggests a positive relationship between death anxiety and stress. This means that as levels of death anxiety increase then levels of stress also increase.

Table 7. Correlations between Death Anxiety and Depression & Anxiety

	Death Anxiety	Depression	Stress
Death Anxiety	1		
Depression	0.41**	1	
Stress	0.43**	0.71**	1

Source: Estimated by authors. **Note:** N=280, ** p<0.01, CRIES = Children's Revised Impact of Event Rating Scale.

Table 8 indicates the results of multiple regression models that predict the variation in dependent variable due to increase in the magnitude of independent variables. Table 8 summarizes the results of a multiple regression analysis that examines the relationship between several predictors (Depression, Stress, and Death Anxiety) and a dependent variable. The model is tested using three different iterations, and the table presents the statistical measures for each model.

Table 8. Results of multiple regression models

Model	R	R Square	ΔR Square	F Change Sign	Sig. F Change
1	0.379[a]	0.144	0.144	46.608	0.000
2	0.491[b]	0.241	0.097	35.468	0.000
3	0.516[c]	0.266	0.025	9.545	0.002
a= Predictors: (Constant), Depression					
b= Predictors: (Constant), Depression, Stress.					
c= Predictors: (Constant), Depression, Stress, Death Anxiety					

In Model 1, the coefficient of determination (R Square) is 0.144, indicating that 14.4% of the variance in the dependent variable is explained by the predictor variable i.e., depression. The change in R Square between Model 1 and the null model (with no predictors) is statistically significant (ΔR Square = 0.144, F (1, N) = 46.608, $p < 0.001$).

Model 2 added the predictor variable Stress in this analysis. The coefficient of determination increased to 0.241, indicating that the addition of Stress accounted for an additional 9.7% of the variance in the dependent variable (ΔR Square = 0.097). The change in R Square from Model 1 to Model 2 is statistically significant F (1, N) = 35.468, $p < 0.001$).

Table 9. ANOVA results for regression models predicting CRIES ANOVA[a]

Model	Source	Sum of Squares	df	Mean Square	F	Sig.
1	Regression	4884.88	1	4884.88	46.608	0.0001[b]
	Residual	29136.74	278	104.80		
	Total	34021.62	279			
2	Regression	8192.21	2	4096.10	43.927	0.0001[c]
	Residual	25829.41	277	93.24		
	Total	34021.62	279			
3	Regression	9055.66	3	3018.553	33.370	0.0001[d]
	Residual	24965.96	276	90.456		
	Total	34021.62	279			

Source: Estimated by authors. **Note:** * $p < 0.001$.
a=Dependent Variable: CRIES
b=Predictors: (Constant), Depression.
c=Predictors: (Constant), Depression, Anxiety
d=Predictors: (Constant), Depression, Anxiety, Stress

In Model 3, the predictor variable death anxiety is included in regression analysis. The coefficient of determination further increased to 0.266, suggesting that the addition of Death Anxiety contributed an additional 2.5% of the explained variance in the dependent variable (ΔR Square = 0.025). The change in R-Square from Model 2 to Model 3 is statistically significant F (1, N) = 9.545, p = 0.002). The results indicate that Depression, Stress, and Death Anxiety collectively account for a significant proportion of the variance in the dependent variable. The addition of each predictor variable resulted in a statistically significant increase in the amount of explained variance in the model.

Table 9 shows the ANOVA results for regression models that are predicting the change in CRIES. The table presents the results of the ANOVA. Each model includes different combinations of predictor variables: For Model 1, the ANOVA reveals a significant regression model ($F(1, 278) = 46.60, p < 0.001$). The predictor variable Depression_Total_DASS accounts for a significant amount of variance in CRIES, as indicated by the significant sum of squares for the regression (4884.884).

Model 2 also demonstrates a significant regression model (F (2, 277) = 43.92, $p < 0.001$). The addition of Stress to the model leads to a significant increase in the explained variance, as reflected by the increased sum of squares for the regression (8192.21). Similarly, Model 3 shows a significant regression model ($F(3, 276) = 33.370, p < 0.001$). The inclusion of Death Anxiety further improves the prediction of CRIES, as indicated by the increased sum of squares for the regression (9055.660).

Table 10. Coefficients for Predicting CRIES

Coefficients[a]
Model Predictor Variable B Std. Error. Beta t Sig.
1 Constant 23.08 1.26 18.28 .001
Depression. 0.517 0.07 0.37 6.82 .001
2 Constant 20.10 1.29 15.56 .001
Depression 0.08 0.10 0.06 0.85 .396
Stress 0.60 0.10 0.44 5.95 .001
3 Constant 18.06 1.43 12.60 .001
Depression 0.01 0.10 0.01 0.13 .894
Stress 0.54 0.10 0.40 5.41 .001
Death Anxiety 0.33. 0.11 0.18 3.09 .002

Source: Estimated by authors. Note: Dependent Variable: CRIES[a]

Table 10 represents the coefficients for three different models with the dependent variable CRIES. Model 1: The constant term in the model is 23.08. For every unit increase in the depression variable, the CRIES variable also increases by 0.51 units. The coefficient is statistically significant ($p < 0.001$). Model 2: The constant term in the model is 20.10. The coefficient for this variable is 0.08, indicating that one unit increase in depression leads to a 0.08 unit increase in CRIES. However, this coefficient is not statistically significant ($p = 0.39$). For every unit increase in the stress variable, the CRIES variable may increase by 0.60 units. The coefficient is statistically significant ($p < 0.001$). Model 3: The constant term in the model is 18.06. The coefficient for this variable is 0.01, which is not statistically significant ($p = 0.894$). For every unit increase in the stress variable, the CRIES variable increases by 0.54 units. The coefficient is statistically significant ($p < 0.001$). For every unit increase in the death anxiety variable, the CRIES variable increases by 0.33 units. While the coefficient is statistically significant ($p = 0.002$).

In Model 1, the depression variable has a significant positive impact on CRIES. However, in Model 2, the effect of depression is not statistically significant, but stress has a significant positive effect. In Model 3, stress and death anxiety have a significant positive effect on CRIES, while the effect of depression is not found statistically significant. Table 11 indicates the results of excluding certain variables from the models predicting the dependent variable "CRIES."

Table 11. The Results of excluding certain variables from the models predicting the dependent variable "CRIES."

Excluded Variable
Model Excluded Variable Beta t Sig. Partial Correlation
1 Anxiety 0.14[b] 1.84 0.066 0.11
Stress 0.44[b] 5.95 0.000 0.33
Death Anxiety 0.23[b] 3.90 0.000 0.22
2 Anxiety 0.02[c] 0.26 0.79 0.01
Death Anxiety 0.18[c] 3.09 0.002 0.18
3 Anxiety -0.008[d] -0.11 0.910 -0.007

Source: Estimated by authors. Dependent Variable: CRIES[a]
Predictors in the Model: (Constant), Depression[b]
Predictors in the Model: (Constant), Depression[c], Stress[c]
Predictors in the Model: (Constant), Depression[d], Stress[d], Death_Anxiety[d]

In Model 1, the variable "anxiety " has a coefficient (Beta) of 0.140, suggesting a positive relationship with "total." However, the coefficient is not statistically significant ($p = 0.066$). The variable "stress" has a coefficient of 0.444, indicating a positive relationship with "CRIES," and it is statistically significant ($p < 0.001$). Similarly, the variable "death anxiety" has a coefficient of 0.237, showing a positive relationship with "CRIES," and it is statistically significant ($p < 0.001$). The partial correlations indicate the strength of the relationships between the excluded variables and the dependent variable, with values of 0.110 for "anxiety," 0.337 for "stress," and 0.228 for "death anxiety."

In Model 2, the variable "anxiety" has a coefficient of 0.020, indicating a negligible positive relationship with "CRIES." However, this coefficient is not statistically significant ($p = 0.791$). The variable "death anxiety" has a coefficient of 0.181, suggesting a positive relationship with "CRIES," and it is statistically significant ($p = 0.002$). The partial correlations for these variables are 0.016 for "anxiety Total DASS" and 0.183 for "death anxiety."

In Model 3, the variable "anxiety" has a coefficient of -0.008, indicating a negligible negative relationship with " CRIES." However, this coefficient is not statistically significant ($p = 0.912$). The partial correlation for "anxiety" is -0.007. Overall, the excluded variables (i.e., "anxiety") do not appear to have significant relationships with the dependent variable "CRIES" across all three models. In Model 1, "stress" and "death anxiety" are significant predictors of "CRIES." In Model 2, only "death anxiety " is a significant predictor. However, in Model 3, none of the excluded variables are significant predictors of " CRIES."

6. MAJOR FINDINGS

The correlation results provide evidence about the relationships between the variables. The results indicate that higher levels of event-related distress (measured by CRIES) are associated with depression, anxiety, and stress among the children. The strong and moderate correlation coefficients suggest substantial connections between depression and anxiety, depression and stress, as well as anxiety and stress. These findings contribute to enhancing our understanding on the interplay between psychological variables in the context of children's responses to distressing events. The results are significant because these highlighted the potential association between experiencing traumatic events and the subsequent development of death anxiety in children. Traumatic events have a profound impact on individuals, and children, in particular, may struggle to cope with the emotional aftermath. The positive correlation between CRIES and death anxiety implies that as children experience higher levels of post-traumatic stress symptoms, their concerns and fears regarding death increase. The results also demonstrate that depression, stress, and death anxiety collectively contribute to the prediction of the dependent variable.

The inclusion of each predictor variable led to a statistically significant increase in the amount of explained variance in the model. This indicates that all three predictors play a key role for predicting the outcome variable. The output of ANOVA indicates that depression, stress, and death anxiety significantly contribute to increased CRIES. Each predictor variable significantly contributes and explains variance, suggesting that these variables play an important role in understanding CRIES scores. Natural disasters have long been recognized as catastrophic events that not only disrupt physical environments but also profoundly affect the mental well-being of individuals and communities (Makwana, 2019). Females tend to have a higher prevalence of PTSD symptoms, death anxiety, and stress compared to males (Kar & Bastia, 2006). These findings provide insights into potential gender differences in mental health outcomes and highlight the importance of considering gender-specific factors in understanding and addressing psychological distress. Children and women face many health issues due to natural disasters. It is also caused to increase mental health issues. They also seek psycho-social rehabilitation and intervention after natural disasters. Post-traumatic stress disorder (PTSD), depression, and anxiety are the most common psychological responses recorded by the respondents. Although, multifaceted relationships are reported between environmental disasters and mental health. The people of Joshimath have faced harsh realities due to natural disasters. Floods, landslides, water crises, and wildfires have a significant impact on livelihood security of people in the survey area.

7. POLICY IMPLICATIONS

Early recognition and counselling can prevent many such prolonged agonies. Researchers have observed that few respondents require psychiatric intervention as well as crisis counselling. However, there are population segments requiring special attention and active rehabilitation and intervention. These include children, frail elderly, people with pre-existing mental illness, and families of those who have died in the disaster. Highly negative emotions are also noted in respondents during the icebreaking. The participants have high negative emotions due to natural disaster. The interactive patterns are so meaningful and full of depth. The other psychological symptoms such as helplessness and loss of social security are also observed in the children. In addition, interventions should aim at mental health promotion and prevention of psychological symptoms among disaster-affected families. It should be included under disaster mental health services by government and non-government organizations in natural disaster-prone areas. It is very crucial to provide Psycho-social rehabilitation to children and women in the study area.

8. SIGNIFICANCE AND INNOVATION OF THE STUDY

This present chapter attempts to estimate the effect of natural disasters on children's psychological health and well-being using primary data. Providing the statistical inference of primary data is an innovation of this chapter. Natural disasters have long been recognized as catastrophic events. They disrupt physical environments as well as profoundly affect the mental well-being of families living in those conditions. These unpredictable cation-like earthquakes, floods, and wildfires, often can harm the psychological health of the directly affected population. The empirical results, therefore, provide crucial policy insights to reduce the numbers and frequency of natural disasters and environmental degradation in Uttarakhand. Further, the chapter also comes with some specific areas in which further research can be conducted by the existing researchers.

9. LIMITATIONS AND SCOPE FOR FUTURE RESEARCH

There may be many other psychological consequences of natural disasters in the study area. While this chapter could not include more variables in the empirical investigation. Future research, therefore, must incorporate other important psychological variables. The findings are completely based on quantitative methods using primary data. Further studies can involve a mixture of both qualitative and quantitative methods to understand the underlying factors on psychological issues of children. This study has a certain number of participants and could not be generalized using such a small sample size. However, the social, emotional, and cognitive, impacts of disaster experience do not appear to be large in aggregate. Hence, primary and secondary data are required to analyses the multiple variables that are responsible for depression, anxiety, and stress in children in further study. Further study can include crucial indicators related to the mental health and well-being of survivors of a natural disaster in India.

REFERENCES

Acierno, R., Ruggiero, K. J., Galea, S., Resnick, H. S., Koenen, K., Roitzsch, J., de Arellano, M., Boyle, J., & Kilpatrick, D. G. (2007). Psychological sequelae resulting from the 2004 Florida hurricanes: Implications for postdisaster intervention. *American Journal of Public Health*, 97(1, Supplement_1), 103–108. DOI: 10.2105/AJPH.2006.087007 PMID: 17413067

Agbola, T., & Ojeleye (2007). Climate change and food crop production in Ibadan, Nigeria. *African Crop Science Conference Proceedings*, 8(1), 1423-1433.

Agyapong, B., Shalaby, R., Eboreime, E., Obuobi-Donkor, G., Owusu, E., Adu, M. K., Mao, W., Oluwasina, F., & Agyapong, V. I. O. (2022). Cumulative trauma from multiple natural disasters increases mental health burden on residents of Fort McMurray. *European Journal of Psychotraumatology*, 13(1), 1–20. DOI: 10.1080/20008198.2022.2059999 PMID: 35599978

Alam, M. Md., Sowar, C., Talib, B., Mokhtar, T. M. E. (2011). Climate change adaptation policy in Malaysia: Issues for agricultural sector. *African Journal of Agricultural Research*, 7(9), 1368–1373.

Ashraf, S. N., & Singh, A. K. (2022). Implications of appropriate technology and farm inputs in the agricultural sector of Gujarat: Empirical analysis based on primary data. *Agricultural Economics Research Review*, 35(2), 1–18. DOI: 10.5958/0974-0279.2022.00031.3

Banholzer, S., Kossin, J., & Donner, S. (2014). The impact of climate change on natural disasters. *Reducing Disaster: Early Warning Systems for Climate Change*, In *Springer eBooks* (pp. 21–49). DOI: 10.1007/978-94-017-8598-3_2

Bastami, F., Mohammadi, R., Piri, Z. A., Valipour, E., Ahmadi, P., Almasian, M., & Motlagh, S. N. (2024). prevalence of post-traumatic stress disorder and its relationship with coping strategies among flood victims: Evidence from Iran. *Journal of Affective Disorders Reports*, 15(1), 10–17. DOI: 10.1016/j.jadr.2023.100704

Bazga, B. (2015). Food security component of sustainable development—Prospects and challenges in the future. *Procedia Economics and Finance*, 32(1), 1075–1082. DOI: 10.1016/S2212-5671(15)01570-1

Bell, J. E., Brown, C. L., Conlon, K., Herring, S., Kunkel, K. E., Lawrimore, J., Luber, G., Schreck, C., Smith, A., & Uejio, C. (2018). Changes in extreme events and the potential impacts on human health. *Journal of the Air & Waste Management Association*, 68(4), 265–287. DOI: 10.1080/10962247.2017.1401017 PMID: 29186670

Benevolenza, M. A., & DeRigne, L. (2018). The impact of climate change and natural disasters on vulnerable populations: A systematic review of literature. *Journal of Human Behavior in the Social Environment*, 29(2), 266–281. DOI: 10.1080/10911359.2018.1527739

Cénat, J. M., & Derivois, D. (2014). Assessment of prevalence and determinants of posttraumatic stress disorder and depression symptoms in adults survivors of earthquake in Haiti after 30 months. *Journal of Affective Disorders*, 159(1), 111–117. DOI: 10.1016/j.jad.2014.02.025 PMID: 24679398

Charuvastra, A., & Cloitre, M. (2008). Social bonds and posttraumatic stress disorder. *Annual Review of Psychology*, 59(1), 301–328. DOI: 10.1146/annurev.psych.58.110405.085650 PMID: 17883334

Chaudhary, M. T., & Piracha, A. (2021). Natural Disasters—Origins, Impacts, management. *Encyclopedia*, 1(4), 1101–1131. DOI: 10.3390/encyclopedia1040084

Cianconi, P., Betrò, S., & Janiri, L. (2020). The impact of climate change on mental health: A systematic descriptive review. *Frontiers in Psychiatry*, 11(1), 74. DOI: 10.3389/fpsyt.2020.00074 PMID: 32210846

Conte, H. R., Weiner, M. B., & Plutchik, R. (1982). Measuring death anxiety: Conceptual, psychometric, and factor-analytic aspects. *Journal of Personality and Social Psychology*, 43(4), 775–785. DOI: 10.1037/0022-3514.43.4.775 PMID: 7175677

Dai, W., Chen, L., Lai, Z., Li, Y., Wang, J., & Liu, A. (2016). The incidence of post-traumatic stress disorder among survivors after earthquakes:a systematic review and meta-analysis. *BMC Psychiatry*, 16(1), 188. DOI: 10.1186/s12888-016-0891-9 PMID: 27267874

Davydzenka, T., Tahmasebi, P., & Shokri, N. (2024). Unveiling the global extent of land subsidence: The sinking crisis. *Geophysical Research Letters*, 51(1), 1–11. DOI: 10.1029/2023GL104497

de Almeida, B. A., & Mostafavi, A. (2016). Resilience of infrastructure systems to sea-level rise in coastal areas: Impacts, adaptation measures, and implementation challenges. *Sustainability (Basel)*, 8(11), 11–15. DOI: 10.3390/su8111115

Faridi, R., & Wadood, S. N. (2010). An econometric assessment of household food security in Bangladesh. *Bangladesh Development Studies*, 33(3), 97–111. https://www.jstor.org/stable/23339854

Fergusson, D. M., Horwood, L. J., Boden, J. M., & Mulder, R. T. (2014). Impact of a major disaster on the mental health of a well-studied cohort. *JAMA Psychiatry*, 71(9), 1025–1031. DOI: 10.1001/jamapsychiatry.2014.652 PMID: 25028897

First, J. M. (2024). Post-traumatic stress and depression following disaster: Examining the mediating role of disaster resilience. *Frontiers in Public Health*, 12(1), 12–19. DOI: 10.3389/fpubh.2024.1272909 PMID: 38299076

Goldmann, E., & Galea, S. (2014). Mental health consequences of disasters. *Annual Review of Public Health*, 35(1), 169–183. DOI: 10.1146/annurev-publhealth-032013-182435 PMID: 24159920

Greg, E. E., Anam, B. E., William, M. F., & Duru, E. J. C. (2011). Climate change, food security and agricultural productivity in African: Issues and policy directions. *International Journal of Humanities and Social Science*, 1(21), 205–223.

Grupe, D. W., & Nitschke, J. B. (2013). Uncertainty and anticipation in anxiety: An integrated neurobiological and psychological perspective. *Nature Reviews. Neuroscience*, 14(7), 488–501. DOI: 10.1038/nrn3524 PMID: 23783199

Hoelterhoff, M., & Chung, M. C. (2013). Death anxiety and well-being; coping with life-threatening events. *Traumatology*, 19(4), 280–291. DOI: 10.1177/1534765613477499

Kar, N., & Bastia, B. K. (2006). Post-traumatic stress disorder, depression and generalised anxiety disorder in adolescents after a natural disaster: A study of comorbidity. *Clinical Practice and Epidemiology in Mental Health*, 2(17), 1–12. DOI: 10.1186/1745-0179-2-17 PMID: 16869979

Keya, T. A., Leela, A., Habib, N., Rashid, M., & Bakthavatchalam, P. (2023). Mental health disorders due to disaster exposure: A systematic review and meta-analysis. *Cureus*, 15(4), 1–7. DOI: 10.7759/cureus.37031 PMID: 37143625

Kumar, A., Ahmad, M. M., & Sharma, P. (2017). Influence of climatic and non-climatic factors on sustainable food security in India: A statistical investigation. *International Journal of Sustainable Agricultural Management and Informatics*, 3(1), 1–100. DOI: 10.1504/IJSAMI.2017.082917

Kumar, S., & Singh, A. K. (2023). Modeling the effects of climate change on agricultural productivity: Evidence from Himachal Pradesh, India. *Asia-Pacific Journal of Regional Science*, 7(2), 521–548. DOI: 10.1007/s41685-023-00291-w

Kuo, H. W., Wu, S. J., Ma, T. C., Chiu, M. C., & Chou, S. Y. (2007). Posttraumatic symptoms were worst among quake victims with injuries following the Chi-chi quake in Taiwan. *Journal of Psychosomatic Research*, 62(4), 495–500. DOI: 10.1016/j.jpsychores.2004.11.012 PMID: 17383502

Lovibond, P. F., & Lovibond, S. H. (1995). The structure of negative emotional states: Comparison of the Depression Anxiety Stress Scales (DASS) with the Beck Depression and Anxiety Inventories. *Behaviour Research and Therapy*, 33(3), 335–343. DOI: 10.1016/0005-7967(94)00075-U PMID: 7726811

Makwana, N. (2019). Disaster and its impact on mental health: A narrative review. *Journal of Family Medicine and Primary Care*, 8(10), 3090–3095. DOI: 10.4103/jfmpc.jfmpc_893_19 PMID: 31742125

Malhi, G. S., Kaur, M., & Kaushik, P. (2021). Impact of climate change on agriculture and its mitigation Strategies: A review. *Sustainability (Basel)*, 13(3), 1318. DOI: 10.3390/su13031318

Mall, M. K., Singh, R., Gupta, A., Srinivasan, G., & Rathore, L. S. (2006). Impact of climate change on Indian agriculture: A review. *Climatic Change*, 78(1), 445–478. DOI: 10.1007/s10584-005-9042-x

McMichael, C., Barnett, J., & McMichael, A. J. (2012). An ill wind? Climate change, migration, and health. *Environmental Health Perspectives*, 120(5), 646–654. DOI: 10.1289/ehp.1104375 PMID: 22266739

Neria, Y., Nandi, A., & Galea, S. (2008). Post-traumatic stress disorder following disasters: A systematic review. *Psychological Medicine*, 38(4), 467–480. DOI: 10.1017/S0033291707001353 PMID: 17803838

Pandey, R. A., Chalise, P., Khadka, S., Chaulagain, B., Maharjan, B., Pandey, J., Nepal, J., & Pandey, C. (2023). Post-traumatic stress disorder and its associated factors among survivors of 2015 earthquake in Nepal. *BMC Psychiatry*, 23(1), 340. DOI: 10.1186/s12888-023-04836-3 PMID: 37189110

Perrin, S., Meiser-Stedman, R., & Smith, P. (2005). The Children's Revised Impact of Event Scale (CRIES): Validity as a Screening Instrument for PTSD. *Behavioural and Cognitive Psychotherapy*, 33(4), 487–498. DOI: 10.1017/S1352465805002419

Prasad, A. S., & Francescutti, L. H. (2017). Natural disasters. *International Encyclopedia of Public Health*, 3(2), 215–222. DOI: 10.1016/B978-0-12-803678-5.00519-1

Rahmani, M., Muzwagi, A., & Pumariega, A. J. (2022). Cultural factors in disaster response among diverse children and youth around the world. *Current Psychiatry Reports. Current Psychiatry Reports*, 24(10), 481–491. DOI: 10.1007/s11920-022-01356-x PMID: 35953637

Saeed, S. A., & Gargano, S. P. (2022). Natural disasters and mental health. *International Review of Psychiatry (Abingdon, England)*, 34(1), 16–25. DOI: 10.1080/09540261.2022.2037524 PMID: 35584023

SAMHSA. (2017). SAMHSA Disaster Technical assistance center supplemental research bulletin greater impact: How disasters affect people of low socioeconomic status. SAMHSA. https://www.samhsa.gov/sites/default/files/dtac/srb-low-ses_2.pdf

Seneviratne, S. I., Nicholls, N., Easterling, D., Goodess, C. M., Kanae, S., Kossin, J., Luo, Y., Marengo, J., McInnes, K., Rahimi, M., Reichstein, M., Sorteberg, A., Vera, C., & Zhang, X. (2012). Changes in climate extremes and their impacts on the natural physical environment. In Field, C. B., V. Barros, T.F. Stocker, D. Qin, D.J. Dokken, K.L. Ebi, M.D. Mastrandrea, K.J. Mach, G.-K. Plattner, S.K. Allen, M. Tignor, and P.M. Midgley (eds.), *Managing the Risks of Extreme Events and Disasters to Advance Climate Change Adaptation. A Special Report of Working Groups I and II of the Intergovernmental Panel on Climate Change (IPCC)* (pp. 109-230). Cambridge University Press, Cambridge, UK, and New York, NY, USA.) DOI: 10.1017/CBO9781139177245.006

Shivanna, K. R. (2022). Climate change and its impact on biodiversity and human welfare. *Proceedings of the Indian National Science Academy. Part A, Physical Sciences*, 88(2), 160–171. DOI: 10.1007/s43538-022-00073-6

Shultz, J. M., Neria, Y., Allen, A., & Espinel, Z. (2013). Psychological impacts of natural disasters. In Bobrowsky P. T. (Edt.), *Encyclopaedia of Natural Hazards. Netherlands*, pp. 779-791. Springer. DOI: 10.1007/978-1-4020-4399-4_279

Singh, A. K., Ashraf, S. N., Sharma, S. K., & Kumar, S. (2023a). Factors affecting farm income and climate adaptation strategies in the farming sector of Gujarat state of India. In Sadhik, B. J., Adewole, J., Yeneneh, A., & Abidoye, L. (Eds.), *Opportunities and Challenges in Climate-Friendly Clean Water and Energy Technologies* (pp. 122–147). IGI Global., DOI: 10.4018/978-1-6684-7303-0.ch006

Singh, A. K., Ashraf, S. N., Sharma, S. K., & Kumar, S. (2023b). Factors affecting farm income and climate adaptation strategies in the farming sector of Gujarat State of India. In *Advances in environmental engineering and green technologies book series* (pp. 122–147). DOI: 10.4018/978-1-6684-7303-0.ch006

Singh, A. K., Jyoti, B., & Mishra, V. K. (2023c). Policy implications from selected countries for a sustainable future of the society through green entrepreneurship and economic development. In Abdelli, M., Ul-Durar, S., & Wasef Hattab, H. (Eds.), *Entrepreneurship Ecosystems and Their Opportunities and Challenges* (pp. 129–162). IGI Global., DOI: 10.4018/978-1-6684-7140-1.ch006

Singh, A. K., Jyoti, B., & Sankaranarayanan, K. G. (2024). A review for analyzing the impact of climate change on agriculture and food security in India. In Samanta, D., & Garg, M. (Eds.), *The Climate Change Crisis and Its Impact on Mental Health* (pp. 227–251). IGI Global., DOI: 10.4018/979-8-3693-3272-6.ch017

Singh, A. K., Kumar, S., & Jyoti, B. (2022). Influence of climate change on agricultural sustainability in India: A state-wise panel data analysis. *Asian Journal of Agriculture*, 6(1). Advance online publication. DOI: 10.13057/asianjagric/g060103

Thompson, N. J., Fiorillo, D., Rothbaum, B. O., Ressler, K. J., & Michopoulos, V. (2018). Coping strategies as mediators in relation to resilience and posttraumatic stress disorder. *Journal of Affective Disorders*, 225(1), 153–159. DOI: 10.1016/j.jad.2017.08.049 PMID: 28837948

Uddin, M. S., Haque, C. E., Khan, M. N., Doberstein, B., & Cox, R. S. (2021). Disasters threaten livelihoods, and people cope, adapt and make transformational changes: Community resilience and livelihoods reconstruction in coastal communities of Bangladesh. *International Journal of Disaster Risk Reduction*, 63(1), 23–34. DOI: 10.1016/j.ijdrr.2021.102444

Wade, T. (2022). Health risks associated with sea level rise. National Collaborating Centre for Environmental Health (NCCEH). https://ncceh.ca/resources/evidence-reviews/health-risks-associated-sea-level-rise

Wang, F., Fang, Y., Deng, H., & Wei, F. (2022). How community medical facilities can promote resilient community constructions under the background of pandemics. *Indoor and Built Environment*, 31(4), 1018–1027. DOI: 10.1177/1420326X211048537 PMID: 35431637

Yehuda, R., Hoge, C. W., McFarlane, A. C., Vermetten, E., Lanius, R. A., Nievergelt, C. M., Hobfoll, S. E., Koenen, K. C., Neylan, T. C., & Hyman, S. E. (2015). Post-traumatic stress disorder. *Nature Reviews. Disease Primers*, 1(1), 15–30. DOI: 10.1038/nrdp.2015.57 PMID: 27189040

Yorulmaz, D. S., & Karadeniz, H. (2021). Effects of disasters on mental health. *Doğ Afet Çev Derg*, 7(2), 392–398. DOI: 10.21324/dacd.786048

Zhou, X., Kang, L., Sun, X., Song, H., Mao, W., Huang, X., Zhang, Y., & Li, J. (2013). Prevalence and risk factors of post-traumatic stress disorder among adult survivors six months after the Wenchuan earthquake. *Comprehensive Psychiatry*, 54(5), 493–499. DOI: 10.1016/j.comppsych.2012.12.010 PMID: 23337407

ADDITIONAL READINGS

Douglas, K. M., Groves, S., Porter, R. J., Jordan, J., Wilson, L., Melzer, T. R., Wise, R. G., Bisson, J. I., & Bell, C. J. (2019). Traumatic imagery following glucocorticoid administration in earthquake-related post-traumatic stress disorder: A preliminary functional magnetic resonance imaging study. *The Australian and New Zealand Journal of Psychiatry*, 53(12), 1167–1178. DOI: 10.1177/0004867419851860 PMID: 31146540

Galea, S., Nandi, A., & Vlahov, D. (2005). The epidemiology of post-traumatic stress disorder after disasters. *Epidemiologic Reviews*, 27(1), 78–91. DOI: 10.1093/epirev/mxi003 PMID: 15958429

Nanduri, A., Vasquez, M., Veluri, S. C., & Ranjbar, N. (2023). Scoping review of PTSD treatments for natural disaster survivors. *Health Psychology Research*, 11(1), 1–17. DOI: 10.52965/001c.89642 PMID: 38089640

Neria, Y., Nandi, A., & Galea, S. (2008). Post-traumatic stress disorder following disasters: A systematic review. *Psychological Medicine*, 38(4), 467–480. DOI: 10.1017/S0033291707001353 PMID: 17803838

Zang, Y., Hunt, N., & Cox, T. (2014). Adapting narrative exposure therapy for Chinese earthquake survivors: A pilot randomised controlled feasibility study. *BMC Psychiatry*, 14(1), 262–268. DOI: 10.1186/s12888-014-0262-3 PMID: 25927297

KEY TERMS

Death Anxiety-: Fear of death or dying. It is usually used to mention to the negative emotions one might have about death.

Natural Disaster Survivors -: Those who have faced and lived through a major natural disaster and experiences it as most difficult situation they've ever encountered.

Psychological Distress –: It can be considered as unpleasant emotions or feelings which can severely affect daily life activities of an Individual.

PTSD-: The Post-traumatic stress disorder (PTSD) can be characterised as a psychological disorder where an individual fail to recover after experiencing or witnessing a terrifying event.

Trauma -: A highly stressful event of an individual's life.

Chapter 6
Air Conditioning for Healthier Living

Shalom Akhai
http://orcid.org/0000-0002-7533-457X
Department of Mechanical Engineering, Maharishi Markandeshwar Engineering College, Maharishi Markandeshwar University, Mullana-Ambala, India

Mahapara Abbass
http://orcid.org/0000-0002-0838-2598
Maharishi Markandeshwar Engineering College, Maharishi Markandeshwar University, Mullana-Ambala, India

ABSTRACT

The chapter explores the role of air conditioning technology in promoting healthier living environments amid climate change. It discusses the impact of climate change on indoor air quality (IAQ) and its implications for human health. It suggests a multifaceted approach, including ventilation, source control, air purification, and moisture regulation. It also proposes climate change adaptation strategies, such as advanced air conditioning systems and early warning systems. The chapter highlights risks associated with traditional cooling methods, particularly for vulnerable populations. Technological advancements like HEPA filters and solar-powered systems are suggested for improving IAQ while minimizing environmental impact. Green construction techniques, photocatalytic oxidation, and green roofs/walls are also discussed. The chapter acknowledges challenges like data privacy and integration complexity but emphasizes the convergence of air conditioning technology, climate change adaptation, and sustainable design for healthier indoor environments.

DOI: 10.4018/979-8-3693-4183-4.ch006

1 NEED FOR AIR CONDITIONING

Indoor air quality (IAQ) is crucial for overall health and well-being, especially in relation to climate change (Saini, Dutta, & Marques, 2020). Contemporary construction techniques can lead to elevated indoor contaminants, causing respiratory issues, sick building syndrome, and long-term health complications (Aziz et al., 2023). To maintain good IAQ, methods include ventilation, source management, air purification, moisture regulation, cleaning, and maintenance (Swamy, 2021). Green construction techniques and adaptation techniques are essential to protect public health and reduce health hazards (Mashallahi et al., 2022). Air conditioning technology can create healthier living environments and reduce risks (Akhai & Khang 2024; Akhai et al., 2021; Akhai et al., 2020). Climate change adaptation is a proactive approach to modify and mitigate the negative effects of shifting climatic conditions, aiming to enhance resilience and reduce vulnerability (Joakim et al., 2021). The necessity for adaptation arises from the global impacts of climate change, such as more frequent and severe heatwaves, changes in rainfall patterns, and rising sea levels Abbass et al. (2022). Adaptation techniques, such as air conditioning and early warning systems, can effectively mitigate health hazards linked to heatwaves (Viguié et al., 2020). Additionally, climate change may intensify pre-existing health problems and introduce novel hazards, such as vector-borne illnesses like malaria and dengue fever, into previously unaffected regions (Alanazi et al., 2022; Fadda, 2020). Furthermore, Indoor air quality can significantly improve cognitive abilities and productivity, leading to improved work performance. It also positively impacts mood and morale in schools and work environments, fostering a more positive atmosphere and better learning conditions (Tanwar & Akhai 2017; Akhai et al., 2017). Cleaner indoor air can reduce the spread of diseases, such as COVID-19, thereby contributing to a healthier environment. Additionally, improving air quality can lead to longer life spans (D'alessandro et al., 2020). By reducing exposure to indoor pollutants, individuals can enjoy better health and potentially longer lives (Akhai et al., 2016; Akhai et al., 2020). Therefore, prioritizing indoor air quality improvements is crucial for creating healthier and more productive indoor environments. These health benefits highlight the importance of prioritizing indoor air quality improvements (Kumar & Akhai 2022).

1.1 Importance of Indoor Air Quality in Health Management

Indoor air quality (IAQ) is crucial for health management, especially in the context of climate change (Sadrizadeh et al. 2022). As individuals spend over 80% of their time indoors, the quality of indoor air can significantly impact their well-being, efficiency, and overall standard of living (Akhai et al., 2016). Indoor air pollution from sources such as combustion appliances, construction materials, and external contaminants can lead to various health conditions, including respiratory ailments, allergies, and cardiovascular disorders (Dehghani et al., 2024). Climate change has further exacerbated indoor air pollution by affecting the spread and intensity of harmful substances. Heatwaves can increase the use of air conditioning systems, which circulate indoor air containing allergens and pathogens, and aggravate respiratory diseases within (Ayejoto et al., 2023; Akhai 2024; Aggarwal et al., 2022). Climate change also affects the frequency and spread of interior allergens, such as mold spores and dust mites, as well as outdoor pollutants like pollen and particulate matter. Fluctuations in temperature and humidity can enhance the formation and spread of allergens, particularly problematic for those with respiratory diseases like asthma and allergies (Syed, Folz, and Ali 2023).

Poor indoor air quality not only has direct health consequences but can negatively impact cognitive function and productivity. Research has shown that being exposed to indoor air pollution can negatively impact cognitive function, reduce concentration, and lead to higher rates of absenteeism in both educational and professional environments. Prioritizing the maintenance of high-quality indoor air is essential for safeguarding physical health, mental well-being, and productivity (Niza et al., 2024).

Air conditioning technology is essential for controlling indoor air quality and fostering healthier living spaces. Contemporary air conditioning systems are specifically engineered to eliminate pollutants, allergens, and toxins, enhancing indoor air quality, and decreasing the likelihood of health problems. Allocating resources towards and advocating for the adoption of effective and environmentally friendly air conditioning systems can actively enhance the physical and mental welfare of individuals and communities amidst the challenges posed by climate change (Tran, Park, and Lee 2020; (Mata et al., 2022).

Table 1 provides a comprehensive look at the various aspects of indoor air quality, detailing how each factor affects health and cognitive function, its relation to climate change, and the role of air conditioning technology in managing these effects. It underscores the multifaceted impact of IAQ and the importance of advanced, eco-friendly air conditioning systems in promoting health and well-being.

Table 1. Importance of IAQ, Health, and Climate Synergies.

Factor	Description	Health Impacts	Cognitive Impacts	Climate Change Effects	Air Conditioning Role
Time Spent Indoors	High percentage of time indoors emphasizes the importance of IAQ.	Poor IAQ can lead to chronic health issues.	Impaired cognitive function and increased absenteeism.	Greater indoor pollutant concentration due to extreme weather.	Regulates temperature and filters air, reducing pollutant exposure.
Pollution Sources	Combustion, construction materials, external contaminants.	Respiratory ailments, allergies, cardiovascular disorders.	Can cause headaches, dizziness, and fatigue.	Heatwaves and cold snaps can intensify indoor pollution levels.	Modern systems can remove a wide range of pollutants.
Health Conditions	Conditions exacerbated by poor IAQ.	Asthma, bronchitis, heart diseases.	Long-term exposure may affect brain health.	Rising temperatures can increase the prevalence of allergens.	Properly maintained systems ensure a healthier indoor environment.
Cognitive Function	Impact of IAQ on brain performance.	Stress and anxiety due to poor air quality.	Decreased concentration, memory, and decision-making abilities.	Fluctuating indoor conditions can stress the body and mind.	Enhances comfort, allowing better focus and productivity.
Productivity	IAQ's effect on work and learning efficiency.	Sick building syndrome reduces overall well-being.	Poor IAQ is linked to lower productivity and performance.	Extreme temperatures can disrupt comfort and efficiency.	AC systems can maintain optimal conditions for peak performance.
Climate Change	The interplay between IAQ and changing climate patterns.	Increased incidence of heat-related illnesses.	Heat stress can impair mental function.	More frequent extreme weather events affect IAQ.	Eco-friendly AC systems can mitigate some effects of climate change.
AC Technology	Advances in AC systems for IAQ management.	Reduces the risk of disease transmission.	Ensures a stable environment for cognitive activities.	AC systems can adapt to changing outdoor conditions.	Smart systems optimize energy use and maintain air quality.

2 HEALTH RISKS ASSOCIATED WITH TRADITIONAL COOLING METHODS

The health risks associated with traditional cooling methods are multifaceted and far-reaching. Rising temperatures and humidity, indoor air quality challenges, and respiratory problems and allergies are just a few of the significant concerns.

2.1 Impact of Rising Temperatures and Humidity

Rising temperatures and humidity pose health risks to vulnerable populations, including the elderly, pre-existing conditions, and low-income communities, with increased air conditioning usage exacerbating these risks (Leap et al., 2024). This includes:

- **Elderly individuals:** Their bodies often have a reduced capacity to regulate temperature, making them more susceptible to heatstroke and dehydration.
- **People with pre-existing medical conditions:** Existing respiratory illnesses, heart disease, and chronic kidney disease can be significantly worsened by heat stress.
- **Low-income communities:** Limited access to air conditioning and proper ventilation can exacerbate the health risks associated with extreme heat.

While air conditioning can provide relief, its increased use due to rising temperatures but creates challenges like:

- **Cost burden:** Air conditioning can be a significant financial strain, particularly for low-income households, further amplifying health disparities.
- **Greenhouse gas emissions:** Increased reliance on air conditioning contributes to greenhouse gas emissions, which in turn, drive further temperature increases.

This highlights the urgent need for multifaceted solutions that address both the immediate health risks and the underlying causes of climate change (Wang et al., 2021).

2.2 Indoor Air Quality Challenges

Traditional air conditioning can negatively affect indoor air quality, causing pollutants and allergens, causing respiratory issues and health problems. Maintaining good indoor air quality is crucial for overall well-being.

2.2.1 Sources of Indoor Air Pollution

Various factors contribute to the degradation of indoor air quality. Gaining a comprehensive understanding of these sources is essential for formulating and implementing effective solutions to reduce or prevent their negative impacts (Vardoulakis et al., 2020; Saini et al., 2021). Below are few crucial points:

- **Combustion Appliances:** Gas stoves, fireplaces, and heaters release harmful pollutants like carbon monoxide, nitrogen dioxide, and particulate matter, impacting respiratory health.
- **Building Materials and Furnishings:** New furniture, carpets, paints, and building materials often off-gas volatile organic compounds (VOCs) for extended periods, causing irritation and respiratory problems.
- **Biological Contaminants:** Mold, mildew, dust mites, and pet dander thrive in damp environments, triggering allergies and asthma attacks.
- **Radon:** This naturally occurring radioactive gas seeps into buildings from the soil and can pose a serious health risk with long-term exposure.
- **Second-hand Smoke:** Exposure to cigarette smoke, even in the absence of a smoker, can lead to a range of health problems, particularly for children and the elderly.
- **Outdoor Air Pollution:** Particulate matter, pollen, and other pollutants from outside can infiltrate buildings through leaks and ventilation systems, impacting indoor air quality.

2.2.2 Climate Change and its Impact on Indoor Air Quality

Climate change has both positive and negative effects on indoor air quality. Extreme heat events not only increase the dependence on air conditioning, which affects energy consumption, but also lead to higher temperatures and humidity (Mansouri et al., 2022; Božić et al., 2019):

- **Promote mold growth:** Damp environments are ideal breeding grounds for mold, which can exacerbate allergies and respiratory issues.
- **Increase outdoor air pollution:** Heatwaves can intensify smog formation, leading to higher levels of pollutants infiltrating buildings.
- **Reduce ventilation effectiveness:** Hotter outside air reduces the effectiveness of natural ventilation, potentially trapping pollutants inside.

2.2.3 Challenges of Achieving Healthy Indoor Air Quality

Maintaining good indoor air quality presents a complex challenge due to several factors (Burroughs and Hansen 2020; Ma et al., 2021):

- **Balancing Comfort and Health:** Airtight buildings designed for energy efficiency can impede natural ventilation, leading to a buildup of pollutants. Strategies are needed to achieve a balance between energy efficiency and adequate ventilation.
- **Cost Considerations:** Implementing effective air filtration systems, maintaining ventilation systems, and addressing moisture issues can be expensive, particularly for low-income households and public buildings.
- **Lack of Awareness:** Many people are unaware of the potential health risks associated with poor indoor air quality and the steps they can take to improve it.
- **Identifying Sources:** Diagnosing the root cause of poor IAQ can be challenging, requiring specialized equipment and expertise.

2.2.4 Beyond Air Conditioning: A Multifaceted Approach

Although air conditioning contributes to temperature regulation, it is not a sufficient solution for ensuring high-quality indoor air. A comprehensive approach is necessary, combining various strategies:

- **Source Control:** Identify and address the source of pollutants whenever possible. This may involve proper ventilation for combustion appliances, using low-VOC building materials and furniture, and controlling moisture levels to prevent mold growth.
- **Ventilation:** Increase ventilation by opening windows regularly, weather permitting, and ensuring proper functioning of mechanical ventilation systems.
- **Air Filtration:** Invest in high-quality air purifiers with HEPA filters to capture particles and allergens.
- **Maintenance:** Regularly change air filters in air conditioning and ventilation systems and maintain proper drainage to prevent moisture buildup.
- **Awareness and Education:** Raise awareness about the importance of indoor air quality and provide resources for individuals to improve their indoor environment.

Thus, ensuring the best possible indoor air quality requires a thorough and varied approach. Through the identification of the sources of indoor air pollution, understanding the complexities resulting from climate change, and implementing a combination of strategies, we may create interior environments that are healthier and better for communities. Participating in this activity not only enhances our physical and mental well-being but also enhances productivity and helps to the advancement of a more eco-conscious future. To address these challenges in a proactive manner, we have the potential to reverse the pattern of indoor air pollution and see enhanced respiratory function (Saran et al., 2020; Asim et al., 2022; Nag & Nag, 2019).

2.3 Respiratory Problems and Allergies

Air conditioning can exacerbate respiratory issues and allergies due to dry air, exacerbating conditions like asthma. Adopting sustainable, evidence-based cooling strategies is crucial for health and well-being in heat-vulnerable settings. Although air conditioning provides relief from extreme heat, it may have both positive and negative effects on the quality of indoor air (Vornanen-Winqvist et al., 2020). The reason is as follows:

- **Dry Air Irritation:** Air conditioning systems remove moisture from the air to create a cooling effect. However, excessively dry air can irritate mucous membranes in the nose, throat, and lungs. This can trigger coughing, sneezing, and exacerbate pre-existing respiratory conditions like asthma.
- **Allergen and Pathogen Proliferation:** Air conditioning systems typically recirculate a significant portion of indoor air. While filters capture some particles, allergens like dust mites, pet dander, and mold spores can remain airborne. Additionally, if not properly maintained, air conditioning systems can harbor and spread bacteria and viruses, further compromising indoor air quality.

These factors can contribute to a range of health problems, including:

- **Asthma Exacerbation:** Dry air and airborne irritants can trigger asthma attacks in susceptible individuals, leading to difficulty breathing, wheezing, and chest tightness.
- **Allergic Rhinitis:** Symptoms like a runny nose, itchy eyes, and congestion can worsen due to the increased circulation of allergens within a closed environment.

- **Increased Susceptibility to Infections:** Dry and irritated mucous membranes can become more susceptible to infections, potentially leading to respiratory illnesses like colds and the flu.

Thus, the reliance on air conditioning, while providing temporary relief from heat, can have unintended consequences for our health. It's crucial to move beyond a singular focus on cooling and adopt a more holistic approach that prioritizes both thermal comfort and healthy indoor air quality. This may involve:

- **Prioritizing Ventilation:** Whenever possible, utilize natural ventilation by opening windows and doors to introduce fresh air and dilute indoor pollutants.
- **Investing in Air Filtration:** High-quality air purifiers with HEPA filters can effectively capture allergens and other airborne irritants. However, regular filter changes are essential for optimal performance.
- **Maintaining Air Conditioning Systems:** Regular cleaning and upkeep of air conditioners, including filter replacement and coil cleaning, helps prevent the growth of mold and bacteria within the system.
- **Exploring Sustainable Cooling Alternatives:** Passive cooling strategies like shading, natural ventilation, and evaporative cooling can be effective in some climates, offering a more environmentally friendly and potentially health-promoting approach.

By acknowledging the drawbacks of traditional cooling methods and exploring alternative solutions, we can create healthier and more sustainable indoor environments for all. This shift is particularly important in heat-vulnerable settings where compromised health can have a significant impact on individuals and communities (Kumar et al., 2023; Hawkins et al., 2020; Raju, Siddharthan, and McCormack 2020, Andersen et al., 2023).

3 TRADITIONAL AIR CONDITIONING SYSTEMS CHALLENGES AND SOLUTIONS

Conventional air conditioning systems encounter difficulties such as high energy use, the release of greenhouse gases, and insufficient filtration and ventilation. Solar-powered systems provide sustainable options, while HEPA filters and UV germicidal irradiation enhance the quality of indoor air. Intelligent ventilation systems and automation enhance air quality while minimising energy use. Sustainable solutions place emphasis on optimising energy efficiency, enhancing indoor air quality, and promoting human health. This is achieved via the implementation of innovative filtration systems, effective ventilation methods, and the use of renewable energy sources. Novel technologies provide several alternatives for a future in cooling that is both sustainable and conducive to good health (Hama Radha 2023, Islam et al., 2023, Ortiz, Itard, &Bluyssen, 2020).

Challenges:

- **Energy Intensity and Greenhouse Gas Emissions:** Traditional air conditioning systems are major consumers of energy, relying heavily on electricity generated by fossil fuels. This heavy energy demand translates into increased greenhouse gas emissions, further accelerating climate change and impacting air quality.
- **Inadequate Filtration and Ventilation:** Many older air conditioning systems lack efficient filtration systems, allowing airborne pollutants, allergens, and even pathogens to recirculate within the building. Additionally, limited ventilation can lead to stagnant air, further compromising indoor air quality.

Solutions:

- **Innovations in Filtration Technology:** Advancements in air filtration offer promising solutions. High-Efficiency Particulate Air (HEPA) filters capture a wide range of airborne particles, including dust mites, pollen, and mold spores, leading to cleaner indoor air. Additionally, Ultraviolet (UV) Germicidal Irradiation technology can effectively inactivate bacteria and viruses within air conditioning systems, reducing the spread of airborne pathogens.
- **Smart Ventilation Systems and Automation:** Technology is revolutionizing ventilation strategies. Smart ventilation systems use sensors to monitor indoor air quality and adjust ventilation rates accordingly. This ensures adequate fresh air exchange while minimizing energy consumption. Automation features like programmable thermostats and occupancy sensors can further optimize cooling systems, reducing energy waste and ensuring comfort when spaces are occupied.

Additional Considerations:

- **Building Design and Efficiency:** Building design plays a crucial role in reducing the demand for air conditioning. Strategies like proper insulation, shading techniques, and natural ventilation can significantly decrease the cooling load on air conditioning systems.

- **Alternative Cooling Methods:** Depending on the climate, exploring alternative cooling methods can be beneficial. Evaporative cooling systems, which utilize the principle of water evaporation to create a cooling effect, can be a more energy-efficient option in arid climates.

By embracing innovative technologies, prioritizing energy efficiency, and exploring alternative cooling methods, we can create a future of comfortable and healthy indoor environments while minimizing our environmental impact. This shift towards sustainable cooling solutions is crucial for mitigating climate change and ensuring the health and well-being of future generations (Ortiz et al., 2020; Alazazmeh & Asif, 2021).

4 INNOVATIONS IN AIR CONDITIONING TECHNOLOGY

Innovative technologies briefed in Table 2 are revolutionizing air conditioning, offering sustainable, healthy, and efficient solutions. Evaporative cooling, a natural approach, uses water evaporation to create a refreshing environment, reducing energy consumption and environmental impact. Photocatalytic oxidation (PCO) uses ultraviolet light and a catalyst to break down pollutants and odors, improving indoor air quality and reducing reliance on chemicals. Green roofs and walls integrate vegetation into building designs, providing natural insulation, reduced energy consumption, and improved air quality. Solar-powered air conditioning uses the sun's abundant energy, reducing carbon footprint, reducing greenhouse gas emissions, and offering lower energy costs. These technologies, along with artificial intelligence and machine learning, are paving the way for a future where cooling homes and workplaces is not only comfortable but also sustainable and healthy for generations to come. These innovative technologies are briefed below:

- **Evaporative Cooling Systems -** Evaporative cooling systems utilize the principle of water evaporation to cool the air. These systems are energy-efficient and environmentally friendly, making them a sustainable alternative to traditional air conditioning methods (Lai et al., 2024, Cengiz et al., 2024).
- **Photocatalytic Oxidation for Air Purification -** Photocatalytic oxidation is a technology used for air purification that eliminates pollutants and odors by breaking them down into harmless byproducts. This innovative method enhances indoor air quality by effectively removing contaminants (Cengiz et al., 2024; Mousavi et al., 2024; Cerrato and Fumo 2024).
- **Green Roofs and Walls for Natural Insulation -** Green roofs and walls involve the use of vegetation to insulate buildings naturally. These systems

provide thermal regulation, reduce energy consumption, and contribute to environmental sustainability by absorbing carbon dioxide and providing habitat for wildlife (Besir & Cuce 2018).

- **Solar-Powered Air Conditioning Systems -** Solar-powered air conditioning systems harness solar energy to provide cooling solutions. By utilizing renewable energy sources, these systems reduce carbon emissions and dependency on traditional power grids, offering a sustainable and cost-effective cooling alternative (Albatayneh et al., 2021; Arif et al., 2021; Haffaf et al, 2021).
- **Integration of Artificial Intelligence and Machine Learning -** The integration of artificial intelligence and machine learning in air conditioning systems enables predictive maintenance, energy optimization, and personalized comfort settings. These intelligent algorithms adjust to user preferences, optimize energy usage, and enhance the overall efficiency and sustainability of air conditioning technology (Haffaf et al., 2021; Rajasekhar et al.; 2020, Tien et al., 2022).

5 CONCLUSION

Air conditioning technology is crucial in managing climate change health risks by providing efficient cooling solutions that prioritize human health and environmental sustainability. By embracing innovative technologies and sustainable practices, we can create healthier living environments that are resilient to climate change challenges. Addressing indoor air quality (IAQ) and climate change adaptation is increasingly important, especially for vulnerable populations. Advancements in air conditioning technology, such as HEPA filtration, solar-powered systems, and smart ventilation, can enhance IAQ, reduce energy consumption, and mitigate climate change impacts. Green construction practices, photocatalytic oxidation, and artificial intelligence can also contribute to healthier indoor environments. A holistic approach, prioritizing health promotion, environmental sustainability, and equitable access to solutions, is needed. Interdisciplinary collaboration is needed to develop comprehensive strategies that safeguard public health, enhance resilience, and advance sustainability goals. In brief:

- Climate change's intensifying impacts necessitate a focus on IAQ and its health implications.
- Challenges include poor IAQ, traditional cooling methods, and climate change's effects on vulnerable populations.

- Opportunities for innovation, collaboration, and sustainable action exist through advancements in air conditioning technology and green construction practices.
- A holistic approach is needed, prioritizing health promotion, environmental sustainability, and equitable access to solutions.
- Education and awareness initiatives are key to fostering a culture of health and environmental stewardship.
- Interdisciplinary collaboration is required for managing IAQ and adapting to climate change.
- The convergence of these areas presents a pathway towards healthier, more sustainable indoor environments.

References

Abbass, K., Qasim, M. Z., Song, H., Murshed, M., Mahmood, H., & Younis, I. (2022). A review of the global climate change impacts, adaptation, and sustainable mitigation measures. *Environmental Science and Pollution Research International*, 29(28), 42539–42559. DOI: 10.1007/s11356-022-19718-6 PMID: 35378646

Aggarwal, P., Rana, M., & Akhai, S. (2022). Briefings on e-waste hazard until COVID era in India. *Materials Today: Proceedings*, 71(2), 389–393. DOI: 10.1016/j.matpr.2022.09.507

Akhai, S. (2024). Trends and Environmental Impact of Paper Consumption: A Prognostic Scenario for the Indian Market by 2030-A Case Study. In *International Conference on Interdisciplinary Approaches in Civil Engineering for Sustainable Development* (Vol. 464, pp. 11-18). Springer Nature Singapore. DOI: 10.1007/978-981-97-0910-6_2

Akhai, S., Bansal, S. A., & Singh, S. (2020). A critical review of thermal insulators from natural materials for energy saving in buildings. *Journal of Critical Reviews*, 7(19), 278–283.

Akhai, S., & Khang, A. (2024). Energy Efficiency and Human Comfort: AI and IoT Integration in Hospital HVAC Systems. *Medical Robotics and AI-Assisted Diagnostics for a High-Tech Healthcare Industry*, 93-108.

Akhai, S., Mala, S., & Jerin, A. A. (2020). Apprehending Air Conditioning Systems in Context to COVID-19 and Human Health: A Brief Communication. *International Journal of Healthcare Education & Medical Informatics (ISSN: 2455-9199), 7*(1&2), 28-30.

Akhai, S., Mala, S., & Jerin, A. A. (2021). Understanding whether air filtration from air conditioners reduces the probability of virus transmission in the environment. *Journal of Advanced Research in Medical Science & Technology (ISSN: 2394-6539), 8*(1), 36-41.

Akhai, S., Singh, V. P., & John, S. (2016). Human performance in industrial design centers with small unit air conditioning systems. *Journal of Advanced Research in Production Industrial Engineering*, 3(2), 5–11.

Akhai, S., Singh, V. P., & John, S. (2016). Investigating Indoor Air Quality for the Split-Type Air Conditioners in an Office Environment and Its Effect on Human Performance. *Journal of Mechanical Civil Engineering*, 13(6), 113–118.

Akhai, S., Thareja, P., & Singh, V. P. (2017). Assessment of Indoor Environment Health Sustenance in Air Conditioned Class Rooms. *Journal of Advanced Research in Civil and Environmental Engineering*, 4(1&2), 1–9. DOI: 10.24321/2393.8307.201701

Alanazi, Z. S. A., Essa, A. M. A., Alalawi, Z. A. M., Albalawi, F. F. M., Almohammadi, Y. M. A., Alnajran, M. A. M., & Alanazi, F. M. F. (2022). Exploring The Nexus Of Global Climate Changes And Infectious Disease Dynamics: A Depth Review Study. *Journal of Namibian Studies: History Politics Culture*, 32, 921–960.

Alazazmeh, A., & Asif, M. (2021). Commercial building retrofitting: Assessment of improvements in energy performance and indoor air quality. *Case Studies in Thermal Engineering*, 26, 100946. DOI: 10.1016/j.csite.2021.100946

Albatayneh, A., Jaradat, M., Al-Omary, M., & Zaquot, M. (2021). Evaluation of coupling PV and air conditioning vs. Solar cooling systems—Case study from Jordan. *Applied Sciences (Basel, Switzerland)*, 11(2), 511. DOI: 10.3390/app11020511

Andersen, Z. J., Vicedo-Cabrera, A. M., Hoffmann, B., & Melén, E. (2023). Climate change and respiratory disease: Clinical guidance for healthcare professionals. *Breathe (Sheffield, England)*, 19(2), 220222. DOI: 10.1183/20734735.0222-2022 PMID: 37492343

Arif, S., Taweekun, J., Ali, H. M., Yanjun, D. A. I., & Ahmed, A. (2021). Feasibility study and economic analysis of grid connected solar powered net zero energy building (NZEB) of shopping mall for two different climates of Pakistan and Thailand. *Case Studies in Thermal Engineering*, 26, 101049. DOI: 10.1016/j.csite.2021.101049

Asim, N., Badiei, M., Mohammad, M., Razali, H., Rajabi, A., Chin Haw, L., & Jameelah Ghazali, M. (2022). Sustainability of heating, ventilation and air-conditioning (HVAC) systems in buildings—An overview. *International Journal of Environmental Research and Public Health*, 19(2), 1016. DOI: 10.3390/ijerph19021016 PMID: 35055838

Ayejoto, D. A., Agbasi, J. C., Nwazelibe, V. E., Egbueri, J. C., & Alao, J. O. (2023). Understanding the connections between climate change, air pollution, and human health in Africa: Insights from a literature review. *Journal of Environmental Science and Health. Part C, Environmental Carcinogenesis & Ecotoxicology Reviews*, 41(3-4), 77–120. PMID: 37880976

Aziz, N., Adman, M. A., Suhaimi, N. S., Misbari, S., Alias, A. R., Abd Aziz, A., & Khan, M. M. H. (2023, February). Indoor air quality (IAQ) and related risk factors for sick building syndrome (SBS) at the office and home: A systematic review. [). IOP Publishing.]. *IOP Conference Series. Earth and Environmental Science*, 1140(1), 012007. DOI: 10.1088/1755-1315/1140/1/012007

Besir, A. B., & Cuce, E. (2018). Green roofs and facades: A comprehensive review. *Renewable & Sustainable Energy Reviews*, 82, 915–939. DOI: 10.1016/j.rser.2017.09.106

Božić, J., Ilić, P., & Ilić, S. (2019). Indoor air quality in the hospital: The influence of heating, ventilating and conditioning systems. *Brazilian Archives of Biology and Technology*, 62, e19180295. DOI: 10.1590/1678-4324-2019180295

Burroughs, H. E., & Hansen, S. J. (2020). *Managing indoor air quality*. River Publishers. DOI: 10.1201/9781003151654

Cengiz, M., Kayri, İ., & Aydın, H. (2024). A collated overview on the evaporative cooling applications for photovoltaic modules. *Renewable & Sustainable Energy Reviews*, 197, 114393. DOI: 10.1016/j.rser.2024.114393

Cerrato, G., & Fumo, N. (2024). Performance of photocatalytic oxidizing air cleaners in different experimental setups–a review. *Energy Reports*, 11, 1412–1419. DOI: 10.1016/j.egyr.2024.01.009

D'alessandro, D., Gola, M., Appolloni, L., Dettori, M., Fara, G. M., Rebecchi, A., & Capolongo, S. (2020). COVID-19 and living space challenge. Well-being and public health recommendations for a healthy, safe, and sustainable housing. *Acta Biomedica*, 91(9-S), 61. PMID: 32701918

Dehghani, M. H., Karri, R. R., Vera, T., & Hassan, S. K. M. (Eds.). (2024). *Health Effects of Indoor Air Pollution: Volume 2: Air Pollution, Human Health, and the Environment*. Elsevier.

Fadda, J. (2020). Climate change: an overview of potential health impacts associated with climate change environmental driving forces. In *Renewable Energy and Sustainable Buildings: Selected Papers from the World Renewable Energy Congress WREC 2018* (pp. 77-119). Springer International Publishing. DOI: 10.1007/978-3-030-18488-9_8

Haffaf, A., Lakdja, F., Abdeslam, D. O., & Meziane, R. (2021). Solar energy for air conditioning of an office building in a case study: Techno-economic feasibility assessment. *Renewable Energy Focus*, 39, 148–162. DOI: 10.1016/j.ref.2021.09.002

Hama Radha, C. (2023). Retrofitting for improving indoor air quality and energy efficiency in the hospital building. *Sustainability (Basel)*, 15(4), 3464. DOI: 10.3390/su15043464

Hawkins, V. R., Marcham, C. L., Springston, J. P., Miller, J. D., Braybrooke, G., Maunder, C., ... & Kollmeyer, B. (2020). The value of IAQ: a review of the scientific evidence supporting the benefits of investing in better indoor air quality.

Islam, S. A., Rahaman, M. O., Hossain, M. F., Mridha, M. S., & Islam, M. J. (2023). *design, fabrication, and performance analysis of a solar powered air purifier* (Doctoral dissertation, Sonargoan University (SU)).

Joakim, E. P., Mortsch, L., & Oulahen, G. (2021). Using vulnerability and resilience concepts to advance climate change adaptation. In *Environmental Hazards and Resilience* (pp. 13–31). Routledge. DOI: 10.4324/9781003171430-1

Kumar, P., & Akhai, S. (2022). Effective energy management in smart buildings using VRV/VRF systems. *Additive Manufacturing in Industry*, 4(0), 27–35. DOI: 10.1201/9781003360001-2

Kumar, P., Singh, A. B., Arora, T., Singh, S., & Singh, R. (2023). Critical review on emerging health effects associated with the indoor air quality and its sustainable management. *The Science of the Total Environment*, 872, 162163. DOI: 10.1016/j.scitotenv.2023.162163 PMID: 36781134

Lai, L., Wang, X., Hu, E., & Ng, K. C. (2024). A vision of dew point evaporative cooling: Opportunities and challenges. *Applied Thermal Engineering*, 244, 122683. DOI: 10.1016/j.applthermaleng.2024.122683

Leap, S. R., Soled, D. R., Sampath, V., & Nadeau, K. C. (2024). Effects of Extreme Weather on Health in Underserved Communities. *Annals of Allergy, Asthma & Immunology*.

Ma, N., Aviv, D., Guo, H., & Braham, W. W. (2021). Measuring the right factors: A review of variables and models for thermal comfort and indoor air quality. *Renewable & Sustainable Energy Reviews*, 135, 110436. DOI: 10.1016/j.rser.2020.110436

Mansouri, A., Wei, W., Alessandrini, J. M., Mandin, C., & Blondeau, P. (2022). Impact of climate change on indoor air quality: A review. *International Journal of Environmental Research and Public Health*, 19(23), 15616. DOI: 10.3390/ijerph192315616 PMID: 36497689

Mashallahi, A., Ardalan, A., Nejati, A., & Ostadtaghizadeh, A. (2022). Climate adaptive hospital: A systematic review of determinants and actions. *Journal of Environmental Health Science & Engineering*, 20(2), 983–1013. DOI: 10.1007/s40201-022-00810-5 PMID: 36406601

Mata, T. M., Martins, A. A., Calheiros, C. S., Villanueva, F., Alonso-Cuevilla, N. P., Gabriel, M. F., & Silva, G. V. (2022). Indoor air quality: A review of cleaning technologies. *Environments (Basel, Switzerland)*, 9(9), 118. DOI: 10.3390/environments9090118

Mousavi, S. M., Pouramini, Z., Babapoor, A., Binazadeh, M., Rahmanian, V., Gholami, A., Omidfar, N., Althomali, R. H., Chiang, W.-H., & Rahman, M. M. (2024). Photocatalysis air purification systems for coronavirus removal: Current technologies and future trends. *Chemosphere*, 353, 141525. DOI: 10.1016/j.chemosphere.2024.141525 PMID: 38395369

Nag, P. K., & Nag, P. K. (2019). Sick building syndrome and other building-related illnesses. *Office Buildings: Health, Safety and Environment*, 53-103.

Niza, I. L., de Souza, M. P., da Luz, I. M., & Broday, E. E. (2024). Sick building syndrome and its impacts on health, well-being and productivity: A systematic literature review. *Indoor and Built Environment*, 33(2), 218–236. DOI: 10.1177/1420326X231191079

Ortiz, M., Itard, L., & Bluyssen, P. M. (2020). Indoor environmental quality related risk factors with energy-efficient retrofitting of housing: A literature review. *Energy and Building*, 221, 110102. DOI: 10.1016/j.enbuild.2020.110102

Ortiz, M., Itard, L., & Bluyssen, P. M. (2020). Indoor environmental quality related risk factors with energy-efficient retrofitting of housing: A literature review. *Energy and Building*, 221, 110102. DOI: 10.1016/j.enbuild.2020.110102

Rajasekhar, B., Tushar, W., Lork, C., Zhou, Y., Yuen, C., Pindoriya, N. M., & Wood, K. L. (2020). A survey of computational intelligence techniques for air-conditioners energy management. *IEEE Transactions on Emerging Topics in Computational Intelligence*, 4(4), 555–570. DOI: 10.1109/TETCI.2020.2991728

Raju, S., Siddharthan, T., & McCormack, M. C. (2020). Indoor air pollution and respiratory health. *Clinics in Chest Medicine*, 41(4), 825–843. DOI: 10.1016/j.ccm.2020.08.014 PMID: 33153698

Sadrizadeh, S., Yao, R., Yuan, F., Awbi, H., Bahnfleth, W., Bi, Y., Cao, G., Croitoru, C., de Dear, R., Haghighat, F., Kumar, P., Malayeri, M., Nasiri, F., Ruud, M., Sadeghian, P., Wargocki, P., Xiong, J., Yu, W., & Li, B. (2022). Indoor air quality and health in schools: A critical review for developing the roadmap for the future school environment. *Journal of Building Engineering*, 57, 104908. DOI: 10.1016/j.jobe.2022.104908

Saini, J., Dutta, M., & Marques, G. (2020). Indoor air quality prediction systems for smart environments: A systematic review. *Journal of Ambient Intelligence and Smart Environments*, 12(5), 433–453. DOI: 10.3233/AIS-200574

Saini, J., Dutta, M., Marques, G., Saini, J., Dutta, M., & Marques, G. (2021). Indoor Air Quality: Impact on Public Health. *Internet of Things for Indoor Air Quality Monitoring*, 1-14.

Saran, S., Gurjar, M., Baronia, A., Sivapurapu, V., Ghosh, P. S., Raju, G. M., & Maurya, I. (2020). Heating, ventilation and air conditioning (HVAC) in intensive care unit. *Critical Care*, 24(1), 1–11. DOI: 10.1186/s13054-020-02907-5 PMID: 32375844

Swamy, G. S. N. V. K. S. N. (2021). Development of an indoor air purification system to improve ventilation and air quality. *Heliyon*, 7(10), e08153. DOI: 10.1016/j.heliyon.2021.e08153 PMID: 34703923

Syed, M., Folz, R. J., & Ali, U. (2023). Environmental factors and their impact on airway diseases: Exploring air pollution, indoor and outdoor allergens, and climate change. *Current Pulmonology Reports*, 12(3), 162–170. DOI: 10.1007/s13665-023-00319-8

Tanwar, N., & Akhai, S. (2017). Survey Analysis for Quality Control Comfort Management in Air Conditioned Classroom. *Journal of Advanced Research in Civil and Environmental Engineering*, 4(1&2), 20–23. DOI: 10.24321/2393.8307.201702

Tien, P. W., Wei, S., Darkwa, J., Wood, C., & Calautit, J. K. (2022). Machine learning and deep learning methods for enhancing building energy efficiency and indoor environmental quality–a review. *Energy and AI*, 10, 100198. DOI: 10.1016/j.egyai.2022.100198

Tran, V. V., Park, D., & Lee, Y. C. (2020). Indoor air pollution, related human diseases, and recent trends in the control and improvement of indoor air quality. *International Journal of Environmental Research and Public Health*, 17(8), 2927. DOI: 10.3390/ijerph17082927 PMID: 32340311

Vardoulakis, S., Giagloglou, E., Steinle, S., Davis, A., Sleeuwenhoek, A., Galea, K. S., Dixon, K., & Crawford, J. O. (2020). Indoor exposure to selected air pollutants in the home environment: A systematic review. *International Journal of Environmental Research and Public Health*, 17(23), 8972. DOI: 10.3390/ijerph17238972 PMID: 33276576

Viguié, V., Lemonsu, A., Hallegatte, S., Beaulant, A. L., Marchadier, C., Masson, V., Pigeon, G., & Salagnac, J. L. (2020). Early adaptation to heat waves and future reduction of air-conditioning energy use in Paris. *Environmental Research Letters*, 15(7), 075006. DOI: 10.1088/1748-9326/ab6a24

Vornanen-Winqvist, C., Järvi, K., Andersson, M. A., Duchaine, C., Létourneau, V., Kedves, O., Kredics, L., Mikkola, R., Kurnitski, J., & Salonen, H. (2020). Exposure to indoor air contaminants in school buildings with and without reported indoor air quality problems. *Environment International*, 141, 105781. DOI: 10.1016/j.envint.2020.105781 PMID: 32417615

Wang, H., Zhao, L., Cao, R., & Zeng, W. (2021). Refrigerant alternative and optimization under the constraint of the greenhouse gas emissions reduction target. *Journal of Cleaner Production*, 296, 126580. DOI: 10.1016/j.jclepro.2021.126580

Chapter 7
Climate Change and Mental Health

Jeffrey Kurebwa
https://orcid.org/0000-0002-8371-8055
Bindura University of Science Education, Zimbabwe

Nyasha Yvonne Kurebwa
Women's University in Africa, Zimbabwe

Valeria Supu
https://orcid.org/0009-0002-0860-5744
Bindura University of Science Education, Zimbabwe

ABSTRACT

Amidst the growing concerns surrounding climate change, the impact on mental health has emerged as a pressing issue that warrants attention. Research has shown that individuals residing in regions vulnerable to extreme weather events, such as Zimbabwe, are particularly susceptible to experiencing mental health challenges as a result of climate-related stressors. The unpredictable weather patterns, recurring droughts, and food insecurities have exacerbated existing mental health conditions and triggered new cases of anxiety, depression, and trauma among the population. As we delve into the intersection between climate change and mental health, it is crucial to understand the multi-faceted ways in which climate-related stressors influence the psychological well-being of individuals, families, and communities. By exploring this complex relationship, we can develop targeted interventions and policies to mitigate the adverse effects of climate change on mental health.

DOI: 10.4018/979-8-3693-4183-4.ch007

INTRODUCTION

The mental well-being of citizens globally is significantly affected by climate change. Exposure to extreme heat has been linked to various mental health issues such as fatigue, anxiety, mood disorders, hostility, and violence. Moreover, individuals with pre-existing mental illness face an increased risk of death during heat waves. Extreme weather events have a direct impact on mental health, leading to conditions such as depression, anxiety, Post-Traumatic Stress Disorder (PTSD), suicidal ideation, substance abuse, and feelings of hopelessness (Hayes et al. 2019). These impacts have been supported by various studies (Hayes & Poland, 2018). The mental health effects of climate change can also be felt when individuals witness large-scale events remotely, such as through the news. Additionally, the awareness of global biodiversity loss can lead to feelings of hopelessness and depression. These factors contribute to the overall mental health challenges faced by people due to climate change.

The recognition of the climate crisis has the potential to cause distress, anxiety, and a sense of impending doom, as noted by Hayes et al. (2019) and Hayes & Poland (2018). Conversely, there have been documented instances of positive psychological growth, leading to feelings of altruism, purpose, and optimism, referred to as 'post-traumatic growth (Cianconi et al. 2020, Hayes et al. 2019). Climate change has unequal impacts on the health of the world's population. Children are disproportionately affected globally due to their developing physiology and longer lifespan, which increases their overall exposure time (Watts et al. 2019). Older adults and individuals with medical comorbidities, such as diabetes, have vulnerable physiology and use prescription medications that put them at risk of heat-related illnesses, such as medications that predispose them to dehydration. (Watss et al. 2019). Climate change has a disproportionate impact on women in various social and cultural settings (Watts et al. 2019, Williams, 2018). Vulnerable populations, including those at risk of homelessness and outdoor workers, are particularly susceptible to heat-related illnesses due to limited shelter and cooling options (Kiefer et al. 2016). The increasing concerns related to climate change have brought attention to the impact on mental health. Studies have indicated that individuals living in regions prone to extreme weather events are at a higher risk of experiencing mental health issues due to climate-related stressors (Inder, 2011). This underscores the need to address the mental health implications of climate change. Understanding the connection between climate change and mental health is essential to address how climate-related stressors impact the psychological well-being of the population. By examining the intricate relationship between these factors, tailored interventions and policies can be formulated to alleviate the negative impact of climate change on mental health (Ojala & Bengtsson, 2019).

The health effects of climate change manifest through both direct and indirect channels. Direct impacts include the rise in frequency and intensity of extreme weather events like heatwaves, droughts, floods, and storms, leading to health issues such as heat-related illnesses, injuries, and fatalities (Ojala & Bengtsson, 2019). Indirectly, climate change can affect health by causing disruptions in livelihoods, loss of land and property, and altering interactions with environmental systems. This can result in changes to the spread and prevalence of diseases transmitted through vectors, water, or food. Furthermore, the consequences of climate change are compounded by their interaction with already fragile social structures, jeopardizing access to sufficient nutrition and clean drinking water. The intricate web of connections between climate change and health underscores the urgent need for comprehensive mitigation and adaptation strategies to safeguard public health (Solomon et al. 2007).

The current obstacles pose a significant danger to human well-being, with the potential to undo the progress achieved in public health over the past fifty years (Palinkas & Wong, 2020). It is crucial to situate the issue of climate change within the context of mental health, as the impact of climate change is becoming more severe in terms of frequency, intensity, duration, and complexity, leading to a rise in climate-related mental health issues worldwide (Ojala & Bengtsson, 2019). Recognizing the mental health implications of climate change enables mental health professionals to identify and predict trends in mental illness, such as Post-Traumatic Stress Disorder (PTSD) in the aftermath of extreme weather events. This awareness is essential in addressing the growing burden of mental health disorders and challenges on a global scale (Bryant et al. 2014).

Recognizing the disproportionate effects of climate change on marginalized communities is crucial for developing effective public health interventions aimed at safeguarding those most vulnerable to mental health issues. Linking mental health outcomes to climate change also provides an opportunity to evaluate, establish, and enhance mental health services (Cianconi, Betro, & Janiri, 2020). The World Health Organization (WHO) introduced a blueprint in 2015 to create climate-resilient healthcare systems, offering guidance for healthcare professionals to anticipate, prevent, and respond to climate change-induced crises, ultimately bolstering the resilience and efficacy of health systems in the face of evolving environmental conditions (WHO, 2015).

The framework fails to address the complexities of mental health systems, such as the widespread lack of mental health infrastructure, funding, and resources on a global scale. However, it does offer valuable guidance for building resilience in mental health systems (Cianconi, Betro, & Janiri, 2020). By highlighting climate change as a factor affecting mental health, researchers, practitioners, and activists not only raise awareness about the wide-ranging impact of climate change on health but also contribute to the improvement of mental health systems. There is now a growing body of research that delves into the current mental health implications of extreme weather events related to climate change. Studies have shown that extreme heat and humidity can lead to an increase in hospital admissions for mood and behavioral disorders, including schizophrenia, mania, and neurotic disorders (Fernandez et al. 2015).

Experts in the field have observed that individuals with impaired thermoregulation, such as those with pre-existing mental health conditions, individuals taking certain prescription medications, and those with substance abuse issues, are more susceptible to heat-related mental health problems (Inder, 2011). Additionally, extreme heat has been associated with an increased risk of wildfires, which can have a direct impact on mental health. Bryant et al. (2014) conducted a study on the psychological effects of the Black Saturday bushfires in Victoria, Australia, and found that communities most vulnerable to the impacts of bushfires experienced higher rates of PTSD, psychological distress, and depression related to the fires (Inder, 2011).

The Transactional Model of Coping

The transactional model of coping provides a framework for understanding how individuals manage intense emotions (Russ, Regalado, & Halfon, 2020). According to researcher Maria Ojala (2019), this model suggests that people employ various coping strategies to deal with stress and negative emotions, and the strategies they choose can have an impact on their behavior and psychological well-being (Ojala & Bengtsson, 2019). This coping model consists of two appraisal stages. The first stage, known as primary appraisal, involves an individual's assessment of whether a situation poses a threat to something they value. If they perceive the situation as a threat, it can lead to the experience of negative emotions. For instance, in the context of climate change, a person might contemplate the potential impact of climate change on the future of their children (Lauren, O'Connel, & Yogman, 2020).

Upon recognizing that climate change poses a risk to the future quality of life for their offspring, individuals may experience a range of emotions such as fear, anger, grief, or hopelessness. In the secondary appraisal phase, individuals strive to manage these emotions stemming from the perceived threat. The transactional coping model delineates two coping strategies: emotion-focused coping and problem-focused coping (Lauren, O'Connel, & Yogman, 2020). Emotion-focused coping involves addressing the emotions themselves, which may entail avoiding situations that trigger negative emotions or attempting to distance oneself from distressing circumstances. On the other hand, problem-focused coping entails directly addressing the situation at hand to alleviate negative emotions (Russ, Regalado, & Halfon, 2020).

Ojala & Bengtsson (2019) suggest that individuals facing a threatening situation have the option to either take direct action or engage in discussions with others to develop a plan. In contrast to the transactional model of coping, which only acknowledges these two approaches, Ojala & Bengtsson (2019) introduce a third method known as meaning-focused coping. This approach emphasizes the use of positive emotions to counteract negative ones, rather than attempting to eliminate negative emotions. According to Ojala & Bengtsson (2019), meaning-focused coping allows positive emotions to serve as buffers, enabling individuals to confront their concerns and promote problem-focused coping and active engagement. Unlike emotion-focused coping strategies such as avoidance or distraction, which can hinder problem-focused coping, meaning-focused coping encourages individuals to take action and reduces the paralyzing effects of negative emotions. Meaning-focused coping proves to be particularly beneficial in situations where the level of threat, such as that posed by climate change, is unalterable (Ojala & Bengtsson, 2019). This approach emphasizes finding significance and purpose in the face of adversity, allowing individuals to better manage and adapt to circumstances beyond their control. By focusing on the meaning behind their experiences, individuals can cultivate resilience and psychological well-being, even in the absence of the ability to mitigate the threat itself (Ojala & Bengtsson, 2019).

Defining Climate Change

Climate is defined as the long-term average weather conditions in a specific region, typically spanning 30 years, and is determined by factors like temperature, precipitation, humidity, and seasonal patterns (Matthews et al. 2021). Climate change, on the other hand, refers to significant alterations in climate that persist for decades or longer. The UN Framework Convention on Climate Change (UNFCCC, 1992) offers a detailed explanation of climate change, attributing it to human activities that impact the global atmosphere and are observed over comparable timeframes alongside natural climate variability.

The Intergovernmental Panel on Climate Change (IPCC, 2014) reported that climate change encompasses various phenomena, including rising temperatures, shifting precipitation patterns, increasing sea levels, intrusion of saltwater, and a heightened likelihood of extreme weather occurrences like floods and droughts (Solomon et al. 2007). Climate change is an escalating worldwide emergency, with its magnitude already substantial and poised to expand further if no action is taken. It gives rise to immediate dangers such as hurricanes, floods, and wildfires, as well as gradual risks like alterations in ecosystems, food and water insecurity, and the erosion of local traditions and heritage (Watts et al., 2018). Climate change stands out as one of the numerous environmental challenges confronting the planet, necessitating urgent attention and concerted efforts to mitigate its impacts and safeguard the well-being of current and future generations. Unsustainable human activities, including deforestation, ecosystem degradation, depletion and loss of biodiversity, and reliance on fossil fuels by economies, are causing water and food insecurity, air pollution, and contamination of land, rivers, and oceans (Vigo et al. 2016). These factors are significantly impacting human health and mental well-being, and exacerbating the climate emergency. The consequences of unsustainable human activities, such as deforestation and reliance on fossil fuels, are leading to water and food insecurity, air pollution, and contamination of land, rivers, and oceans. These effects are measurably impacting human health, and mental well-being, and further worsening the climate emergency (Hayes et al. 2018).

Defining Mental Health

Mental health issues encompass changes in cognition, emotions, or actions and are linked to distress or impaired performance. The World Health Organization (2004) defines mental health as a state of well-being in which the individual realizes their abilities, can handle everyday stress, can work effectively, and can contribute to their community. This definition allows for a range of mental health results. On one end of the spectrum are positive mental health and psychosocial wellness, while on the other end are mental disorders; clearly defined categories of mental illness as recognized by psychiatrists. Within the range of mental health, there exists a middle ground where individuals may experience impairment without meeting the criteria for pathology, yet still have a significant impact on their daily lives. Some of these challenges stem from typical psychological responses to adversity and catastrophic events (Berry, Bowen, & Kjellstrom, 2010). The outcomes of mental health are heavily influenced by both interpersonal and extra-personal factors (Watts et al. 2018).

The impact of current and future climate change on human health is a growing concern that has been highlighted in various reports. According to the 2009 Lancet report, climate change is identified as the most significant global health threat of the 21st century (Brosse et al., 2022). Climate change affects human health directly through changes in temperature, precipitation patterns, and the frequency of extreme weather events. Additionally, there are indirect impacts such as rising sea levels, ocean acidification, flooding, droughts, and wildfires, all of which contribute to the overall health risks posed by climate change. The visible effects of climate change are already being observed, and without effective mitigation strategies, these impacts are expected to exacerbate existing global health challenges and inequalities. It is crucial to address these issues to safeguard public health and well-being in the face of climate change (Brosse et al., 2022).

The World Health Organization (WHO, 2014) projected that climate change would result in around 250,000 additional deaths annually from 2030 to 2050, primarily due to malnutrition, malaria, diarrhea, and heat stress. Climate change has a significant impact on mental health, manifesting in various ways. The increase in extreme weather events such as storms and floods is causing more frequent, intense, and complex disasters, which can lead to mental health issues like post-traumatic stress, depression, and anxiety (Brosse et al., 2022). The gradual effects of climate change, such as rising temperatures, sea levels, and irregular droughts, can disrupt the natural environment. These changes pose threats to food and water supply, alter agricultural and habitation practices, weaken infrastructure, and result in population displacement (Brosse et al., 2022).

Adverse effects on mental health stemming from these occurrences encompass financial strain, strained relationships, and heightened susceptibility to violence and hostility. Conversely, favorable mental health consequences have also been noted, such as individuals coming together to heal, reconstruct, and provide solace to one another in times of adversity (Hayes, 2018). Nevertheless, the impact of climate change on mental health is frequently disregarded, mirroring the general lack of emphasis on mental health in the global context, where mental illnesses are often overlooked as a significant health concern (Vigo et al. 2016). Due to significant demographic, environmental, and socio-political shifts, the global burden of mental disorders has surged worldwide and is projected to increase further.

Nevertheless, even in developed nations, mental health services are frequently neglected in terms of government funding and staffing when compared to the disease burden (Pollack, 2016). The standard of mental health care is consistently lower than that of physical health care. Particularly in low- and middle-income countries (LMIC), access to high-quality mental health services is limited. Merely 14% of low-income countries (LIC) typically provide mental health services at healthcare facilities, with inadequate policies and ineffective community public health interventions often lacking mental health components (Pollack, 2016). The treatment gap for mental health in developing countries is estimated to be approximately 75% (American Public Health Association, 2019).

Individuals with mental health disorders are particularly susceptible to experiencing suffering, disability, human rights violations, stigma, discrimination, impoverishment, and premature death (Pollack, 2016). Financial limitations associated with seeking treatment and the resulting loss of income place individuals with mental health issues and their families in danger of being marginalized (Hanlon et al. 2019). Mental health challenges not only impact patients and their families but also carry significant economic burdens for society as a whole. These burdens include direct costs incurred within the healthcare system, as well as indirect costs stemming from reduced income due to mortality, disability, and decreased productivity resulting from illness-related absences (Patel et al. 2018). In 2010, it was approximated that the combined direct and indirect expenses amounted to a total of $2.5 trillion on a global scale (Trautmann, 2016). The Sustainable Development Goals (SDGs) emphasize the significance of addressing both climate change and mental health, as they are directly linked to SDGs 3 & 13: "Ensure healthy lives and promote wellbeing for all at all ages" and "Take urgent action to combat climate change impacts". Moreover, by adopting an integrated approach to the SDGs, there will be a ripple effect on other goals such as those aimed at reducing poverty and inequality (Patel et al. 2018).

Linking Climate Change to Mental Health

Climate influencers can impact both physical and mental well-being through various channels of exposure. In the realm of mental health, multiple researchers have delved into these exposure pathways, each offering a unique perspective (Benateau et al. 2019, Parker et al. 2008, Viviroli, 2020). Despite some discrepancies among authors, there is a consensus regarding the categorization of climate change effects on mental health into direct and indirect impacts, as well as the differentiation between acute weather occurrences, sub-acute weather events, and enduring environmental alterations (Palinkas & Wong, 2020). The direct consequences of climate change on mental health manifest when individuals face more frequent, severe, and prolonged climate-related calamities and extreme weather conditions. It is evident that extreme events, particularly disasters, are prevalent and contribute significantly to the overall health repercussions of climate change (Parker et al., 2008).

Severe weather conditions can heighten the likelihood of danger, harm, and fatalities, resulting in immediate and traumatic reactions that may escalate the risk of developing PTSDs Hayes et al. 2018). Conversely, less severe weather events can exacerbate existing mental health conditions due to heat exposure and contribute to an uptick in aggressive behaviors (Hayes et al. 2018). Prolonged alterations in the environment can trigger feelings of anxiety, helplessness, and sorrow, leading to a sense of displacement and grief. These emotional responses are highlighted in the US Global Change Research Program's report from 2016. The impact of climate change on mental health can also be indirect, stemming from shifts in social, economic, and environmental factors that influence mental well-being (Martin et al. 2020).

Alterations in natural environments, disrupted availability of food and water, adverse impacts on physical health, changes in agricultural conditions, modified land usage and human settlement, and weakened physical and social infrastructure have the potential to lead to heightened financial and interpersonal stress, increased likelihood of violence and aggression, and the displacement of entire communities (Hayes et al. 2018). Additionally, climate change can affect physical health, which in turn can impact mental health or vice versa, due to the causal and reciprocal relationship between physical and mental health (Berry et al. 2010). The responses of mental health to climate change can be intricate and varied, and may also encompass positive outcomes. These positive outcomes may include psychological growth following a traumatic event, empathy, compassion, altruism, and emotional resilience (Berry et al. 2018). However, the literature documenting these positive effects is limited (Hayes, 2019).

The changing climate poses significant physical public health threats, such as an increase in vector-borne diseases, higher rates of heat-related morbidity and mortality, worsening respiratory conditions due to declining air quality, and an uptick in food and water-borne illnesses from contamination (Alderman et al. 2013). A report by scientists from the IPCC in collaboration with the World Health Organization (WHO) shed light on the health implications of climate change at 1.5 degrees Celsius and 2.0 Celsius, emphasizing that even a slight half-degree rise in temperature could lead to severe physical health consequences, particularly impacting the most vulnerable populations in society (Castleden, 2011).

Ongoing research efforts are continuously enhancing our understanding of the effects of climate change on physical health. These studies have highlighted various consequences such as the escalation of vector-borne, waterborne, and foodborne diseases, as well as the surge in acute and chronic respiratory ailments like asthma and allergies, along with the rise in heat-related and extreme weather-related health issues leading to morbidity and mortality (Bragg & Reser, 2012). Furthermore, global reports on climate change and health are increasingly acknowledging the indirect health impacts associated with these environmental changes. These include health concerns related to food and water safety, malnutrition due to food insecurity, skin cancer like malignant melanoma resulting from ultraviolet (UV) exposure, and chronic kidney disease stemming from dehydration (Cheng & Berry, 2013).

The correlation between mental health and climate change has been demonstrated through various channels, including the worsening of pre-existing mental health conditions after extreme weather occurrences, the rise in new mental health diagnoses in communities facing climate-related stressors, and the heightened susceptibility to physical illnesses leading to increased vulnerability to climate change impacts. Research by (Cianconi, Betrò, & Janiri, 2020, Berry et al. 2010, Panu, 2020) has shed light on the intricate relationship between mental health and climate change, emphasizing the detrimental effects on individuals' well-being and overall health. Notably, heat-related issues have emerged as a significant concern in the context of mental health, with direct connections established between extreme climate events such as heatwaves, floods, and fires, and the exacerbation of mental health conditions among affected populations (Berry et al. 2010).

Individuals with pre-existing mental illnesses are more susceptible to relapse when faced with external social and environmental stressors. Those diagnosed with PTSD, anxiety disorders, or depressive disorders are particularly vulnerable to experiencing worsened symptoms after extreme weather events due to exposure to risk factors associated with their conditions, such as threats to safety, lack of control in threatening situations, and diminished support systems (Berry et al. 2008). Disruption of healthcare services further heightens the likelihood of relapse in mood and substance use disorders. Moreover, extreme weather occurrences have been linked to heightened levels of violence within communities, posing a collective risk for the development or exacerbation of mental health disorders stemming from trauma (WHO, 2017). This interconnected relationship underscores the importance of addressing mental health needs in the aftermath of such events to mitigate adverse outcomes and promote resilience among affected individuals (Willox et al., 2013).

Climate Psychology and Climate Anxiety

Since the beginning of the 21st century, there has been a significant advancement in the field of climate psychology as highlighted by the American Psychological Association (APA, 2010). Ursano, Morganstein, & Cooper (2010) emphasized climate change as a major threat to global health in the 21st century, prompting a response from the psychology community. The 2010 report by the American Psychological Association outlined six crucial areas for psychology to focus on regarding climate change, such as risk perception, psychological and behavioral causes, psychosocial impacts, adaptation strategies, barriers to action, and the role of psychologists in addressing these issues. In parallel, psychotherapists and clinical psychologists are grappling with ecological loss, anxiety, and the impact on their patients and themselves as they confront the challenges posed by the deteriorating biosphere (Rust, 2020).

The focus is on climate anxiety and its impact within the field of climate psychology, acknowledging that anxiety can intersect with a variety of other emotions like grief and rage. The American Psychological Association (2010) defines Eco anxiety as a persistent fear of environmental catastrophe, ranging from mild stress to clinical conditions such as depression, anxiety, post-traumatic stress disorder, and suicide (North, Kawasaki, Spitznagel, & Hongkong, 2004). Additionally, maladaptive coping mechanisms such as intimate partner violence and substance abuse have been associated with this type of anxiety (Flori, Henkin, Floos, Cheely, & Tureck, 2009). Cianconi, Betro, & Janili (2020) have identified various forms of climate anxiety and trauma, including intergenerational effects, particularly in cases where environmental harm results in the loss of a way of life or culture.

Psychological Effects of Climate Change

Scholars have employed a variety of terms to articulate the psychological impacts of climate change, with 'eco-anxiety' being one of the most commonly used. According to the APA (2010), eco-anxiety is characterized as a persistent fear of environmental catastrophe, as well as the apprehension regarding the future for oneself, one offspring, and future generations, as the gradual and seemingly irreversible consequences of climate change continue to unfold (Clayton et al. 2017). Unlike an emotional reaction to directly experiencing the effects of climate change, eco-anxiety is more existential, representing a fear that climate change is robbing prospects (Pihkala, 2018). Preceding the widespread adoption of the term 'eco-anxiety' is the concept of 'solastalgia' which was introduced by (Glenn Albrecht, 2005).

Solastalgia is differentiated from eco-anxiety by its focus on the immediate emotional response to the loss or alteration of one's home environment, rather than a broader apprehension about the future. Albrecht's (2005) definition of solastalgia emphasizes the distress caused by the absence of comfort and the feeling of disconnection from one's familiar surroundings. Unlike eco-anxiety, which is characterized by a more generalized sense of unease about the impending consequences of environmental degradation, solastalgia is rooted in the tangible experience of witnessing one's homeland undergo irreversible transformations due to factors like climate change. The concept of solastalgia underscores the profound emotional impact of environmental changes on individuals, highlighting the deep sense of loss and isolation that can arise from the disruption of one's connection to their home and territory. This distinction emphasizes the personal and immediate nature of solastalgia compared to the more abstract concerns associated with eco-anxiety.

The concept of a person's sense of place is crucial to their identity and the emotional distress caused by their loss (Albrecht, 2005). Additionally, Albrecht introduced the term 'eco paralysis' to illustrate a form of ecological distress where individuals are consumed by their concern for environmental issues. Eco-paralysis is significant as it suggests that inaction is not always a result of apathy. Rather, it stems from an overwhelming care for the environment, leading individuals to be immobilized by their emotions (Pihkala, 2018). Weintrobe (2013), a psychoanalyst specializing in climate change coined the term 'environmental melancholia' to describe the overwhelming emotions people experience in response to environmental threats (Weintrobe, 2013). This concept, along with eco paralysis, illustrates the emotional aspect of climate action, highlighting how the inability to manage distress about environmental issues can result in inaction, ultimately hindering much-needed progress.

Impacts of Climate Change on Mental Health

The mental health of individuals is being significantly impacted by climate change in various ways. Most countries have experienced a rise in temperatures and shifts in weather patterns, leading to an increase in natural disasters such as droughts and floods (Cianconi, Betro, & Janiri, 2020). These events have had devastating consequences for communities, causing physical harm, displacement, and significant mental health implications. The environmental stressors resulting from these climate-related disasters have not only caused physical harm and displacement but have also had a profound impact on the mental well-being of the population (Brosse et al. 2021). The loss of livelihoods, homes, and loved ones has led to psychological distress, contributing to higher rates of anxiety, depression, and PTSD. Additionally, the uncertainty and fear surrounding future climate-related events have created a sense of helplessness and powerlessness among the population, further exacerbating mental health challenges. The multifaceted and far-reaching impacts of climate change on mental health highlight the urgent need for comprehensive support and intervention to address these issues (Cianconi, Betro, & Janiri, 2020).

The necessity of a comprehensive strategy that combines mental health services with climate adaptation measures to enhance resilience and promote the welfare of impacted populations is highlighted in a study by Filho et al. (2016). The impact of climate change on mental health is influenced by various factors such as the local cultural, social, economic, and developmental setting, the geographical spread of exposure, the nature of the meteorological occurrence, the duration and intensity of the event, and the expected short-term and long-term effects on physical health and community welfare (Berry, 2010).

Mental Health

Climate change has both immediate and long-term effects on mental health. Immediate impacts can lead to anxiety, depression, PTSD, and substance abuse, especially following a disaster (Cianconi, Betro, & Janiri, 2020). The risk of psychological trauma and shock is high when individuals experience injury, property damage, or loss after a disaster, which can have lasting effects on mental health. While some acute effects may diminish over time, PTSD can persist as a chronic condition even after the initial shock has subsided (Berry, 2010). Additional long-term consequences documented consist of increased levels of hostility, aggression, and an enduring feeling of mourning over, for instance, locations that hold personal significance (Cunsolo et al. 2012). Further effects outlined in research include feelings of guilt among survivors, climate and ecological distress, heightened psychotic episodes, thoughts of suicide, and actual suicide attempts. Individuals who have experienced severe catastrophes are at a greater risk of taking their own lives. Those with pre-existing mental health issues may experience a more pronounced impact (Cunsolo & Ellis, 2018).

Climate change and the resulting extreme weather events have the potential to give rise to new mental health conditions within affected populations. Responses to these events can vary from immediate distress to long-term psychiatric disorders. Notably, there has been an increase in the diagnosis of conditions such as generalized anxiety, panic disorder, substance use disorders, and major depression following such events (Cianconi, Betrò, & Janiri 2020). The psychological impact of displacement, including the loss of employment, community, and sense of identity, can contribute to the perpetuation of mental illness over an extended period. This is due to the loss of protective social and support factors that are known to play a critical role in reducing the onset and worsening of mental health conditions (Berry, 2010). Social support plays a crucial role in communities that have faced challenges in accessing healthcare and government resources historically. Research has shown that the psychological effects of events like flooding can have long-lasting impacts, with those who have been displaced experiencing a 1.7 times higher risk compared to those who have not been displaced by floods for example (Berry et al. 2018).

Studies have shown that extreme weather events can lead to the development of mental health conditions such as depression, anxiety, and stress-related issues (Berry, 2010). The indirect consequences of climate change on mental health can stem from the destruction of physical and social infrastructure, adverse effects on physical health, shortages of food and water, conflicts, and displacement due to acute, subacute, and chronic changes in climate (Berry, 2009b). Among the various climate hazards, drought stands out as one of the well-documented factors that can have an indirect impact on mental health. Prolonged droughts can disrupt food and water supplies, leading to economic and mental strain on land-based workers, particularly affecting those residing in rural and remote areas (Bunch et al. 2011).

A study conducted on the impact of drought and distress in Australia revealed that rural residents are more affected by drought-related distress compared to their urban counterparts. The study, conducted over seven years, highlighted the heightened vulnerability of rural dwellers to the psychological and emotional effects of droughts, shedding light on the need for targeted support and interventions in these areas (Bunch, 2011). Furthermore, the research also pointed out the increased risks posed by heat to individuals with mental health disorders. Vulnerable patients undergoing medical care and treatment are particularly susceptible to even minor changes in climatic conditions, with hot days significantly elevating the risk of mortality for those suffering from depression and other mental health issues, as opposed to days with lower temperatures (Staffogia et al. 2006). The study also highlighted the heightened risk of heat-related mortality for patients with psychosis, dementia, and substance abuse, emphasizing the need for tailored interventions and support for these vulnerable populations during periods of extreme heat (Page et al. 2018).

Stress-Related Physical Health Problems

Extreme weather events and other climate-related hazards have the potential to cause significant emotional distress (Cianconi et al. 2020). Following an emergency, individuals commonly experience distress, but they are generally able to manage and recover once their basic needs are addressed, and a sense of security and safety is restored (Clayton, 2017). Stress has the potential to weaken the immune system, making individuals more susceptible to air pollution and waterborne diseases (Alderman et al. 2012). Additionally, chronic distress has been associated with sleep disturbances, which can impact physical health and exacerbate mental health issues and psychosocial well-being. Psychological stress has also been found to elevate the risk of developing cardiovascular diseases, autoimmune disorders, and even certain types of cancers (Morey et al. 2015). Furthermore, individuals exposed to climate-related or weather-related disasters often face significant stress and may experience severe mental health implications. Depending on the nature of the disaster, individuals may develop PTSD, depression, and general anxiety simultaneously. While many people can recover from these mental health challenges over time, a notable portion of those affected may go on to experience chronic psychological issues (Cianconi et al. 2020).

Strained Social Relationships

Climate-related hazards have been found to create tension in interpersonal relationships and contribute to instances of intimate partner violence (Hayes, 2018). Additionally, there are other psychosocial impacts such as family separation and disconnection from social support systems (Kousky, 2016). For example, children may have to be temporarily relocated and attend a different school or miss school altogether due to climate-related events. At the community level, the indirect mental health consequences of climate change have been identified as an area that is currently understudied. Bowles (2015) points out that these consequences may include a decrease in community cohesion, the loss of community identity, threats to a sense of continuity, and a sense of belonging as individuals are compelled to move in and out of communities due to environmental stressors. Furthermore, there is a risk of undermining cultural integrity if people are forced to leave their homelands due to climate-related factors. The act of migration poses significant challenges to the identity, sovereignty, and heritage of individuals forced to depart from their native lands. Furthermore, it presents obstacles to the integrity and sustainability of traditional lifestyles. In addition, threats to community well-being may arise, such as a heightened risk of criminal activities, violence, and aggression, as community members grapple with stressors linked to climate change (Bowles, 2015).

Helplessness, Fear, and Grief

Observing the gradual effects of climate change unraveling can result in concerns about what lies ahead, as well as emotions of powerlessness and unease (Albrecht, 2007). Certain individuals encounter emotions of grief, powerlessness, and frustration due to their perceived inability to halt climate change or have an impact (Hayes, 2018). Numerous youths express experiencing heightened distress and a feeling of betrayal and distrust towards the government in light of climate inaction (Hickman et al. 2021).

Increased Risk of Suicidal Behavior

Individuals who have encountered repetitive or severe hazards may face a heightened risk of suicide (Berry et al. 2010). Moreover, Burke et al. (2018) have established a correlation between rising ambient temperatures and an increase in suicide rates across various nations. Although PTSD is commonly associated with severe mental health consequences following acute climate change-related disasters, there is a growing body of evidence pointing towards suicide and suicidal thoughts as additional impacts of extreme weather events. Fernandez et al. (2015) highlight a study conducted in Italy by Chand et al. (2008), which revealed elevated suicide rates in communities located in northern regions with greater climate variability. Dodgen et al. (2016) emphasize the increased danger of homicide-suicides following extreme weather events, citing a twofold rise in these incidents after Hurricane Andrew in 1992 in Miami-Dade County. Additionally, Imperiale & Vanclay (2016) have observed a rise in suicidal thoughts (from 2.8% to 6.4%) and plans to commit suicide (from 1.0% to 2.5%) 18 months after extreme weather events. It is important to note, however, that the overall evidence connecting climate change and suicide remains inconclusive.

SOLUTIONS AND RECOMMENDATIONS

(a) Going forward, policymakers, healthcare providers, and community stakeholders must prioritize mental health in their climate change adaptation strategies.

(b) Countries have the opportunity to enhance the psychological well-being of their population by incorporating mental health initiatives into their climate resilience efforts. Climate change is a significant threat to mental health and emotional well-being, with various pathways through which it can impact individuals.
(c) Policymakers must integrate mental health support systems into programs aimed at climate change adaptation and disaster risk reduction to address the psychological aftermath of environmental emergencies.
(d) Policymakers can enhance the protection of the population's well-being amid environmental challenges by acknowledging the connection between climate change and mental health. It is imperative to prioritize further research and investment in mental health resources to adequately tackle this growing issue. The impact of climate change on the health of individuals and communities is becoming more pronounced, posing significant challenges to health systems worldwide.
(e) Robust public health systems play a crucial role in adaptation efforts, with sufficient resources allocated to mental health care ensuring equitable access to services for those requiring assistance.

FUTURE RESEARCH DIRECTIONS

Further research into the effects of climate change on mental health and mental health-related systems will aid policymakers in creating strong evidence-based mitigation and adaptation policies and plans that could have wide-ranging benefits for society and the environment. Despite being considered the greatest threat to global mental health in the next century, addressing climate change presents a significant opportunity to positively influence mental health for generations to come due to the health co-benefits of transitioning to more sustainable lifestyles.

CONCLUSION

This study highlights the urgent requirement for policymakers to give precedence to mental health interventions amidst the backdrop of climate change. The study has brought to light the notable influence of severe weather occurrences and environmental decline on the mental health of individuals within the nation. More specifically, the research illustrates that heightened exposure to climate-related calamities correlates with increased levels of stress, anxiety, and depression among the Zimbabwean populace. These discoveries carry significant implications for the formulation of policies and the implementation of mental health services.

The mental well-being of people globally is being significantly affected by climate change. The stress, anxiety, and depression experienced by the population are being intensified by extreme weather events, food insecurity, and economic instability. Vulnerable groups, including women, children, and rural communities, are particularly at risk, as indicated. The lack of sufficient mental health services and support further complicates the situation, leaving many individuals without the necessary resources to address the psychological effects of climate change. The connection between climate change and mental health is crucial to address, as extreme weather events can result in severe psychological trauma, leading to increased suicide rates, new cases or exacerbated symptoms of mental illness, and overall deteriorated mental health and emotional well-being in the community. Climate change worsens disparities, making those with mental health conditions more susceptible to the risks posed by extreme heat on their well-being and survival. The distress caused by observing the impacts of climate change on communities, ecosystems, and societies can be overwhelming for a significant number of individuals.

REFERENCES

Albrecht, G. (2005). 'Solastalgia.' A new concept in health and identity. *PAN: philosophy activism nature*, 3, 41-55.

Albrecht, G., Sartore, G., Connor, L., Higginbotham, N., Freeman, S., Kelly, B., Stain, H., Tonna, A., & Pollard, G. (2007). Solastalgia: The distress caused by environmental change. *Australasian Psychiatry*, 15(1_suppl, Suppl 1), S95–S98. DOI: 10.1080/10398560701701288 PMID: 18027145

American Psychological Association (APA). (2010). *Psychology and Global Climate Change: Addressing a Multi-faceted Phenomenon and Set of Challenges*. APA.

American Public Health Association (APHA). (2019). Addressing the Impacts of Climate Change on Mental Health and Well-Being. www.apha.org

Berry, H. L., Bowen, K., & Kjellstrom, T. (2010). Climate change and mental health: A causal pathways framework. *International Journal of Public Health*, 55(2), 123–132. DOI: 10.1007/s00038-009-0112-0 PMID: 20033251

Berry, H. L., Kelly, B. J., Hanigan, I. C., Coates, J. H., McMichael, A. J., Welsh, J. A., & Kjellstrom, T. (2008). *Rural mental health impacts of climate change. Commissioned report for the Garnaut Climate Change Review*. The Australian National University.

Berry, H. L., Waite, T., Dear, K., Capon, A., & Murray, V. (2018). The case for systems thinking about climate change and mental health. *Nature Climate Change*, 8(4), 282–290. DOI: 10.1038/s41558-018-0102-4

Bourque, F., & Cunsolo,W.A. (2014). Climate change: The next challenge for public mental health? *Int Rev Psychiatry.*19;26(4):415–22.

Brosse, M., Benateau, S., Gaudard, A., Stamm, C., & Altermatt, F. (2022). The importance of indirect effects of climate change adaptations on alpine and pre-alpine freshwater systems. *Ecological Solutions and Evidence*, 3(1), e12127. DOI: 10.1002/2688-8319.12127

Bryant, R. A., Waters, E., Gibbs, L., Gallagher, H. C., Pattison, P., Lusher, D., MacDougall, C., Harms, L., Block, K., Snowdon, E., Sinnott, V., Ireton, G., Richardson, J., & Forbes, D. (2014). Psychological outcomes following the Victorian Black Saturday bushfires. *The Australian and New Zealand Journal of Psychiatry*, 48(7), 634–643. DOI: 10.1177/0004867414534476 PMID: 24852323

Burke, M., González, F., Baylis, P., Heft-Neal, S., Baysan, C., Basu, S., & Hsiang, S. (2018). Higher temperatures increase suicide rates in the United States and Mexico. *Nature Climate Change*, 8(8), 723–729. DOI: 10.1038/s41558-018-0222-x

Cianconi, P., Betrò, S., & Janiri, L. (2020). The impact of climate change on mental health: A systematic descriptive review. *Frontiers in Psychiatry*, 11, 1–15. DOI: 10.3389/fpsyt.2020.00074 PMID: 32210846

Clayton, S., Manning, C. M., Krygsman, K., & Speiser, M. (2017). *Mental Health and Our Changing Climate: Impacts, Implications, and Guidance*. American Psychological Association, and ecoAmerica.

Costello, A., Abbas, M., Allen, A., Ball, S., Bell, S., Bellamy, R., Friel, S., Groce, N., Johnson, A., Kett, M., Lee, M., Levy, C., Maslin, M., McCoy, D., McGuire, B., Montgomery, H., Napier, D., Pagel, C., Patel, J., & Patterson, C. (2009). Managing the health effects of climate change. *Lancet*, 373(9676), 1693–1733. DOI: 10.1016/S0140-6736(09)60935-1 PMID: 19447250

Crush, J. (2013). Migration, Urbanization and Food Security in Cities of the Global South: 26–27 November 2012, Cape Town, South Africa.

Cunsolo, W. A., Harper, S., Ford, J., Landman, K., Houle, K., & Edge, V.Rigolet Inuit Community Government. (2012). From this place and of this place: Climate change, health, and place in Rigolet, Nunatsiavut, Canada. *Social Science & Medicine*, 75(3), 538–547. DOI: 10.1016/j.socscimed.2012.03.043 PMID: 22595069

Cusack, L., de Crespigny, C., & Athanasos, P. (2011). Heatwaves and their impact on people with alcohol, drug and mental health conditions: A discussion paper on clinical practice considerations. *Journal of Advanced Nursing*, 67(4), 915–922. DOI: 10.1111/j.1365-2648.2010.05551.x PMID: 21214621

Flory, K., Hankin, B. L., Kloos, B., Cheely, C., & Turecki, G. (2009). Alcohol and cigarette use and misuse among Hurricane Katrina survivors: Psychosocial risk and protective factors. *Substance Use & Misuse*, 44(12), 1711–1724. DOI: 10.3109/10826080902962128 PMID: 19895302

Hanlon, C., Alem, A., Lund, C., Hailemariam, D., Assefa, E., Giorgis, T.W, et al. (2019). Moving towards universal health coverage for mental disorders in Ethiopia. *Int J Ment Health Syst.* 25;13(11).

Hayes, K., Berry, P., & Ebi, K. L. (2019). Factors influencing the mental health consequences of climate change in Canada. *International Journal of Environmental Research and Public Health*, 16(9), 1583. DOI: 10.3390/ijerph16091583 PMID: 31064134

Hayes, K., Blashki, G., Wiseman, J., Burke, S., Reifels, L., & Palinkas, L. A.. (2018). Climate change and mental health: Risks, impacts and priority actions. *International Journal of Mental Health Systems*, 12(1), 1–12. DOI: 10.1186/s13033-018-0210-6 PMID: 29881451

Hayes, K., & Poland, B. (2018). Addressing Mental Health in a Changing Climate: Incorporating Mental Health Indicators into Climate Change and Health Vulnerability and Adaptation Assessment. *International Journal of Environmental Research and Public Health*, 15(9), 1806. DOI: 10.3390/ijerph15091806 PMID: 30131478

Kiefer, M., Rodriguez-Guzman, J., Watson, J., Van Wendel De Joode, B., Mergler, D., & Da Silva, A. S. (2016). Worker health and safety and climate change in the Americas: Issues and research needs. Revista Panamericana de Salud Publica/Pan American. *Journal of Public Health*, 40(3), 192–197.

Kousky, C. (2016). Impacts of natural disasters on children. *The Future of Children*, 26(1), 73–92. DOI: 10.1353/foc.2016.0004

Matthews, J.B., Robin, M., Vincent, van Diemen, R., Fuglestvedt, J.S., Masson-Delmotte, V., Méndez, C., Semenov, S., & Reisinger. (2021). *Annex VII. Glossary: IPCC – Intergovernmental Panel on Climate Change*. IPCC Sixth Assessment Report.

Neria, P., & Schultz, J. M. (2012). Mental health effects of Hurricane Sandy characteristics, potential aftermath, and response. *Journal of the American Medical Association*, 308(24), 2571–2572. DOI: 10.1001/jama.2012.110700 PMID: 23160987

North, C. S., Kawasaki, A., Spitznagel, E. L., & Hong, B. A. (2004). The course of PTSD, major depression, substance abuse, and somatization after a natural disaster. *The Journal of Nervous and Mental Disease*, 293(12), 823–829. DOI: 10.1097/01.nmd.0000146911.52616.22 PMID: 15583503

Ojala, M., & Bengtsson, H. (2019). Young People's Coping Strategies Concerning Climate Change: Relations to Perceived Communication With Parents and Friends and Pro-environmental Behavior. *Environment and Behavior*, 51(8), 907–935. DOI: 10.1177/0013916518763894

Page, L. A., Hajat, S., Kovats, R. S., & Howard, L. M. (2012). Temperature-related deaths in people with psychosis, dementia, and substance misuse. *The Journal of Mental Science*, 200(6), 485–490. PMID: 22661680

Palinkas, L. A., & Wong, M. (2020). Global climate change and mental health. *Current Opinion in Psychology*, 32, 12–16. DOI: 10.1016/j.copsyc.2019.06.023 PMID: 31349129

Patel, V., Saxena, S., Lund, C., Thornicroft, G., Baingana, F., Bolton, P., et al. (2018). *The Lancet Commission on global mental health and sustainable development.* The Lancet: Lancet Publishing Group.

Pihkala, P. (2018). Eco-anxiety, tragedy, and hope: Psychological and spiritual dimensions of climate change. *Zygon*, 53(2), 545–569. DOI: 10.1111/zygo.12407

Pollack, A.A., Weiss, B., Trung, L.T. (2016). Mental health, life functioning, and risk factors among people exposed to frequent natural disasters and chronic poverty in Vietnam. *BJPsych Open,* 2;2(3):221–32.

Russ, S., Regalado, M., & Halfon, N. (2008). Healthcare. *Encyclopaedia of Infant and Early Childhood Development,* pp. 66-80.

Solomon, S., Qin, D., Manning, M., Averyt, K., & Marquis, M. (2007). [*-the physical science basis: Working group I contribution to the fourth assessment report of the IPCC.* UK: Cambridge university press.]. *Climatic Change,* •••, 2007.

Stafoggia, M., Forastiere, F., Agostini, D., Biggeri, A., Bisanti, L., & Cadum, E.. (2006). Vulnerability to Heat-Related Mortality: A Multicity, Population-Based, Case-Crossover Analysis. *Epidemiology (Cambridge, Mass.)*, 17(3), 315–323. DOI: 10.1097/01.ede.0000208477.36665.34 PMID: 16570026

Stephane, H., Mook, B., Laura, B., Marianne, F., Tamaro, K., Ulf, N., Julie, R., David, T., & Adrien, V. (2015). *Shock Waves.* World Bank Publications.

Trautmann, S., Rehm, J., & Wittchen, H. (2016). The economic costs of mental disorders. *EMBO Reports*, 17(9), 1245–1249. DOI: 10.15252/embr.201642951 PMID: 27491723

Vigo, D., Thornicroft, G., & Atun, R. (2016). Estimating the true global burden of mental illness. *The Lancet. Psychiatry*, 3(2), 171–178. DOI: 10.1016/S2215-0366(15)00505-2 PMID: 26851330

Walter, L., Filho, U. M., & Azeiteiro, F. A. (2016). *Climate Change and Health.* Springer.

Watts, N., Adger, W. N., Agnolucci, P., Blackstock, J., Byass, P., Cai, W., Chaytor, S., Colbourn, T., Collins, M., Cooper, A., Cox, P. M., Depledge, J., Drummond, P., Ekins, P., Galaz, V., Grace, D., Graham, H., Grubb, M., Haines, A., & Costello, A. (2015). Lancet Commission: Health and Climate Change: Policy Responses to Protect Public Health. *Lancet*, 386(10006), 1861–1914. DOI: 10.1016/S0140-6736(15)60854-6 PMID: 26111439

Watts, N., Amann, M., Arnell, N., Ayeb-Karlsson, S., Belesova, K., Boykoff, M., Byass, P., Cai, W., Campbell-Lendrum, D., Capstick, S., Chambers, J., Dalin, C., Daly, M., Dasandi, N., Davies, M., Drummond, P., Dubrow, R., Ebi, K. L., Eckelman, M., & Montgomery, H. (2019). The 2019 report of The Lancet Countdown on health and climate change: Ensuring that the health of a child born today is not defined by a changing climate. *Lancet*, 394(10211), 1836–1878. DOI: 10.1016/S0140-6736(19)32596-6 PMID: 31733928

Watts, N., Amann, M., Ayeb-Karlsson, S., Belesova, K., Bouley, T., Boykoff, M., et al. (2018). The Lancet Countdown on health and climate change: from 25 years of inaction to a global transformation for public health. *Lancet*, 10;391(10120):581–630.

Weintrobe, S. (2013). *Engaging with climate change: psychoanalytic and interdisciplinary perspectives*. Routledge.

WHO. (2004). *Promoting Mental Health: Concepts, Emerging Evidence, Practice.* www.who.int

WHO. (2015). *Operational framework for building climate resilient health systems.* www.who.int

Williams, L. (2018). Climate change, colonialism, and women's well-being in Canada: What is to be done? *Canadian Journal of Public Health*, 109(2), 268–271. DOI: 10.17269/s41997-018-0031-z PMID: 29981032

Willox, A. C., Harper, S. L., Ford, J. D., Landman, K., Houle, K., & Edge, V. L. (2012). From this place and of this place: Climate change, sense of place, and health in Nunatsiavut, Canada. *Social Science & Medicine*, 75(3), 538–547. DOI: 10.1016/j.socscimed.2012.03.043 PMID: 22595069

World Health Organization. (WHO) (2017). *Mental health*. 2017. www.who.int

ADDITIONAL READING

Agnew, R. (2012). Dire forecast: A theoretical model of the impact of climate change on crime. *Theoretical Criminology*, 16(1), 21–42. DOI: 10.1177/1362480611416843

Becker, A. E., & Kleinman, A. (2013). Mental health and the global agenda. *The New England Journal of Medicine*, 369(1), 66–73. DOI: 10.1056/NEJMra1110827 PMID: 23822778

Bourque, F., & Cunsolo, W. A. (2014). Climate change: The next challenge for public mental health? *International Review of Psychiatry (Abingdon, England)*, 26(4), 415–422. DOI: 10.3109/09540261.2014.925851 PMID: 25137107

Costello, A., Abbas, M., Allen, A., Ball, S., Bell, S., Bellamy, R., & Lee, M. (2009). Managing the health effects of climate change. *Lancet*, 373(9676), 1693–1733. DOI: 10.1016/S0140-6736(09)60935-1 PMID: 19447250

Doherty, T. J., & Clayton, S. (2011). The psychological impacts of global climate change. *The American Psychologist*, 66(4), 265–276. DOI: 10.1037/a0023141 PMID: 21553952

Giddens, A. (2009). *The politics of climate change*. UK: Cambridge.

Stone, D. A., & Allen, M. R. (2005). The end-to-end attribution problem: From emissions to impacts. *Climatic Change*, 71(3), 303–318. DOI: 10.1007/s10584-005-6778-2

Weissbecker, I. (2011). *Climate change and human well-being: global challenges and opportunities*. Springer. DOI: 10.1007/978-1-4419-9742-5

KEY TERMS AND DEFINITIONS

Climate-: the composite or generalization of weather conditions of a region, as temperature, pressure, humidity, precipitation, sunshine, cloudiness, and winds, throughout the year, averaged over a series of years.

Climate change-: any change in climate over time, whether due to natural variability or as a result of human activity.

Climate change mitigation-: refers to overarching efforts to reduce greenhouse gas

Eco anxiety-: anxiety people face from constantly being surrounded by the wicked and threatening problems associated with a changing climate.

Mental health-: a state of well-being in which the individual realizes his or her abilities, can cope with the normal stresses of life, can work productively and fruitfully, and can contribute to his or her community.

Post-traumatic stress disorder (PTSD)- is a mental health condition that's triggered by a terrifying event: either experiencing it or witnessing it. Symptoms may include flashbacks, nightmares, and severe anxiety, as well as uncontrollable thoughts about the event.

Solastalgia-: homesickness we feel while still at home. It's the pain and longing we feel as we realize the world immediately around us is changing, with our love for that world catalyzing action on its behalf.

Chapter 8
Digital Transformation in Smart Healthcare Canvas of Augmented and Virtual Reality: Prominent Health Gateways in Global Context

Bhupinder Singh
https://orcid.org/0009-0006-4779-2553
Sharda University, India

Christian Kaunert
https://orcid.org/0000-0002-4493-2235
Dublin City University, Ireland

Kamalesh Ravesangar
https://orcid.org/0009-0004-2154-184X
Tunku Abdul Rahman University of Management and Technology, Malaysia

Manmeet Kaur Arora
https://orcid.org/0009-0002-5071-117X
Sharda University, India

Sahil Lal
https://orcid.org/0000-0001-9827-3717
Sharda University, India

DOI: 10.4018/979-8-3693-4183-4.ch008

ABSTRACT

Digital transformation involves using digital technologies to benefit society and the healthcare industry. Healthcare systems must adopt digital technology to develop innovative solutions for enhancing healthcare delivery and addressing medical issues. This transformation encompasses changes related to the internet, digital technologies, and their integration with new therapies and best practices for better health management. Managing the quality of massive data collected can improve patient well-being and reduce service costs. Also, digital technologies will influence medical education, leading experts to devise new training methods. Consequently, practitioners will encounter new opportunities in their field. This chapter focuses on the digital transformation of healthcare with a particular focus on the integration of Augmented Reality (AR) and Virtual Reality (VR) technologies are redefining healthcare delivery, training and patient interaction on a global scale.

1. INTRODUCTION

The healthcare industry is undergoing a significant transformation driven by digital innovation. Among the various technologies propelling this change, AR (Augmented Reality) and VR (Virtual Reality) stand out due to their immersive and interactive capabilities (Sasaki et al., 2024). These technologies offer unprecedented opportunities to enhance medical training, improve patient care, and streamline healthcare operations (Singh et al., 2024). The primary objective of this study is to explore the impact of AR and VR in the smart healthcare landscape (Singh, 2024; Gostin, 2024). The goal is to understand how these technologies are being used globally and to assess their potential for future healthcare delivery (Tulenko et al., 2023). The advent of digital technologies has inspired a paradigm shift in the healthcare sector, fostering the emergence of smart healthcare systems (Hill, 2021).

Digital transformation is a crucial tool for enhancing the effectiveness of preventive and care actions, improving sustainability, and increasing process efficiency (Singh, 2023). Smart infrastructure technology in hospitals uses data from OT and IT systems to create intelligent environments that support the entire patient recovery process and improve management (Singh & Kaunert, 2024). This leads to the development of Smart Hospitals, the hospitals of the future (Kickbusch et al., 2024). To better understand this concept, it can compare it to Smart Cities, where innovation and technological potential are used to transform and improve quality of life (Gostin et al., 2020) (Bodeker & Kronenberg, 2022). In a Smart City, sensors and IoT devices collect data on facilities, traffic systems, security, and supplies, which are then processed and sent to management platforms for real-time visualization, analysis, and operation of city areas or buildings (Bodeker, 2023). In this digitization process, technological enhancement of buildings plays a key role (Magnusson, 2019). For hospitals, this means monitoring centers to regulate performance, optimize resources, and improve staff productivity and customer service (Singh, 2024).

Figure 1. The Landscape of Introduction Split Sections (Source- Original)

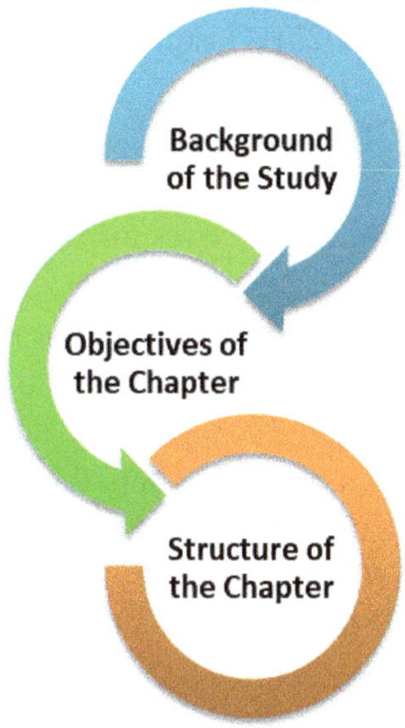

1.1 Background of the Study

Digital transformation is an ongoing process that can unlock opportunities in the healthcare sector, provided there is adequate infrastructure and training (Singh, 2024). It is defined as the use of digital technologies to transform businesses and services (Singh, 2023). The key technologies contributing to digital transformation include IoT digital platforms, cloud computing, and artificial intelligence (Gostin et al., 2024) (Singh, 2024). The sectors such as telecommunications, financial services, and healthcare are significantly impacted by these advancements (Singh, 2024). Digital health can drive innovation in healthcare by enabling patient participation in their healthcare process (Fortanier et al., 2021). Patients can manage their health conditions and make informed decisions by accessing health information through the internet or using digital health applications on mobile devices (Meier et al., 2020). In the coming years, healthcare transformation is expected to prioritize patients who are increasingly taking on the role of "healthcare service consumers: seeking greater control over their health management (Van Lerberghe, 2018) (Singh, 2023). The healthcare industry will evolve to meet the needs and expectations of these consumers, emphasizing personalization, convenience, speed, and immediacy in service delivery (Singh, 2023). New digital technologies will drive healthcare towards digitalization, offering significant benefits to patients and healthcare infrastructure (Sharma & Singh, 2022). These benefits include increased employee productivity, improved operational efficiency and effectiveness of healthcare facilities, and reduced operating costs (Lachat et al., 2023) (Gostin & Sridhar, 2024).

1.2 Objectives of the Chapter

This chapter has the following objectives to-

- explain digital transformation clearly, paying special attention to augmented and virtual reality technology (Marcus et al., 2012).
- analyze the ways that virtual reality (VR) and augmented reality (AR) are being used in healthcare settings throughout the world (Al-Shaar et al., 2017). This covers their application in telemedicine, patient care, surgery planning, and medical education (Gostin, 2022).
- inspect how AR and VR technologies will affect patient outcomes, the effectiveness of healthcare procedures, and the overall cost-effectiveness of the system (Singh, 2022).
- talk about the developments in the field of digital transformation using augmented- virtual reality in smart healthcare (Singh, 2022). This covers prospective developments, new uses, and how these technologies may affect healthcare throughout the world in the future (Singh, 2020).

1.3 Organization/Flow of the Chapter

This chapter deeply dives into the diverse arena on Compass Pointing Exploration of Digital Transformation in Smart Healthcare Canvas of Augmented and Virtual Reality: Projecting Health Gateways in Global Context. Section 2 elaborates the Digital Transformation in Healthcare: Key Drivers of Digital Transformation. Section 3 discusses the Smart Healthcare: Health Gateways in Global Context. Section 4 lays down the Augmented Reality (AR) and Virtual Reality (VR) in Healthcare. Section 5 specifies the Canvas of Augmented and Virtual Reality: Medical Training and Education. Section 6 conveys the Patient Care and Treatment: AR and VR. Section 7 Conclude the Chapter with Future Prospects.

Figure 2. The Flow of this Chapter (Source- Original)

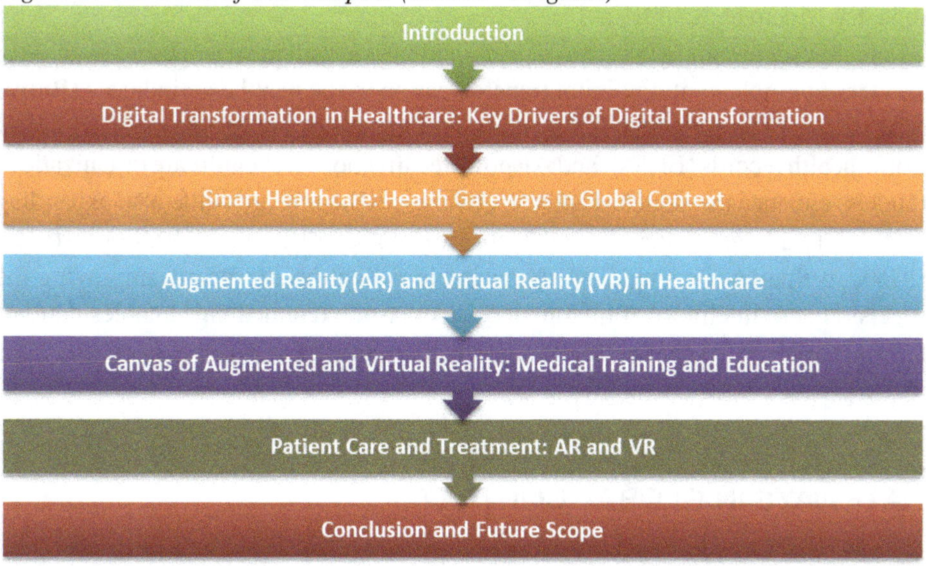

2. DIGITAL TRANSFORMATION IN HEALTHCARE: KEY DRIVERS OF DIGITAL TRANSFORMATION

Digital transformation in healthcare goes beyond mere adoption of technology; it involves prioritizing and fundamentally reshaping healthcare delivery and patient experiences (Singh, 2019). Digital health platforms provide healthcare providers with the opportunity to revolutionize their business processes and cultural practices through technology (Kickbusch et al., 2023). They enable the integration and utilization of data from a wider health ecosystem, which streamlines operations and enhances patient care (Jebril. 2020). This explores the role of digital transformation in healthcare, examines popular use cases, and explores key trends that are shaping the future of healthcare digitalization (Singh, 2023). Digital transformation is revolutionizing healthcare by integrating digital technologies and data-driven processes (Callaway et al., 2018). This approach significantly improves healthcare service delivery, enhances patient outcomes, and streamlines administrative operations, while fostering collaboration among healthcare stakeholders (Singh, 2019). Innovations such as electronic health records (EHRs), telemedicine, advanced data analytics, and the integration of IoT devices and artificial intelligence (AI) are at the core of this transformation (Drager & Sunderland, 2024). Together, these elements drive efficiency, cost-effectiveness, and innovation, marking a new era in the healthcare industry (Cueto et al., 2019).

Healthcare organizations can improve patient care and increase access to medical services by implementing digital transformation in the healthcare industry (Buliya et al., 2017). Through the implementation of technologies like telemedicine, electronic health records (EHRs), and AI-powered diagnostics, health care organizations (HCOs) enhance the standard and effectiveness of healthcare while also expanding its reach to a larger patient base (Talebi Bezmin Abadi et al., 2019). For instance, EHRs give patients a single, easily available patient health record, improving the precision and timeliness of treatment choices, while telemedicine enables patients in remote places to obtain consultations without having to travel (Lopes, 2017).

3. SMART HEALTHCARE: HEALTH GATEWAYS IN GLOBAL CONTEXT

Smart healthcare is a framework that integrates technologies like wearable devices, the Internet of Medical Things (IoMT), advanced machine learning algorithms, and wireless communication to efficiently access health records, connect individuals, resources, and organizations, and intelligently respond to healthcare needs (McCoy et al., 2024). A framework known as "smart healthcare" integrates technologies such as wearables, the Internet of Medical Things (IoMT), sophisticated machine learning algorithms, and wireless communication to connect people, resources, and organizations, retrieve health records quickly, and respond to patients' needs with intelligence. The Internet of Medical Things, or IoMT, is a vital part of smart healthcare (Schneider, 2020). Because illnesses are inherently complex, proper diagnosis frequently requires multimodal medical signals (Folland et al., 2024). The important problem when using multimodal signals is fusing them together well, a topic that the scientific community is becoming more and more interested in (Lgwaran & Okoh, 2019).

Figure 3. The Dimensions of Smart Healthcare- Health Gateways in Global Context (Source- Original)

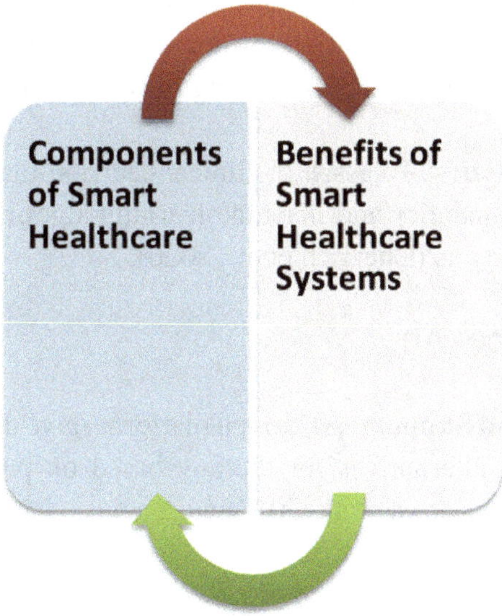

3.1 Components of Smart Healthcare

As information technology has advanced, the concept of smart healthcare has become increasingly prominent. Smart healthcare utilizes next-generation information technologies such as the Internet of Things (IoT), big data, cloud computing, and artificial intelligence to comprehensively transform the traditional medical system, making healthcare more efficient, convenient, and personalized (Ji et al., 2023; Leen et al., 2019). This aims to introduce the concept of smart healthcare. Initially, it outlines the key technologies that underpin smart healthcare and provide an overview of its current applications in several significant fields (Sanyaolu et al., 2019).

IoT (Internet of Things)

Medical Wearables and Gadgets: Internet of Things (IoT) wearables and gadgets gather health data in real time, including vital signs, activity levels, and medication adherence (Shaw et al., 2024).
Remote Monitoring: Early intervention and individualized care are made possible by IoT-enabled continuous patient monitoring outside of conventional clinical settings (Eger et al., 2024).

Large-scale Data and Analytics

Health Data Management: Wearables, medical imaging, electronic health records (EHRs), and other different sources of enormous quantities of health data are all managed and analyzed using big data technology (Erlangga et al., 2019).
Predictive Analytics: To assist in clinical decision-making and preventative treatment, data analytics and machine learning algorithms find patterns and forecast health trends (Chersich et al., 2020).

Artificial Intelligence (AI)

Clinical Decision Support: AI-driven platforms give doctors evidence-based suggestions and alternatives for therapy based on patient information and medical expertise (Dewa et al., 2017).
NLP: Natural language processing is a tool used to organize unstructured medical data such as patient notes and transcripts for analysis and decision assistance (Batalden et al., 2023).

Cloud Processing

Data Storage and Accessibility: Cloud systems provide healthcare organizations scalable storage options that guarantee security and usability (Balarajan et al., 2021).
Telemedicine and Remote Consultation: Cloud-based telemedicine technologies facilitate communication between medical professionals and allow for remote patient consultations (Godman et al., 2020).

3.2 Advantages of Smart Healthcare Systems

The delivery of healthcare has undergone a profound transition thanks to digital revolution, becoming more interactive and patient-centered (Gandhi et al., 2018). Healthcare organizations (HCOs) and technology providers are working together to improve treatment, increase accessibility, and put proactive, economical plans into action (Hassan et al., 2024). Healthcare digitization offers a comprehensive, cheap, and sustainable future, but it's critical to acknowledge that innovation must be balanced with the difficulties it brings (Gan et al., 2020). Although digital transformation brings significant benefits, it also comes with challenges, including data security, interoperability, and the need for healthcare personnel to be upskilled. The effectiveness and long-term viability of digital healthcare projects depend heavily on these variables (Fang et al., 2019).

Enhanced Results for Patients: Data-driven decision-making is encouraged by digital healthcare transformation, which helps hospitals make decisions about admission, treatment, and discharge (Bergwerk et al., 2021). Decision-support algorithms and real-time insights use operational and patient data to make the right suggestions (Tomas et al., 2023). Reducing average hospital stays, lowering readmission rates, and streamlining bed turnover procedures are all possible using predictive analytics for patient and operational flow management. Enhancing hospital resource management leads to better treatment quality and accessibility (Runciman et al., 2017).

Health Information Technology (EHR): Healthcare professionals will always have instant access to vital information thanks to EHRs, which offer a more thorough and user-friendly view of patient records (Li et al., 2018). Both the efficacy of therapies and the accuracy of diagnoses are enhanced by this (Bowman, 2023).

Platforms for Telemedicine: Platforms for telemedicine make virtual consultations possible, improving accessibility, particularly for patients who live far away or have mobility problems (Watts et al., 2023). Better patient outcomes and a more tailored healthcare experience result from this (Chartier, 2024).

4. AUGMENTED REALITY (AR) AND VIRTUAL REALITY (VR) IN HEALTHCARE

Augmented reality (AR) has dual applications in healthcare as it serves both education and training purposes, as well as diagnostics and treatment (Ogugua et al., 2024). AR has the potential to significantly enhance the quality of care. Surgeons, for instance, can use smart glasses and similar devices to monitor vitals and access critical information during surgery, thereby reducing operational risks (Chui et al., 2023). When considering augmenting your healthcare solution with cutting-edge services, it is crucial to understand their practical applications (Malik et al., 2022). Implementing virtual technologies can be complex, so it's essential to carefully weigh the pros and cons. Here are some examples of how augmented reality operates in healthcare (Ghimire et al., 2023). Virtual reality (VR) takes these components to the next level by generating a fully computer-generated simulation of an alternate world (Korzynski et al., 2023). These immersive simulations can create almost any visual or environment imaginable for the user, utilizing specialized equipment such as computers, sensors, headsets, and gloves (Bandi et al., 2023).

Augmented and virtual reality in healthcare is at the center of a spectacular shift being driven by the convergence of technology and healthcare (Zhao et al., 2021). Immersion technologies are not only revolutionizing healthcare delivery, but they are also seeing rapid commercial expansion (Bahroun et al., 2023). Few industries in a world where innovation is always being pushed to its maximum by technology breakthroughs have changed as much as the healthcare sector (Soni, 2023). Virtual reality (VR) and augmented reality (AR) are two cutting-edge technologies that are beginning to change the healthcare industry (Frey & Osborne, 2023). These fully immersive digital experiences are not science fiction anymore; rather, they are a reality that is transforming the way it approach healthcare in general, from diagnosis to treatment and beyond (Cheng et al., 2022). AR and VR in healthcare may improve patient outcomes, medical education, and even our perception of and relationship (Kshetri et al., 2023).

5. MEDICAL TRAINING AND EDUCATION

Because medical students must learn a tremendous quantity of knowledge, their lives are complicated. Learning can be done more effectively with virtual reality (Mondal et al., 2023). Students or interns can practice virtual surgery without running the risk of hurting a patient by playing augmented reality games like Touch Surgery (Rane et al., 2023). The avoiding of the use of cadavers for practical instruction, medical schools can avoid the moral and ethical dilemmas that arise from this training method (Miao et al., 2023). For the following reasons, surgery is where AR and VR in modern medicine are most useful: Before surgery is conducted, doctors or trainees can practice different scenarios by simulating them using virtual reality. Using information from, doctors or students can more accurately evaluate hazards during operations (Leone et al., 2021).

Figure 4. Concepts of Medical Training and Education (Source- Original)

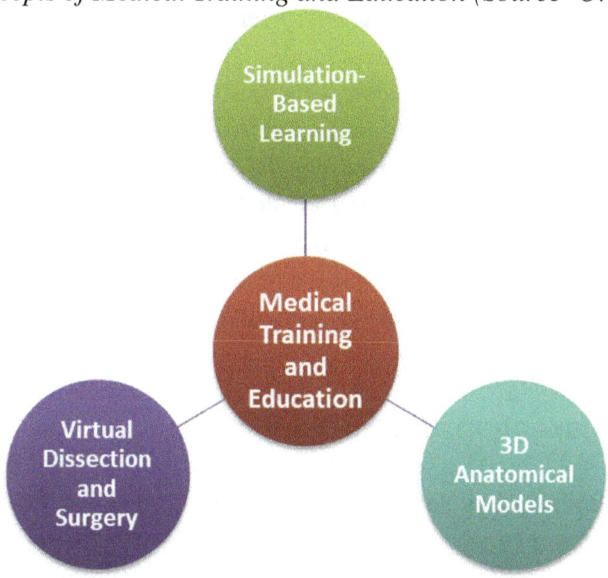

5.1 Simulation-Based Learning

Experiential Learning (EL) is an interdisciplinary educational philosophy and practice that aims to enhance academic learning outcomes, promote student career development, and foster connections within workplace settings through critical thinking (Maddipoti, 2023). With offering intentionally designed, authentic learning experiences and integrated guided reflections, EL opportunities enable students to develop a broad spectrum of knowledge, skills, and values. This approach inspires students to make informed choices about their career paths (Ratten & Jones, 2023). Simulation-based learning is a specific form of experiential learning that provides learners with realistic opportunities to develop and practice their knowledge and skills in a simulated environment (Vyas & Rajendran, 2023).

5.2 3D Anatomical Models

Three primary areas of expertise for Anatomik Modeling are pulmonology (trachea, bronchial tubes, etc.), orthopaedic surgery (pelvis, vertebrae, joints, etc.) and maxillo-facial (jaw, skull, etc.) (Rane, 2023). Scientific or medical data, including MRIs, CT scans, micro-CTs and 3D scans, are used to build these models (Dwivedi et al., 2023). The ability to analyze medical images in three dimensions allows us to create high-quality three-dimensional printouts of intricate diseased anatomical features (Wong et al., 2023). Because every person is different and the human body is so intricate, 3D-printed anatomical models are a great way to get doctors ready for surgery (Rane, 2023). It is better to have a detailed 3D representation of the patient's anatomy than basic 2D images obtained via medical imaging (Doshi & Hauser, 2023). This makes it possible for the surgeons to research, be ready, and even practice a surgical technique beforehand. It may be more efficient to teach future surgeons using 3D-printed anatomical models rather than more conventional dissection techniques (Byrne, 2023). It is a more adaptable and economical method that offers a more accurate depiction of particular diseases (Law et al., 2023) (Lewis & mercer, 2024).

5.3 Virtual Dissection and Surgery

In the rapidly evolving landscape of modern medical education, traditional learning methods are being transformed by cutting-edge technology (Gursoy et al., 2023). That such innovation is the virtual dissection table, which revolutionizes how medical students learn anatomy and surgical techniques (Wamba et al., 2023). It enables students to explore and study human anatomy using highly interactive and detailed virtual models (Harrer, 2023). With high-resolution 3D imaging and intuitive software, this technology replicates the experience of traditional cadaver dissection in a digital format (Gabrielson et al., 2023). The virtual dissection table allows students to repeat procedures as many times as needed, which fosters a deeper understanding and allows them to hone their surgical skills without the limitations of a physical specimen (Budhwar et al., 2023). They can review complex structures repeatedly until they grasp the concepts, without the pressure of keeping up with the rest of the class (Mohamadi et al., 2023). This technology reduces the need for constant procurement of cadavers, thereby cutting costs and minimizing the environmental impact associated with traditional cadaver use (Xu et al., 2024). The virtual dissection table represents a significant advancement in medical education, offering students an enhanced learning experience that is flexible, cost-effective, and environmentally sustainable (Morley et al., 2023).

6. PATIENT CARE AND TREATMENT: AR AND VR

The virtual reality and augmented reality (AR/VR) have the potential to completely with change the healthcare industry by introducing cutting-edge diagnosis and treatment options and by altering the way that care is provided (Bryan-Kinns et al., 2023). Their capacity to deliver both conventional and wholly original material in incredibly realistic and immersive ways, adapted to many therapeutic contexts and available remotely, is what gives them potential for use in diagnosis and therapy (Thirunavukarasu et al., 2023). AR/VR can be used by doctors, patients, and caretakers to carry out or prepare for certain treatments or procedures (Li et al., 2021). With superimposing or fusing simulated digital images with the user's actual view usually seen through a smartphone camera, head-mounted display (HMD) or heads-up display (HUD) augmented reality (AR) improves the real-world experience (Sezgin, 2023). Mixed reality, sometimes known as merged reality, is the term used to describe the interaction between digital visuals and the real world, which is frequently controlled by the user (Raj et al., 2023). Via entirely replacing the user's vision of the actual world with a simulated, interactive environment, virtual reality (VR) offers an immersive virtual world experience that frequently calls for a headset (Minseen et al., 2023).

The term "extended reality" (XR) refers to both virtual reality and augmented reality techniques (Hadi et al., 2023) (Dhoni & Kumar, 2023). It shall be concentrating on augmented reality and virtual reality for the purposes of this webpage. Augmented and virtual reality is already being utilized in the real world to treat patients (Stasevych & Zvarych, 2023). An augmented reality device that helps a surgeon with their technique by superimposing medical pictures on a patient during surgery and VR therapy for veterans of the armed forces treating post-traumatic stress disorder (PTSD) (Joshi et al., 2024). Virtual reality (VR) rehabilitation treatment is used to help patients recuperating from strokes or other physical disabilities by simulating real-life settings and improving bodily functioning (Sikka et al., 2020). With improving medical processes, offering therapy for mental health issues, and boosting the results of physical rehabilitation, these technologies are revolutionizing the healthcare industry (Music, 2022).

Figure 5. Aspects on Patient Care and Treatment- AR and VR (Source- Original)

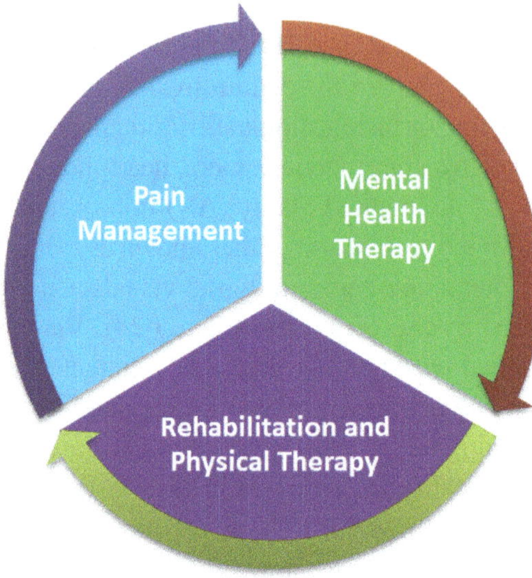

Due to advancements in science and technology, conventional biotechnology-based medicine is gradually digitizing and becoming more informationized in today's era of informatization (Knagg et al., 2024). A new generation of information technology is integrated into smart healthcare as a result of this shift (Pool et al., 2024). Smart healthcare is a multifaceted, all-encompassing shift rather than just a straightforward technology breakthrough (Goldsteen et al., 2024). This change is noticeable in a number of important areas as Changes in the Medical Model: A move from patient-centered treatment to disease-centered care; Transitioning from clinical to regional medical informatization involves changes in the process of informatics construction; Changes in Medical Management: A shift from general to individualized care; Concepts of Prevention and Treatment Have Changed: Preventive healthcare has replaced the focus on treating diseases (Slaunwhite et al., 2024). These changes aim to meet individual healthcare needs while enhancing medical care efficiency, greatly improving the overall medical and health service experience (Fuse Brown & Hall, 2024).

7. CONCLUSION AND FUTURE SCOPE

The healthcare providers are increasingly focusing on virtual care, vertical integration, and retail-based health services to meet evolving demands influenced by changing patient demographics and treatment approaches. Artificial intelligence

(AI) and machine learning (ML) technologies offer significant advantages in both medical and business fields. AI/ML is employed to enhance resource management, operational efficiency, and to develop personalized treatment plans and predictive models for patient care. Cloud platforms and software-as-a-service applications enable healthcare providers to innovate, experiment with new ideas, and formulate data-driven strategies to customize patient care and accelerate innovation. Despite significant advancements in medical care, access and affordability remain challenges in highly developed healthcare systems, often dictated by geographical location. Approximately half of the global population lacks access to basic medical services, highlighting the need for healthcare systems reform to ensure fairness and improve healthcare access. This transformation includes decentralizing services to bring healthcare closer to patients, establishing interconnected healthcare systems to support equity, and embracing digital transformation. In the future, healthcare providers will evolve into integrated care systems rather than traditional hospitals, reflecting the ongoing transformation of the healthcare system.

References

Al-Shaar, L., Vercammen, K., Lu, C., Richardson, S., Tamez, M., & Mattei, J. (2017). Health effects and public health concerns of energy drink consumption in the United States: A mini-review. *Frontiers in Public Health*, 5, 286776. DOI: 10.3389/fpubh.2017.00225 PMID: 28913331

Bahroun, Z., Anane, C., Ahmed, V., & Zacca, A. (2023). Transforming education: A comprehensive review of generative artificial intelligence in educational settings through bibliometric and content analysis. *Sustainability (Basel)*, 15(17), 12983. DOI: 10.3390/su151712983

Balarajan, Y., Selvaraj, S., & Subramanian, S. V. (2021). Health care and equity in India. *Lancet*, 377(9764), 505–515. DOI: 10.1016/S0140-6736(10)61894-6 PMID: 21227492

Bandi, A., Adapa, P. V. S. R., & Kuchi, Y. E. V. P. K. (2023). The power of generative ai: A review of requirements, models, input–output formats, evaluation metrics, and challenges. *Future Internet*, 15(8), 260. DOI: 10.3390/fi15080260

Batalden, M., Batalden, P., Margolis, P., Seid, M., Armstrong, G., Opipari-Arrigan, L., & Hartung, H. (2023). Coproduction of healthcare service. *BMJ Quality & Safety*. PMID: 26376674

Bergwerk, M., Gonen, T., Lustig, Y., Amit, S., Lipsitch, M., Cohen, C., Mandelboim, M., Levin, E. G., Rubin, C., Indenbaum, V., Tal, I., Zavitan, M., Zuckerman, N., Bar-Chaim, A., Kreiss, Y., & Regev-Yochay, G. (2021). Covid-19 breakthrough infections in vaccinated health care workers. *The New England Journal of Medicine*, 385(16), 1474–1484. DOI: 10.1056/NEJMoa2109072 PMID: 34320281

Bodeker, G. (2023). *WHO global atlas of traditional, complementary, and alternative medicine* (Vol. 1). World Health Organization.

Bodeker, G., & Kronenberg, F. (2022). A public health agenda for traditional, complementary, and alternative medicine. *American Journal of Public Health*, 92(10), 1582–1591. DOI: 10.2105/AJPH.92.10.1582 PMID: 12356597

Bowman, S. (2023). Impact of electronic health record systems on information integrity: Quality and safety implications. *Perspectives in Health Information Management*, 10(Fall). PMID: 24159271

Bryan-Kinns, N., Banar, B., Ford, C., Reed, C. N., Zhang, Y., Colton, S., & Armitage, J. (2023). Exploring XAI for the arts: Explaining latent space in generative music. *arXiv preprint arXiv:2308.05496.*

Budhwar, P., Chowdhury, S., Wood, G., Aguinis, H., Bamber, G. J., Beltran, J. R., Boselie, P., Lee Cooke, F., Decker, S., DeNisi, A., Dey, P. K., Guest, D., Knoblich, A. J., Malik, A., Paauwe, J., Papagiannidis, S., Patel, C., Pereira, V., Ren, S., & Varma, A. (2023). Human resource management in the age of generative artificial intelligence: Perspectives and research directions on ChatGPT. *Human Resource Management Journal*, 33(3), 606–659. DOI: 10.1111/1748-8583.12524

Buliva, E., Elhakim, M., Tran Minh, N. N., Elkholy, A., Mala, P., Abubakar, A., & Malik, S. M. M. R. (2017). Emerging and reemerging diseases in the World Health Organization (WHO) Eastern Mediterranean Region—Progress, challenges, and WHO initiatives. *Frontiers in Public Health*, 5, 276. DOI: 10.3389/fpubh.2017.00276 PMID: 29098145

Byrne, M. (2023). The disruptive impacts of next generation generative artificial intelligence. *CIN: Computers, Informatics. Nursing*, 41(7), 479–481. PMID: 37417716

Callaway, M. V., Connor, S. R., & Foley, K. M. (2018). World Health Organization public health model: A roadmap for palliative care development. *Journal of Pain and Symptom Management*, 55(2), S6–S13. DOI: 10.1016/j.jpainsymman.2017.03.030 PMID: 28801003

Chartier, Y. (Ed.). (2024). *Safe management of wastes from health-care activities*. World Health Organization.

Cheng, Z., Lee, D., & Tambe, P. (2022). Innovae: Generative ai for understanding patents and innovation. *Available atSSRN* 3868599. DOI: 10.2139/ssrn.3868599

Chersich, M. F., Gray, G., Fairlie, L., Eichbaum, Q., Mayhew, S., Allwood, B., English, R., Scorgie, F., Luchters, S., Simpson, G., Haghighi, M. M., Pham, M. D., & Rees, H. (2020). COVID-19 in Africa: Care and protection for frontline healthcare workers. *Globalization and Health*, 16(1), 1–6. DOI: 10.1186/s12992-020-00574-3 PMID: 32414379

Chui, M., Hazan, E., Roberts, R., Singla, A., & Smaje, K. (2023). The economic potential of generative AI.

Cueto, M., Brown, T. M., & Fee, E. (2019). *The world health organization: A history*. Cambridge University Press. DOI: 10.1017/9781108692878

Dewa, C. S., Loong, D., Bonato, S., & Trojanowski, L. (2017). The relationship between physician burnout and quality of healthcare in terms of safety and acceptability: A systematic review. *BMJ Open*, 7(6), e015141. DOI: 10.1136/bmjopen-2016-015141 PMID: 28637730

Dhoni, P., & Kumar, R. (2023). Synergizing generative ai and cybersecurity: Roles of generative ai entities, companies, agencies, and government in enhancing cybersecurity. *Authorea Preprints*.

Doshi, A. R., & Hauser, O. (2023). Generative artificial intelligence enhances creativity. *Available atSSRN*. DOI: 10.2139/ssrn.4535536

Drager, N., & Sunderland, L. (2024). Public health in a globalising world: The perspective from the World Health Organization. In *Governing Global Health* (pp. 67–78). Routledge.

Dwivedi, Y. K., Kshetri, N., Hughes, L., Slade, E. L., Jeyaraj, A., Kar, A. K., Baabdullah, A. M., Koohang, A., Raghavan, V., Ahuja, M., Albanna, H., Albashrawi, M. A., Al-Busaidi, A. S., Balakrishnan, J., Barlette, Y., Basu, S., Bose, I., Brooks, L., Buhalis, D., & Wright, R. (2023). "So what if ChatGPT wrote it?" Multidisciplinary perspectives on opportunities, challenges and implications of generative conversational AI for research, practice and policy. *International Journal of Information Management*, 71, 102642. DOI: 10.1016/j.ijinfomgt.2023.102642

Eger, H., Chacko, S., El-Gamal, S., Gerlinger, T., Kaasch, A., Meudec, M., Munshi, S., Naghipour, A., Rhule, E., Sandhya, Y. K., & Uribe, O. L. (2024). Towards a Feminist Global Health Policy: Power, intersectionality, and transformation. *PLOS Global Public Health*, 4(3), e0002959. DOI: 10.1371/journal.pgph.0002959 PMID: 38451969

Erlangga, D., Suhrcke, M., Ali, S., & Bloor, K. (2019). The impact of public health insurance on health care utilisation, financial protection and health status in low-and middle-income countries: A systematic review. *PLoS One*, 14(8), e0219731. DOI: 10.1371/journal.pone.0219731 PMID: 31461458

Fang, H., Eggleston, K., Hanson, K., & Wu, M. (2019). Enhancing financial protection under China's social health insurance to achieve universal health coverage. *bmj, 365*.

Folland, S., Goodman, A. C., & Stano, M. (2024). *The economics of health and health care: Pearson new international edition*. Routledge.

Fortanier, F., Kolk, A., & Pinkse, J. (2021). Harmonization in CSR reporting: MNEs and global CSR standards. *MIR. Management International Review*, 51(5), 665–696. DOI: 10.1007/s11575-011-0089-9

Frey, C. B., & Osborne, M. (2023). Generative AI and the Future of Work: A Reappraisal. *The Brown Journal of World Affairs*, •••, 1–12.

Fuse Brown, E. C., & Hall, M. A. (2024). Private Equity and the Corporatization of Health Care. *Stanford Law Review*, •••, 76.

Gabrielson, A. T., Odisho, A. Y., & Canes, D. (2023). Harnessing generative artificial intelligence to improve efficiency among urologists: Welcome ChatGPT. *The Journal of Urology*, 209(5), 827–829. DOI: 10.1097/JU.0000000000003383 PMID: 36795957

Gan, W. H., Lim, J. W., & Koh, D. (2020). Preventing intra-hospital infection and transmission of coronavirus disease 2019 in health-care workers. *Safety and Health at Work*, 11(2), 241–243. DOI: 10.1016/j.shaw.2020.03.001 PMID: 32292622

Gandhi, T. K., Kaplan, G. S., Leape, L., Berwick, D. M., Edgman-Levitan, S., Edmondson, A., Meyer, G. S., Michaels, D., Morath, J. M., Vincent, C., & Wachter, R. (2018). Transforming concepts in patient safety: A progress report. *BMJ Quality & Safety*, 27(12), 1019–1026. DOI: 10.1136/bmjqs-2017-007756 PMID: 30018115

Ghimire, P., Kim, K., & Acharya, M. (2023). Generative ai in the construction industry: Opportunities & challenges. *arXiv preprint arXiv:2310.04427.*

Godman, B., Haque, M., Islam, S., Iqbal, S., Urmi, U. L., Kamal, Z. M., Shuvo, S. A., Rahman, A., Kamal, M., Haque, M., Jahan, I., Islam, M. Z., Hossain, M. M., Munzur-E-Murshid, , Kumar, S., Charan, J., Bhatt, R., Dutta, S., Abhayanand, J. P., & Sefah, I. (2020). Rapid assessment of price instability and paucity of medicines and protection for COVID-19 across Asia: Findings and public health implications for the future. *Frontiers in Public Health*, 8, 585832. DOI: 10.3389/fpubh.2020.585832 PMID: 33381485

Goldsteen, R. L., Goldsteen, K., & Dwelle, T. (2024). Introduction to public health: promises and practices.

Gostin, L. O. (2022). A framework convention on global health: Health for all, justice for all. *Journal of the American Medical Association*, 307(19), 2087–2092. PMID: 22665108

Gostin, L. O. (2024). A framework convention on global health: Health for all, justice for all. *Journal of the American Medical Association*, 307(19), 2087–2092. PMID: 22665108

Gostin, L. O., Habibi, R., & Meier, B. M. (2020). Has global health law risen to meet the COVID-19 challenge? Revisiting the International Health Regulations to prepare for future threats. *The Journal of Law, Medicine & Ethics*, 48(2), 376–381. DOI: 10.1177/1073110520935354 PMID: 32631189

Gostin, L. O., & Sridhar, D. (2024). Global health and the law. *The New England Journal of Medicine*, 370(18), 1732–1740. DOI: 10.1056/NEJMra1314094 PMID: 24785208

Gostin, L. O., Sridhar, D., & Hougendobler, D. (2024). The normative authority of the World Health Organization. *Public Health*, 129(7), 854–863. DOI: 10.1016/j.puhe.2015.05.002 PMID: 26100341

Gursoy, D., Li, Y., & Song, H. (2023). ChatGPT and the hospitality and tourism industry: An overview of current trends and future research directions. *Journal of Hospitality Marketing & Management*, 32(5), 579–592. DOI: 10.1080/19368623.2023.2211993

Hadi, M. U., Qureshi, R., Shah, A., Irfan, M., Zafar, A., Shaikh, M. B., & Mirjalili, S. (2023). A survey on large language models: Applications, challenges, limitations, and practical usage. *Authorea Preprints*. DOI: 10.36227/techrxiv.23589741.v1

Harrer, S. (2023). Attention is not all you need: The complicated case of ethically using large language models in healthcare and medicine. *EBioMedicine*, 90, 90. DOI: 10.1016/j.ebiom.2023.104512

Hassan, I., Chisty, A., & Bui, T. (2024). Structural and Social Determinants of Health. In *Leading an Academic Medical Practice* (pp. 343–355). Springer International Publishing.

Hill, P. S. (2021). Understanding global health governance as a complex adaptive system. *Global Public Health: An International Journal for Research, Policy and Practice*, 6(6), 593–605. DOI: 10.1080/17441691003762108 PMID: 20432102

Igwaran, A., & Okoh, A. I. (2019). Human campylobacteriosis: A public health concern of global importance. *Heliyon*, 5(11), e02814. DOI: 10.1016/j.heliyon.2019.e02814 PMID: 31763476

Jebril, N. (2020). World Health Organization declared a pandemic public health menace: a systematic review of the coronavirus disease 2019 "COVID-19". *Available atSSRN* 3566298. DOI: 10.2139/ssrn.3566298

Ji, X., Chun, S. A., Wei, Z., & Geller, J. (2023). Twitter sentiment classification for measuring public health concerns. *Social Network Analysis and Mining*, 5, 1–25. PMID: 32226558

Joshi, G., Jain, A., Araveeti, S. R., Adhikari, S., Garg, H., & Bhandari, M. (2024). FDA-Approved Artificial Intelligence and Machine Learning (AI/ML)-Enabled Medical Devices: An Updated Landscape. *Electronics (Basel)*, 13(3), 498. DOI: 10.3390/electronics13030498

Kickbusch, I., Brindley, C., & World Health Organization. (2023). *Health in the post-2015 development agenda: an analysis of the UN-led thematic consultations, High-Level Panel report and sustainable development debate in the context of health*. World Health Organization.

10. Kickbusch, I., Hein, W., & Silberschmidt, G. (2024). Addressing global health governance challenges through a new mechanism: The proposal for a Committee C of the World Health Assembly. *The Journal of Law, Medicine & Ethics*, 38(3), 550–563. DOI: 10.1111/j.1748-720X.2010.00511.x PMID: 20880240

Knagg, R., Dorey, J., Evans, R., & Hitchman, J. (2024). Sustainability in healthcare: Patient and public perspectives. *Anaesthesia*, 79(3), 278–283. DOI: 10.1111/anae.16159 PMID: 38205532

Korzynski, P., Mazurek, G., Altmann, A., Ejdys, J., Kazlauskaite, R., Paliszkiewicz, J., & Ziemba, E. (2023). *Generative artificial intelligence as a new context for management theories: analysis of ChatGPT*. Central European Management Journal.

Kshetri, N., Dwivedi, Y. K., Davenport, T. H., & Panteli, N. (2023). Generative artificial intelligence in marketing: Applications, opportunities, challenges, and research agenda. *International Journal of Information Management*, •••, 102716.

Lachat, C., Van Camp, J., De Henauw, S., Matthys, C., Larondelle, Y., Remaut–De Winter, A. M., & Kolsteren, P. (2023). A concise overview of national nutrition action plans in the European Union Member States. *Public Health Nutrition*, 8(3), 266–274. DOI: 10.1079/PHN2004691 PMID: 15918923

Law, R., Lin, K. J., Ye, H., & Fong, D. K. C. (2023). Artificial intelligence research in hospitality: A state-of-the-art review and future directions. *International Journal of Contemporary Hospitality Management.*

Leen, J. L., & Juurlink, D. N. (2019). Carfentanil: A narrative review of its pharmacology and public health concerns. *Canadian Journal of Anaesthesia*, 66(4), 414–421. DOI: 10.1007/s12630-019-01294-y PMID: 30666589

Leone, D., Schiavone, F., Appio, F. P., & Chiao, B. (2021). How does artificial intelligence enable and enhance value co-creation in industrial markets? An exploratory case study in the healthcare ecosystem. *Journal of Business Research*, 129, 849–859. DOI: 10.1016/j.jbusres.2020.11.008

Lewis, M., & Mercer, E. (2024). The AI Roadmap–charting our future in healthcare communications. *Current Medical Research and Opinion*, 40(1), 5–7. DOI: 10.1080/03007995.2023.2261737 PMID: 37791596

Li, X., Jiang, Y., Rodriguez-Andina, J. J., Luo, H., Yin, S., & Kaynak, O. (2021). When medical images meet generative adversarial network: Recent development and research opportunities. *Discover Artificial Intelligence*, 1(1), 1–20. DOI: 10.1007/s44163-021-00006-0

Li, Y., Wang, H., Jin, X. R., Li, X., Pender, M., Song, C. P., Tang, S.-L., Cao, J., Wu, H., & Wang, Y. G. (2018). Experiences and challenges in the health protection of medical teams in the Chinese Ebola treatment center, Liberia: A qualitative study. *Infectious Diseases of Poverty*, 7(1), 1–12. DOI: 10.1186/s40249-018-0468-6 PMID: 30134982

Lopes, M. B. S. (2017). The 2017 World Health Organization classification of tumors of the pituitary gland: A summary. *Acta Neuropathologica*, 134(4), 521–535. DOI: 10.1007/s00401-017-1769-8 PMID: 28821944

Maddipoti, A. (2023). *Pathway Forward for Responsible Generative AI Implementation in Healthcare* (Doctoral dissertation).

Magnusson, R. S. (2019). Rethinking global health challenges: Towards a 'global compact' for reducing the burden of chronic disease. *Public Health*, 123(3), 265–274. DOI: 10.1016/j.puhe.2008.12.023 PMID: 19278695

Malik, S., Tyagi, A. K., & Mahajan, S. (2022). Architecture, Generative Model, and Deep Reinforcement Learning for IoT Applications: Deep Learning Perspective. *Artificial Intelligence-based Internet of Things Systems*, 243-265.

Marcus, M., Yasamy, M. T., van Ommeren, M. V., Chisholm, D., & Saxena, S. (2012). Depression: A global public health concern.

McCoy, D., Kapilashrami, A., Kumar, R., Rhule, E., & Khosla, R. (2024). Developing an agenda for the decolonization of global health. *Bulletin of the World Health Organization*, 102(2), 130–136. DOI: 10.2471/BLT.23.289949 PMID: 38313156

Meier, B. M., Cinà, M. M., & Gostin, L. O. (2020). Advancing human rights through global health governance. *Foundations of global health and human rights*.

Miao, H., Li, C., & Wang, J. (2023). A future of smarter digital health empowered by generative pretrained transformer. *Journal of Medical Internet Research*, 25, e49963. DOI: 10.2196/49963 PMID: 37751243

Minssen, T., Vayena, E., & Cohen, I. G. (2023). The challenges for regulating medical use of ChatGPT and other large language models. *Journal of the American Medical Association*, 330(4), 315. DOI: 10.1001/jama.2023.9651 PMID: 37410482

Mohamadi, S., Mujtaba, G., Le, N., Doretto, G., & Adjeroh, D. A. (2023). ChatGPT in the age of generative AI and large language models: a concise survey. *arXiv preprint arXiv:2307.04251*.

Mondal, S., Das, S., & Vrana, V. G. (2023). How to bell the cat? A theoretical review of generative artificial intelligence towards digital disruption in all walks of life. *Technologies*, 11(2), 44. DOI: 10.3390/technologies11020044

Morley, J., DeVito, N. J., & Zhang, J. (2023). Generative AI for medical research. *bmj, 382*.

Music, T. (2022). Protecting patients, protecting healthcare workers: A review of the role of influenza vaccination. *International Nursing Review*, 59(2), 161–167. DOI: 10.1111/j.1466-7657.2011.00961.x PMID: 22591085

Ogugua, J. O., Anyanwu, E. C., Olorunsogo, T., Maduka, C. P., & Ayo-Farai, O. (2024). Ethics and strategy in vaccination: A review of public health policies and practices. *International Journal of Science and Research Archive*, 11(1), 883–895. DOI: 10.30574/ijsra.2024.11.1.0141

Pool, J., Akhlaghpour, S., Fatehi, F., & Burton-Jones, A. (2024). A systematic analysis of failures in protecting personal health data: A scoping review. *International Journal of Information Management*, 74, 102719. DOI: 10.1016/j.ijinfomgt.2023.102719

Raj, R., Singh, A., Kumar, V., & Verma, P. (2023). Analyzing the potential benefits and use cases of ChatGPT as a tool for improving the efficiency and effectiveness of business operations. *BenchCouncil Transactions on Benchmarks. Standards and Evaluations*, 3(3), 100140.

Rane, N. (2023). Roles and challenges of ChatGPT and similar generative artificial intelligence for achieving the sustainable development goals (SDGs). *Available atSSRN* 4603244. DOI: 10.2139/ssrn.4603244

Rane, N., Choudhary, S., & Rane, J. (2023). Integrating Building Information Modelling (BIM) with ChatGPT, Bard, and similar generative artificial intelligence in the architecture, engineering, and construction industry: applications, a novel framework, challenges, and future scope. *Bard, and similar generative artificial intelligence in the architecture, engineering, and construction industry: applications, a novel framework, challenges, and future scope (November 22, 2023).*

Rane, N. L. (2023). Multidisciplinary collaboration: key players in successful implementation of ChatGPT and similar generative artificial intelligence in manufacturing, finance, retail, transportation, and construction industry.

Ratten, V., & Jones, P. (2023). Generative artificial intelligence (ChatGPT): Implications for management educators. *International Journal of Management Education*, 21(3), 100857. DOI: 10.1016/j.ijme.2023.100857

Runciman, B., Merry, A., & Walton, M. (2017). *Safety and ethics in healthcare: a guide to getting it right.* CRC Press.

Sanyaolu, A., Okorie, C., Qi, X., Locke, J., & Rehman, S. (2019). Childhood and adolescent obesity in the United States: a public health concern. *Global pediatric health, 6*, 2333794X19891305.

Sasaki, N., Rosenberg, M., Shin, J. H., Kunisawa, S., & Imanaka, Y. (2024). Hidden Populations for Healthcare Financial Protection in the Super-Aging Society: Closing the Gap Between Policy and Practice. *Clinical Social Work Journal*, 52(3), 1–12. DOI: 10.1007/s10615-023-00914-x PMID: 39188583

Schneider, M. J. (2020). *Introduction to public health.* Jones & Bartlett Learning.

Sezgin, E. (2023). Artificial intelligence in healthcare: Complementing, not replacing, doctors and healthcare providers. *Digital Health*, 9, 20552076231186520. DOI: 10.1177/20552076231186520 PMID: 37426593

Sharma, A., & Singh, B. (2022). Measuring Impact of E-commerce on Small Scale Business: A Systematic Review. *Journal of Corporate Governance and International Business Law*, 5(1).

Shaw, F. E., Asomugha, C. N., Conway, P. H., & Rein, A. S. (2024). The Patient Protection and Affordable Care Act: Opportunities for prevention and public health. *Lancet*, 384(9937), 75–82. DOI: 10.1016/S0140-6736(14)60259-2 PMID: 24993913

Sikka, R., Morath, J. M., & Leape, L. (2020). The quadruple aim: Care, health, cost and meaning in work. *BMJ Quality & Safety*, 24(10), 608–610. DOI: 10.1136/bmjqs-2015-004160 PMID: 26038586

Singh, B. (2019). Affordability of Medicines, Public Health and TRIPS Regime: A Comparative Analysis. *Indian Journal of Health and Medical Law*, 2(1), 1–7.

Singh, B. (2023). Blockchain Technology in Renovating Healthcare: Legal and Future Perspectives. In *Revolutionizing Healthcare Through Artificial Intelligence and Internet of Things Applications* (pp. 177-186). IGI Global.

Singh, B. (2023). Unleashing Alternative Dispute Resolution (ADR) in Resolving Complex Legal-Technical Issues Arising in Cyberspace Lensing E-Commerce and Intellectual Property: Proliferation of E-Commerce Digital Economy. *Revista Brasileira de Alternative Dispute Resolution-Brazilian Journal of Alternative Dispute Resolution-RBADR*, 5(10), 81–105. DOI: 10.52028/rbadr.v5i10.ART04.Ind

Singh, B. (2023). Tele-Health Monitoring Lensing Deep Neural Learning Structure: Ambient Patient Wellness via Wearable Devices for Real-Time Alerts and Interventions. *Indian Journal of Health and Medical Law*, 6(2), 12–16.

Singh, B. (2023). Blockchain Technology in Renovating Healthcare: Legal and Future Perspectives. In *Revolutionizing Healthcare Through Artificial Intelligence and Internet of Things Applications* (pp. 177-186). IGI Global.

Singh, B. (2023). Federated Learning for Envision Future Trajectory Smart Transport System for Climate Preservation and Smart Green Planet: Insights into Global Governance and SDG-9 (Industry, Innovation and Infrastructure). *National Journal of Environmental Law*, 6(2), 6–17.

Singh, B. (2023). Blockchain Technology in Renovating Healthcare: Legal and Future Perspectives. In Kaushik, K., Dahiya, S., Aggarwal, S., & Dwivedi, A. (Eds.), *Revolutionizing Healthcare Through Artificial Intelligence and Internet of Things Applications* (pp. 177–186). IGI Global., DOI: 10.4018/978-1-6684-5422-0.ch012

Singh, B. (2023). Blockchain Technology in Renovating Healthcare: Legal and Future Perspectives. In *Revolutionizing Healthcare Through Artificial Intelligence and Internet of Things Applications* (pp. 177-186). IGI Global.

Singh, B. (2024). Featuring Consumer Choices of Consumable Products for Health Benefits: Evolving Issues from Tort and Product Liabilities. *Journal of Law of Torts and Consumer Protection Law*, 7(1).

Singh, B. (2024). Green Infrastructure in Real Estate Landscapes: Pillars of Sustainable Development and Vision for Tomorrow. *National Journal of Real Estate Law*, 7(1), 4–8.

Singh, B. (2024). Cherish Growth, Advancement and Tax Structure: Addressing Social and Economic Prospects. *Journal of Taxation and Regulatory Framework*, 7(1), 7–10.

Singh, B. (2024). Legal Dynamics Lensing Metaverse Crafted for Videogame Industry and E-Sports: Phenomenological Exploration Catalyst Complexity and Future. *Journal of Intellectual Property Rights Law*, 7(1), 8–14.

Singh, B. (2024). Lensing Legal Dynamics for Examining Responsibility and Deliberation of Generative AI-Tethered Technological Privacy Concerns: Infringements and Use of Personal Data by Nefarious Actors. In Ara, A., & Ara, A. (Eds.), *Exploring the Ethical Implications of Generative AI* (pp. 146–167). IGI Global., DOI: 10.4018/979-8-3693-1565-1.ch009

Singh, B. (2024). Social Cognition of Incarcerated Women and Children: Addressing Exposure to Infectious Diseases and Legal Outcomes. In Reddy, K. (Ed.), *Principles and Clinical Interventions in Social Cognition* (pp. 236–251). IGI Global., DOI: 10.4018/979-8-3693-1265-0.ch014

Singh, B. (2024). Evolutionary Global Neuroscience for Cognition and Brain Health: Strengthening Innovation in Brain Science. In Prabhakar, P. (Ed.), *Biomedical Research Developments for Improved Healthcare* (pp. 246–272). IGI Global., DOI: 10.4018/979-8-3693-1922-2.ch012

Singh, B., Jain, V., Kaunert, C., Dutta, P. K., & Singh, G. (2024). Privacy Matters: Espousing Blockchain and Artificial Intelligence (AI) for Consumer Data Protection on E-Commerce Platforms in Ethical Marketing. In Saluja, S., Nayyar, V., Rojhe, K., & Sharma, S. (Eds.), *Ethical Marketing Through Data Governance Standards and Effective Technology* (pp. 167–184). IGI Global., DOI: 10.4018/979-8-3693-2215-4.ch015

Singh, B., & Kaunert, C. (2024). Computational Thinking for Innovative Solutions and Problem-Solving Techniques: Transforming Conventional Education to Futuristic Interdisciplinary Higher Education. In Fonkam, M., & Vajjhala, N. (Eds.), *Revolutionizing Curricula Through Computational Thinking, Logic, and Problem Solving* (pp. 60–82). IGI Global., DOI: 10.4018/979-8-3693-1974-1.ch004

Slaunwhite, A., Min, J. E., Palis, H., Urbanoski, K., Pauly, B., Barker, B., ... & Nosyk, B. (2024). Effect of Risk Mitigation Guidance opioid and stimulant dispensations on mortality and acute care visits during dual public health emergencies: retrospective cohort study. *bmj, 384*.

Soni, V. (2023). Impact of Generative AI on Small and Medium Enterprises' Revenue Growth: The Moderating Role of Human, Technological, and Market Factors. *Reviews of Contemporary Business Analytics*, 6(1), 133–153.

Stasevych, M., & Zvarych, V. (2023). Innovative robotic technologies and artificial intelligence in pharmacy and medicine: Paving the way for the future of health care—a review. *Big Data and Cognitive Computing*, 7(3), 147. DOI: 10.3390/bdcc7030147

Talebi Bezmin Abadi, A., Rizvanov, A. A., Haertlé, T., & Blatt, N. L. (2019). World Health Organization report: Current crisis of antibiotic resistance. *BioNanoScience*, 9(4), 778–788. DOI: 10.1007/s12668-019-00658-4

Thirunavukarasu, A. J., Ting, D. S. J., Elangovan, K., Gutierrez, L., Tan, T. F., & Ting, D. S. W. (2023). Large language models in medicine. *Nature Medicine*, 29(8), 1930–1940. DOI: 10.1038/s41591-023-02448-8 PMID: 37460753

Tomas, M. E., Kundrapu, S., Thota, P., Sunkesula, V. C., Cadnum, J. L., Mana, T. S. C., Jencson, A., O'Donnell, M., Zabarsky, T. F., Hecker, M. T., Ray, A. J., Wilson, B. M., & Donskey, C. J. (2023). Contamination of health care personnel during removal of personal protective equipment. *JAMA Internal Medicine*, 175(12), 1904–1910. DOI: 10.1001/jamainternmed.2015.4535 PMID: 26457544

Tulenko, K., Mgedal, S., Afzal, M. M., Frymus, D., Oshin, A., Pate, M., & Zodpey, S. (2023). Community health workers for universal health-care coverage: From fragmentation to synergy. *Bulletin of the World Health Organization*, 91(11), 847–852. DOI: 10.2471/BLT.13.118745 PMID: 24347709

Van Lerberghe, W. (2018). [*primary health care: now more than ever*. World Health Organization.]. *World Health Report*, •••, 2008.

Vyas, B., & Rajendran, R. M. (2023). Generative Adversarial Networks for Anomaly Detection in Medical Images. *International Journal of Multidisciplinary Innovation and Research Methodology, ISSN: 2960-2068, 2*(4), 52-58.

Wach, K., Duong, C. D., Ejdys, J., Kazlauskaitė, R., Korzynski, P., Mazurek, G., Paliszkiewicz, J., & Ziemba, E. (2023). The dark side of generative artificial intelligence: A critical analysis of controversies and risks of ChatGPT. *Entrepreneurial Business and Economics Review*, 11(2), 7–30. DOI: 10.15678/EBER.2023.110201

Wamba, S. F., Queiroz, M. M., Jabbour, C. J. C., & Shi, C. V. (2023). Are both generative AI and ChatGPT game changers for 21st-Century operations and supply chain excellence? *International Journal of Production Economics*, 265, 109015. DOI: 10.1016/j.ijpe.2023.109015

Watts, N., Adger, W. N., Agnolucci, P., Blackstock, J., Byass, P., Cai, W., Chaytor, S., Colbourn, T., Collins, M., Cooper, A., Cox, P. M., Depledge, J., Drummond, P., Ekins, P., Galaz, V., Grace, D., Graham, H., Grubb, M., Haines, A., & Costello, A. (2023). Health and climate change: Policy responses to protect public health. *Lancet*, 386(10006), 1861–1914. DOI: 10.1016/S0140-6736(15)60854-6 PMID: 26111439

Wong, I. A., Lian, Q. L., & Sun, D. (2023). Autonomous travel decision-making: An early glimpse into ChatGPT and generative AI. *Journal of Hospitality and Tourism Management*, 56, 253–263. DOI: 10.1016/j.jhtm.2023.06.022

Xu, M., Du, H., Niyato, D., Kang, J., Xiong, Z., Mao, S., Han, Z., Jamalipour, A., Kim, D. I., Shen, X., Leung, V. C. M., & Poor, H. V. (2024). Unleashing the power of edge-cloud generative ai in mobile networks: A survey of aigc services. *IEEE Communications Surveys and Tutorials*, 26(2), 1127–1170. DOI: 10.1109/COMST.2024.3353265

Zhao, C. W., Yang, J., & Li, J. (2021). Generation of hospital emergency department layouts based on generative adversarial networks. *Journal of Building Engineering*, 43, 102539. DOI: 10.1016/j.jobe.2021.102539

Chapter 9
Climate Change and Household Food Security

Jeffrey Kurebwa
https://orcid.org/0000-0002-8371-8055
Bindura University of Science Education, Zimbabwe

Nyasha Yvonne Kurebwa
Women's University in Africa, Zimbabwe

ABSTRACT

The purpose of this study is to provide an overview of the impact of climate change on household food security in Africa. It relies on the Integral Theory to understand climate change and food security. The four pillars of food security which are availability, access, utilization, and stability are discussed. Determinants of food security and climate factors affecting food security are further analyzed. The study concludes that there is a need to intensify agriculture and engage in diversification, conservation, and/or climate-smart practices to improve food security. The study recommends the need to improve agricultural extension, storage technology, food banks, and spiritual prayers to promote food security.

INTRODUCTION

Climate change has a significant impact on food security. With global warming exceeding 2oC, many countries face increased drying and severe food security risks, particularly affecting Indigenous farmers and herders in rural communities (Batterbury, 1996; Parkes et al. 2018). As a result, smallholder dryland farmers and marginal rural areas are at the highest risk due to the sensitivity of their production system and the limited adaptive capacity to address these risks (Molotoks et al.

DOI: 10.4018/979-8-3693-4183-4.ch009

2021; Shukla et al. 2021). Fluctuating precipitation patterns are having a significant impact on agricultural productivity and livelihoods by disrupting natural cycles and traditional planting seasons, which are specific periods of the year when conditions for crop cultivation are ideal (Connolly-Boutin and Smit, 2016; Leisner, 2020; Wing et al. 2021). Severe weather events, such as high temperatures, droughts, and floods, have a detrimental effect on agricultural activities, leading to consequences for food supply, quality, safety, and stability (Bhowmik et al. 2021; Cammarano et al. 2020; Gitz et al. 2016). For instance, tropical cyclones and resulting floods in countries like Zimbabwe, Malawi, and Mozambique have displaced numerous individuals and trapped larger populations in severe food crises. The COVID-19 pandemic has worsened food insecurity in African countries, highlighting the vulnerability of supply chains to external shocks. In addition, climate change presents significant challenges to the agricultural sector in Africa, with limited resources for adaptation and a reliance on climate-sensitive economic sectors. As a result, farming households and communities are implementing strategies to mitigate these short- and long-term risks.

Africa has been facing an increase in aridity, particularly in areas like the Sahel, and is projected to encounter a significant rise in temperatures in dry and semi-arid regions (Khan and Akhtar, 2015). Goal 2 of the Sustainable Development Goals (SDGs) aims to eradicate all hunger and malnutrition by 2030. However, a quarter of the global population lacks access to safe and nutritious food. Ironically, food insecurity is widespread among food producers, especially smallholder farmers in Africa. The high prevalence of food insecurity among smallholder farmers is linked to climate change as well as other biophysical and socio-economic factors. Climate change and food security pose significant challenges to the livelihoods of people, particularly smallholder farmers.

Rain-fed agriculture is the primary source of income for a large number of people in Africa (Derbile et al. 2016). Rural households mainly rely on crop farming, livestock rearing, or both for their subsistence and commercial needs. These activities account for approximately 75% of the livelihoods of many communities (World Food Programme, 2012). However, smallholder farmers face various challenges such as poverty, agro-infrastructural issues, customary land tenure systems, institutional constraints, and inadequate social safety nets (Nyantakyi-Frimpong, 2020; Yaro, 2010). Furthermore, climate change is exacerbating the vulnerability of agricultural systems due to resource limitations, which in turn restrict adaptive capacity (Abdul-Razak and Kruse, 2017; Owusu et al. 2021). The susceptibility of agriculture to climate change is currently recognized as a significant factor contributing to food insecurity and malnutrition in Africa (Darfour and Rosentrater, 2016; Wood et al. 2019).

The Integral Theory to Climate Change and Food Security

The challenges posed by climate change necessitate proactive and comprehensive response strategies across all levels. The Integral theory (IT) framework, which employs the all-quadrant-all level (AQAL) model, is an innovative and inclusive approach that enables the examination of diverse viewpoints to offer a thorough comprehension of climate change risks and local adaptation possibilities through various epistemological perspectives and methodological techniques. Despite criticisms regarding the potential for conflicting viewpoints within the IT model, it is widely recognized as a valuable analytical tool for grasping social realities across various disciplines, contexts, and scopes. The IT model focuses on the perspectives of individuals and groups at various levels, within the wider context of both internal and external aspects of reality. These viewpoints work together to understand how social systems demonstrate resilience at both individual and community levels amidst socio-ecological changes (Folke, 2006). Therefore, IT highlights the importance of individual cognition and awareness, shared values and belief systems, and scientific observations in the management of socio-ecological resources (O'Brien, 2010). The internal aspect of the IT model delves into the intangible realms like knowledge, comprehension, mindsets, and motivations that are assessed on both an individual and collective basis.

The exterior aspects delineate the concrete realm encompassing climate, natural resources, and production levels that can be quantitatively measured. These interactions, both internal and external, are rooted in various methodological approaches and fields of study that aid in comprehending and addressing socio-ecological and environmental transformations. In contrast, the IT model acknowledges the 'human dimension' of individuals and groups in grasping socio-ecological system contexts, while also encompassing 'behavioral and system dimensions' to comprehend adaptation processes in reaction to socio-ecological shifts. The IT model has been employed in research on climate change and food security to capture perspectives from farming households and communities regarding the impacts of climate change and potential adaptation strategies.

This method has proved to be effective in research due to its thoroughness in addressing intricate individual and collective issues in society (Hochachka, 2005; McLaren & Kelleher, 2005). Africa stands out as one of the most susceptible continents to climate change because of its high vulnerability to climate-related hazards (Leal Filho et al. 2017; Niang et al. 2014; Owusu et al. 2021), as well as the fragility of its production and livelihood systems. Although the contribution of Sub-Saharan African (SSA) countries to global warming is relatively small compared to developed nations, the region is expected to face disproportionate consequences of climate change risks due to its heightened exposure and vulnerability.

Agriculture is expected to be the sector most vulnerable to climate change, as agricultural systems heavily rely on specific climatic conditions (Pereira, 2017; Shukla et al. 2021; Zeleke et al. 2021). While climate change may have the potential to increase agricultural productivity in certain countries, it is anticipated that productivity in the majority of African countries will decrease (Gibon et al. 2018; IPCC, 2018), primarily due to significant constraints in adaptive capacity (Herslund et al. 2016). The limited adaptive capacity of many developing countries in Africa is influenced by various factors including extreme poverty, ongoing agro-infrastructural challenges, governance constraints, and inadequate social safety nets (Chepkoech et al., 2020).

Climate Change

The Earth's climate system has undergone changes spanning from decades to millennia, and more significant changes are anticipated in the future (Dessler & Parson, 2019; Lemery et al. 2021; IPCC, 2018). Climate change is defined as "the alteration in the climate's state that can be observed through changes in its properties' average and/or variability, persisting for an extended period, typically decades or longer" (IPCC, 2014: p5). Ongoing changes in the global climate, caused by both natural and human forces, are considered inevitable and undeniable in the 21st century (Chen et al. 2021). Global climate models (e.g. General Circulation Model) predict continued climate change in the future, although specific projections often come with high levels of uncertainty.

During the modern historical era, the earth's climate system began to undergo gradual warming trends in the 20th Century (IPCC, 2014). Nevertheless, the rates of global warming have accelerated in the 21st Century (Osborn et al. 2021; World Meteorological Organization, 2018). It is acknowledged that the warming trends will not be uniformly distributed, with significant warming anticipated on land, particularly in arid and semi-arid regions linked to changes in precipitation and extreme weather events (IPCC, 2019). Biophysical and socio-ecological systems in dryland areas will be especially vulnerable to the risks associated with the pronounced warming trend in the Earth's climate system.

Climate change can result from natural processes or human activities. Natural processes such as volcanic eruptions, internal climate variability, and solar radiation can contribute to climate change. On the other hand, human activities like the emission of greenhouse gases and changes in land use can also impact the climate. These climate changes directly affect global food systems and accessibility (Ray et al. 2019). Food security is achieved when individuals have access to safe and nutritious food that meets their dietary requirements and preferences (Global Strategic Framework, 2012). Unfortunately, in 2017, widespread hunger and famine led to severe malnutrition and poor health globally (Food and Agricultural Organization, 2017).

Malnutrition encompasses disparities in energy intake, whether it be an excess, deficiency, or imbalance (FAO, 2018). The intricate global food systems are currently facing challenges in meeting the growing demand for food due to the expanding world population, and are at risk from the impact of climate change (Chase, et al. 2019, Stevanovic, et al. 2016). Without a timely solution, this threat is projected to jeopardize the fundamental aspects of food security, including availability, accessibility, utilization, and stability.

Climate Shocks

Poor individuals experience a variety of shocks that have a direct or indirect impact on their livelihoods and food security (Von Braun, 2009). The concept of shock is broadly defined in different ways in various literature. For example, Dercon et al. (2005, p. 5) describe shocks as "adverse events that result in a loss of household income, a decrease in consumption, and/or a loss of productive assets." According to Dercon et al. (2005), a specific shock is considered serious if it leads to a significant reduction in asset holdings, a substantial decrease in household income, or a significant reduction in consumption. Shocks at the household level can be either idiosyncratic or covariate.

Unique shocks, known as idiosyncratic shocks, target specific households within a community, while covariate shocks impact multiple individuals in the same area simultaneously, as explained by Von Braun (2009). For example, natural disasters like rainfall shocks and droughts fall under the category of covariate shocks since they affect all households in a village and potentially those nearby. Vulnerability to idiosyncratic risks is often attributed to factors such as social standing, occupation, and geographic location, according to Von Braun (2009). Dercon et al. (2005) categorized shocks into various groups including climatic, economic, political, crime, and health-related shocks. Among climatic shocks, idiosyncratic events like floods, excessive rainfall, crop pests/diseases, and livestock pests/diseases are the most common occurrences.

In certain instances, pests and diseases that impact crops or livestock can be a combination of idiosyncratic and covariate shocks. The impact of climate shock is particularly harsh in developing nations, as a significant portion of the impoverished population relies on agriculture for sustenance and income. Von Braun (2009) highlights that climate shock encompasses a rise in extreme weather occurrences like droughts and floods, leading to reduced agricultural output and subsequently lower food availability. In addition to well-known examples of climatic shocks such as droughts and floods, Dercon et al. (2005) also identify erosion, frost occurrences, and pestilence as factors that can adversely affect crops or livestock due to climate shocks.

Food Security

Food insecurity, a significant global threat linked to climate change, goes beyond the mere availability of food. It encompasses the accessibility and sustainable utilization of sufficient, safe, and nutritious food for all individuals at all times, as defined by the FAO in 2009. The burden of climate change on agriculture in Africa, compounded by socioeconomic factors, plays a crucial role in driving food insecurity in the region, as highlighted by various reports from the FAO, IPCC, and other organizations. Despite the considerable attention given to addressing food insecurity on a global scale, the stark reality remains that approximately 2 billion people worldwide continue to face food insecurity, as reported by the FAO and other relevant entities in 2020.

Food security refers to the state in which every person has consistent physical and financial access to an adequate, safe, and nutritious food supply, allowing them to fulfill their dietary requirements and preferences for a healthy and active lifestyle (World Bank Group, 2023). Climate change can directly affect food security through its impact on agricultural practices, which is particularly crucial for the world's impoverished (FAO, 2002). For example, Africa relies heavily on agriculture for food production and sustenance (Sirba and Chimdessa, 2021). Nevertheless, this region is confronted with a substantial threat to agricultural productivity and food security as a result of the consequences of climate change (van Ittersum et al. 2016). Consequently, significant changes in agricultural practices are necessary due to the challenges posed by climate change, leading to food insecurity among farmers. It is imperative to develop strategies for agricultural adaptation and mitigation to reduce the growing risks to food security, as highlighted by the FAO in 2002.

The Four Pillars of Food Security

The idea of food security has progressed, acknowledging its multifaceted nature (Connolly-Boutin and Smit, 2016) and encompassing the four key components of food security; food access, availability, food use, and stability (FAO, 2006). This section delves into the pillars of food security and their evolution.

Availability

The study focuses on how climate change is projected to have a direct impact on food security, specifically by affecting crop productivity. This impact is particularly significant for societies that heavily depend on subsistence farming to fulfill their dietary needs (Wlokas, 2008b). Food security analysis commonly highlights national indicators related to food availability, including crop production and food production indices, livestock ownership indices, and national food balance sheets (Barrett, 2010; Renzaho and Mellor, 2010). The emphasis on availability is due to the relative ease of obtaining productivity estimates on different scales. At first, the focus of food security was mainly on food production, as seen in the conversations held during the 1974 World Food Conference (Anderson and Cook, 1999; Maxwell, 1996; Baro and Deubel, 2006). Nevertheless, researchers quickly realized that simply ensuring the availability of food did not automatically guarantee households' access to it (Jones et al. 2013).

Access

The ability to obtain food is contingent upon the ability to produce it, a process that can be impeded by various obstacles like high food costs or limited financial means (Thompson et al. 2010b). Factors that impact access to food encompass income levels, availability of resources, physical and social surroundings, food prices, as well as governmental and trade regulations (Renzaho and Mellor, 2010). Commonly utilized indicators to gauge access to food include food consumption patterns, monitoring of food prices, income levels, and assets (Webb et al. 2006; Baiphethi and Jacobs, 2009; Renzaho and Mellor, 2010). Amartya Sen's (1982) research on entitlements and access has been a pivotal contribution to the field of food security studies, underscoring the significance of food access (Connolly-Boutin and Smit, 2016). Sen emphasized that despite the presence of enough food, individuals living in poverty may still struggle to access it due to elevated food costs and other barriers, which ultimately hinder their ability to acquire food even though it is essential for their household's well-being (Sen, 1982). This change in perspective prompted a redefined understanding of food security during the 1983 World Food Summit, placing importance on both the availability and accessibility of essential food items (FAO, 1983).

Utilization

The notion of food security has evolved to encompass issues concerning the unequal distribution or insufficient consumption of food (Jones et al. 2013). The utilization aspect of food security takes into account how food is distributed within households, the nutritional value of the food, and the capacity of individuals in the household to absorb and process nutrients (Jones et al. 2013). Studies have indicated that women's choices regarding spending in households tend to prioritize investments in the health, nutrition, and education of children. The behaviors related to acquiring food within households play a critical role in translating both physical and economic access to food into food security (Jones et al. 2013). Consequently, there has been a shift towards embracing 'collective' methods to model decision-making within households, acknowledging that household dynamics have an impact on food security outcomes (Haddad et al. 1997).

Stability

In the 1990s, nutrition research primarily emphasized caloric sufficiency rather than overall diet quality (Jonsson, 2010). This approach led to the identification of a third component, 'utilization'. As a result, it became apparent within the framework of food security that while physical and economic access to food and food acquisition are important, they alone are not enough to guarantee food security within households (Jones et al. 2013). The FAO's definition of food security, as outlined in 1996, underscores the significance of acquiring culturally and socially appropriate foods through acceptable means, introducing a fourth element that is frequently disregarded: the constancy and reliability of food security across time (Jones et al. 2013). Various factors such as weather fluctuations, alterations in food availability, price fluctuations, and governmental regulations can influence food security (Schmidhuber and Tubiello, 2007). For example, the variability in crop production can disproportionately impact the most impoverished individuals in Africa (Bruinsma, 2003). Moreover, economic, environmental, and political variables play a role in food stability issues, evident in the repercussions of food price crises on poverty levels (Nellemann et al. 2009).

Household Food Security

Various interpretations of Household Food Security (HFS) exist, but they all share a common understanding that the fundamental characteristic of HFS is consistent access to a sufficient amount of food. Maxwell and Frankenberger (1992) emphasize this aspect in their definition of HFS, highlighting the importance of secure access to food at all times. On the other hand, scholars like Bonti-Ankomah (2001) provide a slightly different perspective on Household Food Security, defining it as the access of all households to an adequate, safe, and nutritious food supply for maintaining a healthy and productive life. This definition underscores the importance of not only having enough food but also ensuring its quality and nutritional value. Furthermore, the International Fund for Agricultural Development (IFAD, 1992) describes Household Food Security as the ability of a household to obtain a consistent and sustainable supply of sufficient food. This definition takes into account factors such as the household's current food inventory, past food availability, and potential future food supply, which is influenced by the resources available to the household, including capital like land, labor, and time.

The Household Food Security (HFS) concept goes beyond just food production within a household, as it is intricately connected to the overall livelihood strategies of households. This connection can be quite complex, as households may prioritize different objectives, such as asset preservation, over immediate food needs, leading them to go hungry to achieve other goals. Interestingly, it has been observed that households with higher levels of crop production may experience poorer food security outcomes. For instance, smallholders who shift towards cash-oriented maize production tend to have higher rates of child malnutrition and a less diverse diet compared to households that rely on traditional subsistence crops. This highlights the paradoxical nature of the relationship between crop production and food security.

Household food security (HFS) may have a stronger correlation with income levels, cash flows, and cash management within a household, rather than just production levels. In certain climate zones, diversifying away from sole reliance on food-crop production could offer better opportunities for utilizing available resources to enhance food self-sufficiency. This suggests that a comprehensive approach is necessary for developing interventions to tackle household food-security concerns, one that shifts focus from solely nutrition and food production to a more inclusive view of food systems and sustainable livelihoods. The dynamic nature of household food security is also a point of interest. This is based on the understanding that a household is a dynamic unit where many factors are to be considered before food security can be attained.

Determinants of Food Security

Jacobs (2009) posited that developing a universal set of food security goals is a challenging and intricate endeavor. This complexity arises from the various factors that influence the food security situation of a household or its individuals. When examining household food security, it becomes imperative to elucidate the determinants that impact the overall food security status, which encompasses a wide range of variables. The makeup of households can differ based on the number of members, age distribution, and gender composition. Nord and Hopwood (2007) explore the significance of household composition in determining the food security situation of children within the household (Jacobs, 2009). These include:

(a) Wealth and livelihood strategy- This consists of various incomes (wages, social grants) and assets (land, livestock).
(b) Geographic location- This refers to the rural and urban locations, whether the settlement is largely formal or informal, and the distance from the nearest or frequently used food markets.
(c) Institutions- markets, the state, social capital/networks.
(d) Time- the food security condition could be transitory or chronic.
(e) Risk- shocks that are related to the weather, health, as well as commodity price movements. Risk has evolved into a major cross-cutter as it affects all elements of food insecurity (Webb et al. 2006). It includes disruptions such as climatic fluctuations, social conflicts, and other crises that make households more vulnerable to food shortages (Jacobs, 2009).

Climate Factors Affecting Food Security

Several climatic factors affect food security as discussed below.

Droughts

Droughts are the result of prolonged periods of insufficient precipitation, causing water shortages in specific areas (Wang et al. 2016). These occurrences have global impacts, affecting communities through crop failures and subsequent famines, leading to malnutrition and other health issues. The consequences of droughts have been particularly significant in Africa, with an estimated 642 drought events occurring worldwide between 1900 and 2013 (Masih et al. 2014). The complexity of drought as a phenomenon is evident in its varying intensity, onset, cessation, duration, and geographical coverage. Reports have indicated the occurrence of megadroughts in both America and China. The expansion of dry and arid areas throughout the 20th century has been linked to the drying trend observed since the 1970s, with projections suggesting a further increase in these conditions throughout the 21st century (Masih et al. 2014).

Floods

Floods represent the overflow of a large body of water that inundates the land, posing a significant threat to various countries. In Thailand, the 2011 floods had a devastating impact on agriculture, industries, and the overall economy, resulting in the destruction of farmland, displacement of millions of people, and substantial economic losses. Similarly, in Sri Lanka, heavy rainfall in 2017 led to floods and landslides, causing food insecurity for around 900,000 individuals in the southern western regions. According to Week and Wizor (2020), floods are a common disaster with enormous destructive potential, as evidenced by the significant economic damage and human impact in both Thailand and Sri Lanka. Haraguchhi and Upmanu (2015) estimated the economic losses from the Thai floods to be USD 46.5 billion, while FAO (2017) highlighted the food insecurity affecting a large population in Sri Lanka due to the 2017 floods and landslides.

Post-harvest Losses

Post-harvest losses are defined as the losses that occur from the time of harvest until the produce reaches the consumer, as stated by Magray et al. (2017). It is crucial to address the reduction of post-harvest food losses to ensure global food security, given that over one-third of food is wasted post-harvest, according to Kumar and Kalita (2017). In the Asian Pacific regions, the spoilage of fresh farm produce is primarily influenced by the hot and humid climatic conditions. For example, in Bangladesh, post-harvest losses contribute to 10-20% of total losses, as highlighted by Majumder et al. (2016). Similarly, a study conducted in Peshawar Valley, Pakistan, revealed that approximately 20% of the total tomato crop harvested is lost post-harvest, as reported by Mujib et al. (2007).

The Impact of Climate Change on Food Security

The potential consequences of climate change on food security encompass not only direct impacts on nutrition, such as changes in the quantity and quality of food consumption but also indirect effects on livelihoods, including shifts in employment opportunities and the cost of obtaining sufficient nutrition. Climate change can influence both of these aspects. When considering the impact of climate change on food security, it is crucial to take into account both biophysical and social vulnerability. Social vulnerability involves an examination of the demographic, social, economic, and other characteristics of a population that influence their exposure to risk and their capacity to respond to and cope with adverse events. A social vulnerability perspective is indispensable for comprehending why certain individuals, households, or communities may experience varying impacts from climate change, even if they are located in the same geographic area.

The Nature of Impacts

The impact of climate change on food insecurity primarily stems from its biophysical influence on crop, livestock, and farming system productivity. Variations in temperature and precipitation levels, as well as heightened variability, result in fluctuations in both the quantity and quality of food produced. Consequently, this has a ripple effect on the income of food producers and the affordability of food for both rural and urban consumers. Foreseen rises in climate variability are expected to further exacerbate fluctuations in agricultural production, leading to increased price and income instability. Therefore, effective risk management by all stakeholders within the food supply chain, ranging from individuals and households to entire nations, will become increasingly crucial.

The Magnitude of Impacts

The extent of influence on systems, communities, households, and individuals is significantly determined by the level of exposure and susceptibility to the impacts of climate change, which varies across different regions and populations. Certain regions may experience more severe consequences from shifts in temperature and precipitation patterns, while others may initially benefit. The impact will be directly experienced in rural areas and indirectly in urban areas through increased prices and greater variability. These effects can be direct, such as local alterations in precipitation and temperature, as well as indirect, through biophysical changes in other areas that affect local livelihoods, for example, by altering global prices. Certain demographic segments will experience a disproportionate impact, either due to their reliance on agricultural activities for financial sustenance or because they allocate a larger portion of their earnings towards sustenance.

The Ability to Respond

Climate change resilience encompasses various aspects, shifting from the examination of plants and animals to agricultural systems, individuals, households, communities, and nations. Plants and animals exhibit specific susceptibilities to climate variations, while agricultural systems can enhance their resilience through modified practices, adjusted planting schedules, and diversification of varieties or species. Human resilience is contingent upon the available resources, whether physical, social, or financial, within an individual or household. Those in poverty are at heightened risk of food insecurity due to climate change, whereas the affluent have the means to secure food, at least in the short run.

Food insecurity persists even in the wealthiest nations, and the decision to pursue a development trajectory that exacerbates inequality or leads to environmental degradation could increase vulnerability to food insecurity due to climate change in the future (Pieter et al. 2011). Rural farming households in Africa are increasingly aware of the significant impacts of climate change on food security. Recent studies have highlighted that farmers identify various climate change risks, including rising temperatures, unpredictable rainfall patterns, and extreme weather events like prolonged droughts, floods, and wildfires (Guodaar et al. 2021). These effects contribute to worsening food security situations across many African nations. The impact of climate change on food production can be significant, as it affects both plant and animal physiology, leading to a reduction in agricultural labor productivity. This has been well-documented in various studies, highlighting the detrimental effects of heat extremes on outdoor labor, particularly in the agricultural sector. The decline in crop production due to climate change and variability is a pressing issue affecting both developed and developing countries. This not only impacts the global food supply but also contributes to the energy crisis, posing a serious challenge for nations worldwide.

Nevertheless, the most significant impact is being experienced in less affluent developing nations, where a projected global warming increase in average temperature of 2.5° C could result in a 1.3% decrease in per capita income (Richard, 2018). In the regions of Kazakhstan, Kyrgyzstan, and Uzbekistan in central Asia, a study spanning two decades examined meteorological data. The analysis revealed that an increase in precipitation above normal levels was correlated with higher crop yields throughout the year, except in spring, during which excessive rainfall led to reduced yields. The decline in yield during spring was attributed to delays in crop planting and field operations, as well as widespread flooding that created favorable conditions for the spread of plant diseases such as yellow rust (Mirzabaev, 2013).

The Role of Social Networks in Improving Food Security

The role of social networks in managing environmental risks and governing natural resources has been highlighted by various researchers (Alexander et al. 2015; Bodin and Crona, 2009). These networks contribute to sustained collective action, particularly in enhancing food security (Lee et al. 2018; Martin et al. 2004; Nosratabadi et al. 2020). However, there is still a lack of understanding regarding the impact of social networks on food security and risk perceptions in dryland rural areas of Africa, especially in the context of climate change (Dzanja et al. 2013; Thamaga-Chitja and Tamako, 2017). In this context, social networks refer to the relationships and connections between individuals and groups with shared interests, aimed at achieving mutual benefits (Chaudhury et al. 2017). The absence of knowledge in this area is crucial for gaining a thorough comprehension of how rural areas, often lacking well-established and efficient formal organizations, can harness social connections to enhance robust local food networks (Lowitt et al. 2014). Consequently, this research posits that social ties will be instrumental in addressing food insecurity within rural households and communities.

SOLUTIONS AND RECOMMENDATIONS

The effects of climate change on food security have sparked ongoing conversations regarding effective strategies to enhance food security. Researchers worldwide are emphasizing the importance of intensifying agriculture and implementing diversification, conservation, and climate-smart practices to address this issue. In Africa, experts have highlighted the significance of enhancing agricultural extension services, utilizing advanced storage technology, establishing food banks, and incorporating spiritual practices to bolster food security. Additionally, on-farm initiatives, communal food sharing, seasonal migration, and empowering women are identified as crucial approaches to enhance household food security.

It is of great significance to note that evidence indicates that the combination of enhanced technology and farmers' adaptive practices can help mitigate the adverse effects of climate change to some extent. This is because technological advancements can contribute to the increase or maintenance of crop yields, reduction of crop prices, and higher consumption levels (Islam et al. 2016). The role of education in rural communities cannot be overstated, as various research has demonstrated that rural households with higher levels of education are less vulnerable to the impacts of climate change on agriculture. Farmers with higher education levels are more inclined to employ climate adaptation measures such as cultivating drought-resistant crops, diversifying their crop selection, opting for early maturing crops, and seeking alternative sources of income (Gebre et al. 2023).

Households that possess higher levels of education and financial resources are better equipped to diversify their sources of income to mitigate the effects of climate change on agriculture (Gentle and Maraseni, 2012). Studies conducted in countries like Pakistan (Khan et al. 2020), Vietnam (Huong et al. 2019), and Ethiopia (Sertse et al. 2021) have shown that increased education levels can help minimize the negative consequences of climate-related events on agricultural activities. These case studies serve as examples of how education can play a crucial role in lessening the impact of climate change on rural livelihoods and potentially influencing migration patterns in rural areas.

It is essential to provide support to farmers to encourage the adoption of practices that will strengthen their ability to withstand challenges and ensure food security, while also contributing to long-term climate benefits. Creating an environment that enables these changes to take place, including the provision of services and institutions to assist farmers, such as extension services, is crucial. Livelihood diversification and collective decision-making offer promising methods for enhancing food security and building resilience to climate change within smallholder communities. By diversifying livelihoods and making decisions collectively, resources from various activities can be pooled to help households mitigate risks. Strategies such as farm and agricultural diversification involve smallholder farmers engaging in a range of primary agricultural activities to diversify risk and reduce vulnerability to climate change impacts, including diversifying crop and livestock production.

The diversification of farm livelihoods does not solely involve growing different types of crops or raising various livestock but also encompasses the integration of a wide range of farm management techniques. In Africa, smallholder farmers are familiar with the concept of diversifying crops and livestock, with some countries like Zimbabwe now commonly incorporating high-value crops alongside traditional subsistence crops. It is crucial to merge efforts aimed at enhancing food security with those addressing climate change. Policies and initiatives aimed at mitigating the impacts of climate change should complement, rather than operate independently from, those focused on ensuring sustainable food security. Climate change poses a significant threat to food security, and interventions aimed at bolstering the resilience of food systems are likely to also aid in adapting to climate change challenges.

Redirecting efforts to increase expenditure solely on adaptation towards increasing overall expenditures on sustainable food security, with a specific focus on the unique and uncertain threats from climate change, is crucial. This approach should involve public, private, and other sectors, and should prioritize the needs and contributions of the disadvantaged. It is essential to address climate change with explicit attention to the needs of the disadvantaged. This principle should be applied to all vulnerable groups, with a particular emphasis on the role of women as agricultural decision-makers. Women are integral to the planning, design, and implementation of policies and programs aimed at addressing the challenges posed by climate change to food security. In many development discussions, women are often seen as recipients rather than contributors. It is crucial to advocate for women's involvement and leadership in making decisions about food security, especially when addressing adaptation and mitigation needs. Additionally, it is important to acknowledge that women are not a uniform group when considering gender dynamics, as inequities are influenced by factors such as class, culture, and socio-economic conditions.

FUTURE RESEARCH DIRECTION

The study offers valuable insights for policymakers regarding the necessity of implementing a comprehensive strategy for evaluating and addressing climate change risks. It highlights that the climate change strategies and adaptation policies in many African nations have not adequately emphasized the significance of integrating knowledge. The findings of this research present an opening for governments, development professionals, and policymakers to incorporate diverse viewpoints and knowledge systems into forthcoming climate change and food security policies, aiming for sustainable results. Due to the intricate nature of climate risks in many African nations, governments at the local, regional, and national levels must implement adaptation strategies that are multi-dimensional to enhance the resilience of agricultural systems in rural communities with limited resources. This research could prove highly beneficial for African countries seeking to establish collective and communal social structures to assist households in mitigating their exposure to food security threats.

CONCLUSION

The agricultural systems globally are being negatively impacted by climate change, leading to repercussions on the livelihoods, food security, and poverty rates of households, especially in rural communities. Due to the significant effects of climate change on agricultural systems and the opportunities for livelihoods within socio-ecological systems, various entities such as governments, non-governmental organizations, foreign donors, and researchers are promoting and assisting farmers in adapting to climate change by enhancing agriculture through the utilization of modern techniques like modern seed varieties to enhance food security across the continent. Contemporary strategies play a crucial role in managing environmental threats, however, the issue of food insecurity persists as a significant obstacle in Africa, particularly with the escalating risks and extremes brought about by climate change. Despite the implementation of modernization policies and innovations in numerous African nations, the challenges related to food security have not been adequately resolved. As a result, farmers in various rural areas are leveraging their traditional knowledge and social connections as valuable social assets to effectively tackle the risks associated with climate change and food security.

REFERENCES

Afful, D. B. (2016). Public Extension Agents Need For New Competences: Evidence from a Climate Variability Study in Limpopo province of South Africa. *Journal of Agricultural Extension*, 44(2), 59–70.

Alexander, S. M., Armitage, D., & Charles, A. (2015). Social networks and transitions to co-management in Jamaican marine reserves and small-scale fisheries. *Global Environmental Change*, 35, 213–225. DOI: 10.1016/j.gloenvcha.2015.09.001

Amenuveve, V., (2010). Northern Floods Claim 17 Lives: But Residents Still Adamant. *Daily Graphic*, Monday, 13, September, 2010.

Anderson, M. D., & Cook, J. T. (1999). Community food security: Practice in need of theory? *Agriculture and Human Values*, 16(2), 141–150. DOI: 10.1023/A:1007580809588

Antwi-Agyei, P., & Nyantakyi-Frimpong, H. (2021). Evidence of climate change coping and adaptation practices by smallholder farmers in northern Ghana. *Sustainability (Basel)*, 13(3), 1308. DOI: 10.3390/su13031308

Barrett, C. B. (2010). Measuring Food Insecurity. *Science*, 327(5967), 825–828. DOI: 10.1126/science.1182768 PMID: 20150491

Batterbury, S. P. J. (2008). Anthropology and global warming: The need for environmental engagement. *The Australian Journal of Anthropology*, 19(1), 62–65. DOI: 10.1111/j.1835-9310.2008.tb00108.x

Bhowmik, J., Irfanullah, H. M., & Selim, S. A. (2021). Empirical evidence from Bangladesh of assessing climate hazard-related loss and damage and the state of adaptive capacity to address them. *Climate Risk Management*, 31, 100273. DOI: 10.1016/j.crm.2021.100273

Bodin, Ö., & Crona, B. I. (2009). The role of social networks in natural resource governance: What relational patterns make a difference? *Global Environmental Change*, 19(3), 366–374. DOI: 10.1016/j.gloenvcha.2009.05.002

Bonti-Ankomah, S. (2001). *Addressing food insecurity in South Africa, The National Institute for Economic Policy.* Paper presented at the SARPN conference on Land Reform and Poverty Alleviation in Southern Africa. Pretoria.

Bruinsma, J. (Ed.). (2003). *World agriculture: towards 2015/2030.* FAO: Earthscan.

Cammarano, D., Valdivia, R. O., Beletse, Y. G., Durand, W., Crespo, O., Tesfuhuney, W. A., Jones, M. R., Walker, S., Mpuisang, T. N., Nhemachena, C., Ruane, A. C., Mutter, C., Rosenzweig, C., & Antle, J. (2020). Integrated assessment of climate change impacts on crop productivity and income of commercial maize farms in northeast South Africa. *Food Security*, 12(3), 659–678. DOI: 10.1007/s12571-020-01023-0

Chase, S., Kimberly, F., & Christian, M. (2019). Climate Change and Food Security: A Test of U.S leadership in a Fragile world. Center for strategic and international studies. Brief Report). https://www.csis.org/analysis/climate

Chaudhury, A. S., Thornton, T. F., Helfgott, A., Ventresca, M. J., & Sova, C. (2017). Ties that bind: Local networks, communities and adaptive capacity in rural Ghana. *Journal of Rural Studies*, 53, 214–228. DOI: 10.1016/j.jrurstud.2017.05.010

Chen, J., Li, X., Martel, J. L., Brissette, F. P., Zhang, X. J., & Frei, A. (2021). The relative importance of internal climate variability versus anthropogenic climate change in global climate change. *Journal of Climate*, 34(2), 465–478. DOI: 10.1175/JCLI-D-20-0424.1

Chepkoech, W., Mungai, N. W., Stöber, S., & Lotze-Campen, H. (2020). Understanding the adaptive capacity of smallholder African Indigenous vegetable farmers to climate change in Kenya. *Climate Risk Management*, 27, 100204. DOI: 10.1016/j.crm.2019.100204

Connolly-Boutin, L., & Smit, B. (2016). Climate change, food security, and livelihoods in sub-Saharan Africa. *Regional Environmental Change*, 16(2), 385–399. DOI: 10.1007/s10113-015-0761-x

Dari, L., Addo, A., & Dzisi, K. A. (2016). Pesticide use in the production of Tomato (Solanum lycopersicum L.) in some areas of Northern Ghana. *African Journal of Agricultural Research*, 11(5), 352–355. DOI: 10.5897/AJAR2015.10325

de Lima, C. Z., Buzan, J. R., Moore, F. C., Uris, L. C., Baldos, M. H., & Hertel, T. W. (2021). Heat stress on agricultural workers exacerbates crop impacts of climate change. *Environmental Research Letters*, 16(4), 044020. DOI: 10.1088/1748-9326/abeb9f

De Waal, A. (1989). *Famine That Kills: Darfur, Sudan, 1984–1985*. Clarendon Press.

Dercon, S. (2004). Growth and shocks: Evidence from rural Ethiopia. *Journal of Development Economics*, 74(2), 309–329. DOI: 10.1016/j.jdeveco.2004.01.001

Dessler, A. E., & Parson, E. A. (2019). *The science and politics of global climate change: A guide to the debate*. Cambridge University Press. DOI: 10.1017/9781316832158

Dzanja, J. L., Christie, M., Fazey, I., & Hyde, A. G. (2013). The role of social capital on rural food security: The case study of Dowa and Lilongwe Districts in Central Malawi. *Access International Journal of Agricultural Sciences*, 1(4), 46–56.

Esbjörn-Hargens, S., & Zimmerman, M. E. (2009). *Integral ecology: Uniting multiple perspectives on the natural world*. Random House/ Integral Books.

FAO. (1983). World food security: a reappraisal of the concepts and approaches. Director General's Report. www.fao.org

FAO. (2002). The State of Food Insecurity in the World 2001. Food and Agricultural Organization, Rome. www.fao.org

FAO. (2017). Extreme Drought and Floods Threaten Food Security in Sri Lanka. Food and Agriculture Organization of the United Nations 2017. www.fao.org

FAO. (2018). The Future of Food and Agricultural Alternative Pathways to 2050. Food and Agricultural Organization of United Nations. Rome Italy. www.fao.org

Folke, C. (2006). Resilience: The emergence of a perspective for social-ecological systems analyses. *Global Environmental Change*, 16(3), 253–267. DOI: 10.1016/j.gloenvcha.2006.04.002

Food and Agricultural Organization. Economic Commission for Africa, & African Union. (2020). 2019 Africa Regional Overview of Food Security and Nutrition: Containing the damage of economic slowdowns and downturns to food security in Africa. www.fao.org

Frankenburger, T. R. (1992). Indicators and Data Collection Methods for Assessing Household Food Security. In Maxwell, S., & Frankenburger, T. (Eds.), *Household food security: Concepts, indicators, and measurement: A technical review*. UNICEF and IFAD.

Gebre, G. G., Amekawa, Y., Fikadu, A. A., & Rahut, D. B. (2023). Farmers' use of climate change adaptation strategies and their impacts on food security in Kenya. *Climate Risk Management*, 40, 100495. DOI: 10.1016/j.crm.2023.100495 PMID: 37283879

Gentle, P., & Maraseni, T. N. (2012). Climate change, poverty and livelihoods: Adaptation practices by rural mountain communities in Nepal. *Environmental Science & Policy*, 21, 24–34. DOI: 10.1016/j.envsci.2012.03.007

Gibon, F., Pellarin, T., Román-Cascón, C., Alhassane, A., Traoré, S., Kerr, Y., Seen, D. L., & Baron, C. (2018). Millet yield estimates in the Sahel using satellite-derived soil moisture time series. *Agricultural and Forest Meteorology*, 262, 100–109. DOI: 10.1016/j.agrformet.2018.07.001

Gitz, V., Meybeck, A., Lipper, L., Young, C. D., & Braatz, S. (2016). Climate change and food security: risks and responses. Rome: Food and Agriculture Organization of the United Nations (FAO) Report.

Guodaar, L., Bardsley, D. K., & Suh, J. (2021). Integrating local perceptions with scientific evidence to understand climate change variability in northern Ghana: A mixed-methods approach. *Applied Geography (Sevenoaks, England)*, 130(102440), 102440. DOI: 10.1016/j.apgeog.2021.102440

Herslund, L. B., Jalayer, F., Jean-Baptiste, N., Jørgensen, G., Kabisch, S., Kombe, W., Lindley, S., Nyed, P. K., Pauleit, S., Printz, A., & Vedeld, T. (2016). A multi-dimensional assessment of urban vulnerability to climate change in Sub-Saharan Africa. *Natural Hazards*, 82(2), 149–172. DOI: 10.1007/s11069-015-1856-x

Hesselberg, J., & Yaro, J. A. (2006). An assessment of the extent and causes of food insecurity in northern Ghana using a livelihood vulnerability framework. *GeoJournal*, 67(1), 41–55. DOI: 10.1007/s10708-006-9007-2

Huong, N. T. L., Bo, Y. S., & Fahad, S. (2019). The economic impact of climate change on agriculture using the Ricardian approach: A case of northwest Vietnam. *Journal of the Saudi Society of Agricultural Sciences*, 18(4), 449–457. DOI: 10.1016/j.jssas.2018.02.006

IFAD. (2009). Research and Innovation for Smallholder Farmers in the Context of Climate Change. Discussion paper prepared for the Round Table organized during the Thirty-second session of IFAD's Governing Council, 18 February 2009. http://www.ifad.org

Islam, S., Cenacchi, N., Sulser, T. B., Gbegbelegbe, S., Hareau, G., Kleinwechter, U., Mason-D'Croz, D., Nedumaran, S., Robertson, R., Robinson, S., & Wiebe, K. (2016). Structural approaches to modeling the impact of climate change and adaptation technologies on crop yields and food security. *Global Food Security*, 10(September), 63–70. DOI: 10.1016/j.gfs.2016.08.003

Jacobs, P. T., Altman, M., & Hart, T. (2009). The Status of Household Food Security Targets in South Africa. *Agrekon*, 48(4), 410–433. DOI: 10.1080/03031853.2009.9523834

Jones, A. D., Ngure, F. M., Pelto, G., & Young, S. L. (2013). What are we assessing when we measure food security? A compendium and review of current metrics. *Advances in Nutrition*, 4(5), 481–505. DOI: 10.3945/an.113.004119 PMID: 24038241

Jonsson, U. (2010). The rise and fall of paradigms in world food and nutrition policy. *World Nutrition,* 1(3).

Kansiime, M. K., Tambo, J. A., Mugambi, I., Bundi, M., Kara, A., & Owuor, C. (2021). COVID-19 implications on household income and food security in Kenya and Uganda: Findings from a rapid assessment. *World Development*, 137, 105199. DOI: 10.1016/j.worlddev.2020.105199 PMID: 32982018

Khan, I., Lei, H., Shah, I. A., Ali, I., Khan, I., Muhammad, I., Huo, X., & Javed, T. (2020). Farm households' risk perception, attitude and adaptation strategies in dealing with climate change: Promise and perils from rural Pakistan. *Land Use Policy*, 91, 104395. DOI: 10.1016/j.landusepol.2019.104395

Leal Filho, W., Simane, B., Kalangu, J., Wuta, M., Munishi, P., & Musiyiwa, K. (2017). *Climate change adaptation in Africa.* Springer International Publishing. DOI: 10.1007/978-3-319-49520-0

Lee, G. O., Surkan, P. J., Zelner, J., Olórtegui, M. P., Yori, P. P., Ambikapathi, R., Caulfield, L. E., Gilman, R. H., & Kosek, M. N. (2018). Social connectedness is associated with food security among peri-urban Peruvian Amazonian communities. *SSM - Population Health*, 4, 254–262. DOI: 10.1016/j.ssmph.2018.02.004 PMID: 29854909

Leisner, C. P. (2020). Climate change impacts food security-focus on perennial cropping systems and nutritional value. *Plant Science*, 293, 110412. DOI: 10.1016/j.plantsci.2020.110412 PMID: 32081261

Lemery, J., Knowlton, K., & Sorenson, C. (2021). *Global climate change and human health: from Science to Practice*. John Wiley & Sons.

Magray, M., Hussain, K., Aijax, M., Singh, P., & Nabi, A. (2017). Post-Harvest Management of Vegetable Crops: Introduction. *Journal of Pharmacology and Phytochemistry*, 12(23). Pp12-24.

Margaret, G. R. (2018). Climate Change and Individual. *The American Journal of Comparative Law*, 66(suppl_1), 345–378. DOI: 10.1093/ajcl/avy018

Martin, K. S., Rogers, B. L., Cook, J. T., & Joseph, H. M. (2004). Social capital is associated with a decreased risk of hunger. *Social Science & Medicine*, 58(12), 2645–2654. DOI: 10.1016/j.socscimed.2003.09.026 PMID: 15081212

McLaren, K., & Kelleher, D. (2005). New horizons in organizational development. Au Courant. *Canadian Council for International Cooperation*, 13, 6–9.

Mirzabaev, A. (2013). Impact of Weather Variability and Climate Change on Agricultural Resources in Central Asia. *Zeitschrift für Ausländische Landwirtschaft*, 52(3), 237–252.

Molotoks, A., Smith, P., & Dawson, T. P. (2021). Impacts of land use, population, and climate change on global food security. *Food and Energy Security*, 10(1), e261. DOI: 10.1002/fes3.261

Mujib, R. B., Nausahad, K., & Inayatullah, J. (2007). Post-Harvest Losses in Tomato Crop. A case of Peshawar valley. *Sarhad Journal of Agriculture*, 23(4), 1281.

Musah, B. A., Mumuni, E., Abayomi, O., & Jibrel, M. B. (2013). Effects of floods on the livelihoods and food security of households in the Tolon/kumbumgu district of the Northern region of Ghana. *American Journal of Research Communication*, 1(8), 160–171.

Niang, I., Ruppel, O. C., Abdrabo, M. A., Essel, A., Lennard, C., Padgham, J., & Urquhart, P. (2014). Africa. In Barros, V. R. (Eds.), *Climate Change 2014: Impacts, Adaptation, and Vulnerability* (pp. 1199–1265). University of Cambridge Press.

Nosratabadi, S., Khazami, N., Abdallah, M. B., Lackner, Z., Band, S., Mosavi, A., & Mako, C. (2020). Social capital contributions to food security: A comprehensive literature review. *Foods*, 9(11), 1650. DOI: 10.3390/foods9111650 PMID: 33198127

O'Brien, K. (2010). *Responding to climate change. Integral Theory in Action: Applied, Theoretical, and Constructive Perspectives on the AQAL Model.* State University of New York Press.

O'Brien, K., & Hochachka, G. (2010). Integral adaptation to climate change. *Journal of Integral Theory & Practice*, 5(1), 89–102.

O'Brien, K., & Sygna, L. (2013). Responding to climate change: The three spheres of transformation. Proceedings of Transformation in a Changing Climate, 19-21 June 2013, Oslo, Norway: University of Oslo (pp.16-23).

Osborn, T. J., Jones, P. D., Lister, D. H., Morice, C. P., Simpson, I. R., Winn, J. P., Hogan, E., & Harris, I. C. (2021). Land Surface Air Temperature Variations Across the Globe Updated to 2019: The CRUTEM5 Data Set. *Journal of Geophysical Research. Atmospheres*, 126(2), e2019JD032352. DOI: 10.1029/2019JD032352

Owusu, V., Ma, W., Emuah, D., & Renwick, A. (2021). Perceptions and vulnerability of farming households to climate change in three agroecological zones of Ghana. *Journal of Cleaner Production*, 293, 12. DOI: 10.1016/j.jclepro.2021.126154

Parsons, L. A., Shindell, D., Tigchelaar, M., Zhang, Y., & Spector, J. T. (2021). Increased labor losses and decreased adaptation potential in a warmer world. *Nature Communications*, 12(1), 7286. DOI: 10.1038/s41467-021-27328-y PMID: 34907184

Paulson, D. S. (2008). Wilber's integral philosophy: A summary and critique. *Journal of Humanistic Psychology*, 48(3), 364–388. DOI: 10.1177/0022167807309748

Pereira, L. (2017). *Climate change impacts on agriculture across Africa*. Oxford Research Encyclopedia of Environmental Science.

Pickson, R. B., & Boateng, E. (2022). Climate change: A friend or foe to food security in Africa? *Environment, Development and Sustainability*, 24(3), 1–26. DOI: 10.1007/s10668-021-01621-8 PMID: 34276245

Ray, D.K., West, P.C., Clark, M., Gerber, S.S., Alexander, P.V., & Chattejee, S. (2019). Climate Change has likely Affected Global Food Production. *Plose One* 14(5) e 0217148.

Renzaho, A. M., & Mellor, D. (2010). Food security measurement in cultural pluralism: Missing the point or conceptual misunderstanding? *Nutrition (Burbank, Los Angeles County, Calif.)*, 26(1), 1–9. DOI: 10.1016/j.nut.2009.05.001 PMID: 19804955

Schmidhuber, J., & Tubiello, F. N. (2007). Global food security under climate change. *National Academy of Sciences*, 104(50).

Sen, A. (1982). *Poverty and famines: an essay on entitlement and deprivation*. Oxford University Press.

Sertse, S. F., Khan, N. A., Shah, A. A., Liu, Y., & Naqvi, S. A. A. (2021). Farm households' perceptions and adaptation strategies to climate change risks and their determinants: Evidence from Raya Azebo district, Ethiopia. *International Journal of Disaster Risk Reduction*, 60, 102255. DOI: 10.1016/j.ijdrr.2021.102255

Sirba, H. Y., & Chimdessa, T. B. (2021). Review of Impact of Climate Change on Food Security in Africa. *Int. J. Res. Innov. Earth Sci*, 8, 2394–1375.

Stevanovic, M., Popp, A., Lotze-canpen, H., & Dietrich, J. (2016). The Impact of High-end Climate Change on Agricultural Welfare. *Science Advances*, 12(14), 16–34. DOI: 10.1126/sciadv.1501452 PMID: 27574700

Thamaga-Chitja, J. M., & Tamako, N. (2017). Does social capital play a role in climate change adaptation among smallholder farmers for improving food security and livelihoods? *Journal of Consumer Sciences*, 2, 16–27.

Tigchelaar, M., Cheung, W. W. L., Mohammed, E. Y., Phillips, M. J., Payne, H. J., Selig, E. R., Wabnitz, C. C. C., Oyinlola, M. A., Frölicher, T. L., Gephart, J. A., Golden, C. D., Allison, E. H., Bennett, A., Cao, L., Fanzo, J., Halpern, B. S., Lam, V. W. Y., Micheli, F., Naylor, R. L., & Troell, M. (2021). Compound climate risks threaten aquatic food system benefits. *Nature Food*, 2(9), 673–682. DOI: 10.1038/s43016-021-00368-9 PMID: 37117477

Van Ittersum, M. K., Van Bussel, L. G., Wolf, J., Grassini, P., Van Wart, J., Guilpart, N., ... & Cassman, K. G. (2016). Can sub-Saharan Africa feed itself?. *National Academy of Sciences,* 113(52), 14964-14969.

Webb, P., Coates, J., Frongillo, E. A., Rogers, B., Swindale, A., & Bilinsky, P. (2006). Measuring Household Food Insecurity: Why It's So Important and Yet So Difficult to Do. *The Journal of Nutrition*, 136(5), 1404S–1408S. DOI: 10.1093/jn/136.5.1404S PMID: 16614437

Wing, I. S., De Cian, E., & Mistry, M. N. (2021). Global vulnerability of crop yields to climate change. *Journal of Environmental Economics and Management*, 109, 102462. DOI: 10.1016/j.jeem.2021.102462

Wlokas, H. L. (2008b). The impacts of climate change on food security and health in Southern Africa. *Journal of Energy in Southern Africa*, 19(4), 12–20. DOI: 10.17159/2413-3051/2008/v19i4a3334

Wolfson, J. A., & Leung, C. W. (2020). Food insecurity and COVID-19: Disparities in early effects for US adults. *Nutrients*, 12(6), 1648. DOI: 10.3390/nu12061648 PMID: 32498323

World Bank Group. (2023). What is food security? In World Bank. www.worldbank.org

World Meteorological Organisation (WMO). (2021). *State of the Global Climate in 2020.* World Meteorological Organization.

Wudil, A. H., Usman, M., Rosak-Szyrocka, J., Pilař, L., & Boye, M. (2022). Reversing Years for Global Food Security: A Review of the Food Security Situation in Sub-Saharan Africa (SSA). *International Journal of Environmental Research and Public Health*, 19(22), 14836. DOI: 10.3390/ijerph192214836 PMID: 36429555

Zander, K. K., Baggen, H. S., & Garnett, S. T. (2023). Topic modeling the mobility response to heat and drought. *Climatic Change*, 176(4), 42. DOI: 10.1007/s10584-023-03524-1

Zeleke, T., Beyene, F., Deressa, T., Yousuf, J., & Kebede, T. (2021). Vulnerability of Smallholder Farmers to Climate Change-Induced Shocks in East Hararghe Zone, Ethiopia. *Sustainability (Basel)*, 13(4), 2162. DOI: 10.3390/su13042162

ADDITIONAL READING

Daum, T., & Birner, R. (2017). The neglected governance challenges of agricultural mechanization in Africa – Insights from Ghana. *Food Security*, 9(5), 959–979. DOI: 10.1007/s12571-017-0716-9

Leisner, C. P. (2020). Climate change impacts food security on perennial cropping systems and nutritional value. *Plant Science*, 293, 110412. DOI: 10.1016/j.plantsci.2020.110412 PMID: 32081261

Lesk, C., Anderson, W., Rigden, A., Coast, O., Jägermeyr, J., McDermid, S., Davis, K. F., & Konar, M. (2022). Compound heat and moisture extreme impacts on global crop yields under climate change. *Nature Reviews. Earth & Environment*, 3(12), 872–889. DOI: 10.1038/s43017-022-00368-8

Mugambiwa, S. S. (2018). Adaptation measures to sustain indigenous practices and the use of indigenous knowledge systems to adapt to climate change in Mutoko rural district of Zimbabwe. *Jàambá*, 10(1), 1–9. DOI: 10.4102/jamba.v10i1.388 PMID: 29955251

Owusu, V., Ma, W., Emuah, D., & Renwick, A. (2021). Perceptions and vulnerability of farming households to climate change in three agroecological zones of Ghana. *Journal of Cleaner Production*, 293, 126154. DOI: 10.1016/j.jclepro.2021.126154

Thornton, P. K., van de Steeg, J., Notenbaert, A., & Herrero, M. (2009). The impacts of climate change on livestock and livestock systems in developing countries: A review of what we know and what we need to know. *Agricultural Systems*, 101(3), 113–127. DOI: 10.1016/j.agsy.2009.05.002

World Food Programme (WFP). (2012). *Comprehensive food security and vulnerability analysis: focus on northern Ghana*. United Nations World Food Programme.

KEY TERMS AND DEFINITIONS

Adaptation: The process through which societies make themselves better able to cope with an uncertain future.

Climate Change: Permanent long-term shifts in climate, such as temperature and other weather variables.

Climate Change Adaptation: Adjusting to current or anticipated climate-related consequences.

Climate Resilience: The ability of households to anticipate, prepare, and absorb climate stress without losing their function and structure.

Food Insecurity: A situation, where an economic and social condition at a household level is a limiting factor to accessing adequate food.

Food Security: The condition in which all individuals have constant physical and economic access to sufficient, safe, and nutritious food, enabling them to meet their dietary needs and preferences for a healthy and active life.

Livelihood Diversification: The practice by which households build a diverse range of activities and social support abilities to improve and sustain their living standards.

Conclusion

As countries confront the intricate challenges posed by climate change, it becomes increasingly evident that the associated health risks are both severe and extensive. The findings articulated throughout this book highlight the critical necessity for cohesive strategies that place health at the forefront of climate initiatives. The repercussions of climate change, ranging from elevated temperatures to severe weather phenomena, extend beyond environmental concerns; they represent substantial dangers to the health and welfare of communities across the globe. To effectively tackle these issues, it is essential to embrace a proactive stance that prioritizes resilience, prevention, and adaptation. This approach necessitates the reduction of greenhouse gas emissions and the enhancement of health systems to ensure they are strong and adaptable to the shifting realities of climate-induced health threats. Public health experts, policymakers, and communities must work in unison to develop and execute strategies that safeguard at-risk populations and improve health outcomes amidst the challenges posed by climate change.

The interplay between climate change and health necessitates a thorough examination of the social determinants that influence health outcomes, emphasizing the importance of equitable and inclusive responses. By focusing on the specific needs of marginalized populations, countries can foster a more just and resilient society capable of effectively addressing the challenges posed by climate change. Managing the health implications of climate change transcends mere survival; it presents a unique opportunity to reform countries' systems and practices in pursuit of a healthier and more sustainable future. By integrating health considerations into climate strategies, a positive feedback loop can be established where enhanced health outcomes drive more robust climate action, ultimately benefiting the planet as a whole. As the world progresses, it is essential to embrace a shared vision that acknowledges the vital connection between climate and health, enabling countries to pave the way for a resilient future where both humanity and the environment flourish together.

The ramifications of climate change significantly impact human development and overall well-being. It triggers health crises, affecting both communicable and non-communicable diseases in the short and long term, while also influencing mental health and general wellness. Furthermore, it intensifies adverse effects stemming from various social determinants of health. The stability of societies, the health of populations, and the ability of health systems to cope with increasingly intricate and widespread health challenges are all influenced by climate change. The potential health benefits derived from enhanced climate governance aimed at adaptation and mitigation are often overlooked. Urgent and comprehensive action is required to tackle the rapidly escalating health risks posed by climate change, which threaten economic and social stability. Additionally, the connection between climate change and the transgression of other planetary boundaries presents even more significant threats to health security, warranting serious consideration.

The management of health impacts resulting from climate change necessitates contributions from all levels of government and civil society, as well as collaboration across various academic fields and innovative forms of international cooperation that have previously been difficult to achieve. Engaging local communities in the processes of monitoring, dialogue, advocacy, and support for adaptation strategies will be essential. To effectively mitigate the negative health consequences of climate change, a comprehensive and multidisciplinary strategy must encompass at least three tiers of action. Initially, it is imperative to implement policies aimed at reducing carbon emissions and enhancing carbon bio-sequestration to decelerate global warming and ultimately stabilize temperature levels. Subsequently, measures should be taken to address the connections between climate change and disease. It is also crucial to establish robust public health systems capable of managing the adverse health outcomes associated with these changes.

The impact of climate change will be most pronounced on individuals who have limited access to global resources and have played a minimal role in its emergence. In the absence of effective mitigation and adaptation strategies, health disparities are likely to widen, particularly due to adverse effects on the social determinants of health within the most impoverished communities. A significant challenge faced globally is the capacity of health systems to adequately address both the direct and indirect health consequences of climate change, a situation that is particularly acute in numerous low- and middle-income nations that grapple with disorganized, inefficient, and underfunded health infrastructures. To address these challenges, many countries will need to allocate additional investments and resources towards strengthening their health systems. Furthermore, the health risks posed by climate change underscore the essential need for enhanced governance, population-centered planning, and the effective management of limited resources.

An effective strategy to mitigate the negative impacts of climate change necessitates action on at least three distinct levels. Initially, it is essential to implement policies aimed at decreasing carbon emissions, which will help to decelerate global warming and ultimately stabilize temperature fluctuations. Subsequently, it is crucial to address the connections between climate change and its detrimental effects on health. Robust public health systems must be established to manage the adverse health outcomes that arise. This book presents numerous pressing and complex questions for a wide array of stakeholders, including policymakers, civil servants, scholars, health practitioners, non-governmental organizations, advocacy groups, and local communities. Climate change poses a significant threat to global health in the 21st century, necessitating the emergence of a new public health movement that is both multidisciplinary and multisectoral, fostering coordinated thought and action among governments, international organizations, NGOs, and academic entities. Any adaptation efforts must be complemented by a commitment to primary mitigation, specifically the reduction of greenhouse gas emissions.

Recognition by governments and electorates of the significant health consequences of climate change should facilitate advocacy and political initiatives aimed at both mitigation and adaptation efforts. It is imperative to empower impoverished nations, as well as local governments and communities worldwide, to comprehend the implications of climate change and to take proactive measures. Health professionals and academic scholars play a crucial catalytic role in this process. Interdisciplinary teams from higher education institutions can effectively engage with community leaders, civil society organizations, and students in these critical discussions. Furthermore, there is a pressing need for innovative financing mechanisms that foster global connections between developed and developing nations, promoting a reciprocal dialogue. Developed countries can contribute by enhancing the capacity for high-quality research and data collection in developing nations while developing countries can provide insights that help developed nations better understand sustainability and low-carbon lifestyles.

Climate change must be woven into the fabric of contemporary discourse and considered in all governance initiatives. Advocacy movements must prioritize the health impacts of climate change, ensuring that these issues are prominently featured on the agendas of research and development funders, philanthropists, academic journals, scientific conferences, professional gatherings, and educational curricula at all levels. Academics are encouraged to spearhead advocacy efforts within their respective domains of influence. The prevalent fragmentation within health systems, often neglecting long-term sustainability, should transition towards the establishment of integrated, population-centered, and grassroots health planning. Health systems ought to function not merely as conduits for clinical service delivery but also as a robust foundation for an effective public health response to the myriad health threats posed by climate change. Achieving this goal will necessitate a greater focus on the organizational and managerial shortcomings within health ministries, including those at subnational levels. Furthermore, long-term strategies and investments are essential to enhance the clinical and managerial capacities of health systems. In certain nations, it will also be crucial to regulate and organize the private sector to better align its resources with public health interests. Currently, many countries lack a coherent, long-term, and sustainable development strategy for their health systems, and this situation must be addressed.

Climate change is currently occurring and has a direct impact on fundamental health necessities such as access to clean air and water, adequate nutrition, and proper housing. This phenomenon introduces new obstacles and financial burdens in the management of infectious diseases, particularly those that are sensitive to variations in temperature and precipitation, such as cholera and various diarrheal illnesses, as well as vector-borne diseases like malaria, dengue, and schistosomiasis. Changes in the environment serve as significant risk factors for the displacement of populations and their health. The implications of climate change pose a threat to the advancements made by the global public health sector in combating numerous diseases, while simultaneously complicating the efforts of disaster risk management professionals in addressing natural, biological, and social emergencies.

Enhancing health systems is essential for effectively predicting, detecting, preparing for, and responding to climate-related risks and disasters. This can be achieved by establishing climate-informed surveillance and early-warning systems, expanding the capacity of the health workforce in the context of climate and health, and ensuring that healthcare infrastructure is resilient to climate impacts. Additionally, it is important to transition health systems towards low-carbon, high-quality service delivery, which can be facilitated by utilizing clean, renewable energy sources for infrastructure and transportation, as well as adopting low-carbon medicines and equipment. This transition is vital, as the health sector not only faces significant vulnerabilities due to climate change but also plays a substantial role in exacerbating the issue, contributing approximately 5% of global greenhouse gas emissions. To effectively tackle the underlying causes of climate change and its health implications, it is necessary to collaborate across various sectors to enhance initiatives in areas such as One Health, water, sanitation, and hygiene, as well as energy efficiency.

Immediate and decisive measures are essential to prepare for and effectively address the health threats posed by climate change. In addition to strengthening existing health protection initiatives, public health organizations and institutions must enhance their ability to recognize and link evolving disease patterns to climate change. They should also utilize climate forecasts to predict potential health outcomes that may shift under various climate scenarios. The accumulation of greenhouse gases in the atmosphere continues to rise at an alarming pace, leading to increased surface temperatures and climate instability. The health consequences of climate change may manifest as heightened rates of morbidity and mortality associated with deteriorating cardiopulmonary conditions, exacerbated allergies, and an elevated risk of infectious diseases and mental health disorders, including anxiety, depression, and post-traumatic stress disorder resulting from extreme weather events. Family physicians should counsel patients on reducing their exposure to air pollution, which is intensified by rising temperatures.

The utilization of self-monitoring devices, the monitoring of local weather conditions, and the awareness of events such as wildfires can serve as important indicators for patients regarding poor air quality. Particularly vulnerable individuals should refrain from engaging in strenuous outdoor activities and either remain indoors or don protective N95 masks when air quality levels are deemed hazardous. Healthcare providers can educate patients on the early signs of heat-related illnesses and recommend appropriate hydration and cooling strategies during periods of high temperatures. Additionally, physicians must recognize the signs and symptoms associated with vector-borne diseases to facilitate prompt treatment and mitigate their transmission. Furthermore, when advising patients, physicians should consider the health benefits associated with active transportation and plant-based diets to climate. By integrating counseling, public health principles, and advocacy into their practice, physicians can significantly contribute to raising awareness about climate change and influencing related policies.

Compilation of References

10. Kickbusch, I., Hein, W., & Silberschmidt, G. (2024). Addressing global health governance challenges through a new mechanism: The proposal for a Committee C of the World Health Assembly. *The Journal of Law, Medicine & Ethics*, 38(3), 550–563. DOI: 10.1111/j.1748-720X.2010.00511.x PMID: 20880240

Abbass, K., Qasim, M. Z., Song, H., Murshed, M., Mahmood, H., & Younis, I. (2022). A review of the global climate change impacts, adaptation, and sustainable mitigation measures. *Environmental Science and Pollution Research International*, 29(28), 42539–42559. DOI: 10.1007/s11356-022-19718-6 PMID: 35378646

Abedin, M. A., Habiba, U., & Shaw, R. (2013). Gender and climate change: Impacts and coping mechanisms of women and special vulnerable groups. *Climate change adaptation actions in Bangladesh,* pp.165-184.

Acierno, R., Ruggiero, K. J., Galea, S., Resnick, H. S., Koenen, K., Roitzsch, J., de Arellano, M., Boyle, J., & Kilpatrick, D. G. (2007). Psychological sequelae resulting from the 2004 Florida hurricanes: Implications for postdisaster intervention. *American Journal of Public Health*, 97(1, Supplement_1), 103–108. DOI: 10.2105/AJPH.2006.087007 PMID: 17413067

Adger, W. N., Hughes, T. P., Folke, C., Carpenter, S. R., & Rockström, J. (2005). Social-Ecological Resilience to Coastal Disasters. *Science*, 309(5737), 1036–1039. DOI: 10.1126/science.1112122 PMID: 16099974

Afful, D. B. (2016). Public Extension Agents Need For New Competences: Evidence from a Climate Variability Study in Limpopo province of South Africa. *Journal of Agricultural Extension*, 44(2), 59–70.

Agbola, T., & Ojeleye (2007). Climate change and food crop production in Ibadan, Nigeria. *African Crop Science Conference Proceedings*, 8(1), 1423-1433.

Aggarwal, P., Rana, M., & Akhai, S. (2022). Briefings on e-waste hazard until COVID era in India. *Materials Today: Proceedings*, 71(2), 389–393. DOI: 10.1016/j.matpr.2022.09.507

Aguilar, L., Araujo, A., & Quesada-Aguilar, A. (2007). Gender and climate change. *IUCN (International Union for the Conservation of Nature) Fact Sheet,* http://www. genderandenvironment. org/admin/admin_biblioteca/documentos/Fact*sheet%20ClimateChange. pdf.*

Agyapong, B., Shalaby, R., Eboreime, E., Obuobi-Donkor, G., Owusu, E., Adu, M. K., Mao, W., Oluwasina, F., & Agyapong, V. I. O. (2022). Cumulative trauma from multiple natural disasters increases mental health burden on residents of Fort McMurray. *European Journal of Psychotraumatology*, 13(1), 1–20. DOI: 10.1080/20008198.2022.2059999 PMID: 35599978

Ahmad, N. (2012). Gender and Climate Change in Bangladesh: The Role of Institutions in Reducing Gender Gaps in Adaptation Program. *Social Development Papers of World Bank, 126*(67820). https://hdl.handle.net/10986/27416

Ahmed, S., & Fajber, E. (2009). Engendering adaptation to climate variability in Gujarat, India. *Gender and Development*, 17(1), 33–50. DOI: 10.1080/13552070802696896

Akhai, S. (2024). Trends and Environmental Impact of Paper Consumption: A Prognostic Scenario for the Indian Market by 2030-A Case Study. In *International Conference on Interdisciplinary Approaches in Civil Engineering for Sustainable Development* (Vol. 464, pp. 11-18). Springer Nature Singapore. DOI: 10.1007/978-981-97-0910-6_2

Akhai, S., & Khang, A. (2024). Energy Efficiency and Human Comfort: AI and IoT Integration in Hospital HVAC Systems. *Medical Robotics and AI-Assisted Diagnostics for a High-Tech Healthcare Industry*, 93-108.

Akhai, S., Mala, S., & Jerin, A. A. (2020). Apprehending Air Conditioning Systems in Context to COVID-19 and Human Health: A Brief Communication. *International Journal of Healthcare Education & Medical Informatics (ISSN: 2455-9199), 7*(1&2), 28-30.

Akhai, S., Mala, S., & Jerin, A. A. (2021). Understanding whether air filtration from air conditioners reduces the probability of virus transmission in the environment. *Journal of Advanced Research in Medical Science & Technology (ISSN: 2394-6539), 8*(1), 36-41.

Akhai, S., Bansal, S. A., & Singh, S. (2020). A critical review of thermal insulators from natural materials for energy saving in buildings. *Journal of Critical Reviews*, 7(19), 278–283.

Akhai, S., Singh, V. P., & John, S. (2016). Human performance in industrial design centers with small unit air conditioning systems. *Journal of Advanced Research in Production Industrial Engineering*, 3(2), 5–11.

Akhai, S., Singh, V. P., & John, S. (2016). Investigating Indoor Air Quality for the Split-Type Air Conditioners in an Office Environment and Its Effect on Human Performance. *Journal of Mechanical Civil Engineering*, 13(6), 113–118.

Akhai, S., Thareja, P., & Singh, V. P. (2017). Assessment of Indoor Environment Health Sustenance in Air Conditioned Class Rooms. *Journal of Advanced Research in Civil and Environmental Engineering*, 4(1&2), 1–9. DOI: 10.24321/2393.8307.201701

Alam, K., & Rahman, M. H. (2014). Women in natural disasters: A case study from the southern coastal region of Bangladesh. *International Journal of Disaster Risk Reduction*, 8, 68–82. DOI: 10.1016/j.ijdrr.2014.01.003

Alam, M. Md., Sowar, C., Talib, B., Mokhtar, T. M. E. (2011). Climate change adaptation policy in Malaysia: Issues for agricultural sector. *African Journal of Agricultural Research*, 7(9), 1368–1373.

Alanazi, Z. S. A., Essa, A. M. A., Alalawi, Z. A. M., Albalawi, F. F. M., Almohammadi, Y. M. A., Alnajran, M. A. M., & Alanazi, F. M. F. (2022). Exploring The Nexus Of Global Climate Changes And Infectious Disease Dynamics: A Depth Review Study. *Journal of Namibian Studies: History Politics Culture*, 32, 921–960.

Alazazmeh, A., & Asif, M. (2021). Commercial building retrofitting: Assessment of improvements in energy performance and indoor air quality. *Case Studies in Thermal Engineering*, 26, 100946. DOI: 10.1016/j.csite.2021.100946

Albatayneh, A., Jaradat, M., Al-Omary, M., & Zaquot, M. (2021). Evaluation of coupling PV and air conditioning vs. Solar cooling systems—Case study from Jordan. *Applied Sciences (Basel, Switzerland)*, 11(2), 511. DOI: 10.3390/app11020511

Albrecht, G. (2005). 'Solastalgia.' A new concept in health and identity. *PAN: philosophy activism nature*, 3, 41-55.

Albrecht, G. L., Ravaud, J. F., & Stiker, H. J. (2001). L'émergence des disability studies: état des lieux et perspectives. The emergence of disability studies: state of places and of perspectives. *Sciences Sociales et Sante*, 19(4), 43–73. DOI: 10.3406/sosan.2001.1535

Albrecht, G., Sartore, G., Connor, L., Higginbotham, N., Freeman, S., Kelly, B., Stain, H., Tonna, A., & Pollard, G. (2007). Solastalgia: The distress caused by environmental change. *Australasian Psychiatry*, 15(1_suppl, Suppl 1), S95–S98. DOI: 10.1080/10398560701701288 PMID: 18027145

Alexander, D. (2013). Resilience and Disaster Risk Reduction: An Etymological Journey. *Natural Hazards and Earth System Sciences*, •••, 13.

Alexander, S. M., Armitage, D., & Charles, A. (2015). Social networks and transitions to co-management in Jamaican marine reserves and small-scale fisheries. *Global Environmental Change*, 35, 213–225. DOI: 10.1016/j.gloenvcha.2015.09.001

Al-Shaar, L., Vercammen, K., Lu, C., Richardson, S., Tamez, M., & Mattei, J. (2017). Health effects and public health concerns of energy drink consumption in the United States: A mini-review. *Frontiers in Public Health*, 5, 286776. DOI: 10.3389/fpubh.2017.00225 PMID: 28913331

Alston, M. (2014). Gender mainstreaming and climate change. *Women's Studies International Forum, 47*(Part B), 287–294. DOI: 10.1016/j.wsif.2013.01.016

Amenuveve, V., (2010). Northern Floods Claim 17 Lives: But Residents Still Adamant. *Daily Graphic*, Monday, 13, September, 2010.

American Psychological Association (APA). (2010). *Psychology and Global Climate Change: Addressing a Multi-faceted Phenomenon and Set of Challenges*. APA.

American Public Health Association (APHA). (2019). Addressing the Impacts of Climate Change on Mental Health and Well-Being. www.apha.org

Andersen, Z. J., Vicedo-Cabrera, A. M., Hoffmann, B., & Melén, E. (2023). Climate change and respiratory disease: Clinical guidance for healthcare professionals. *Breathe (Sheffield, England)*, 19(2), 220222. DOI: 10.1183/20734735.0222-2022 PMID: 37492343

Anderson, A. (2010). *Combating climate change through quality education*. Brookings Global Economy and Development., https://www.brookings.edu/articles/combating-climate-change-through-quality-education

Anderson, M. D., & Cook, J. T. (1999). Community food security: Practice in need of theory? *Agriculture and Human Values*, 16(2), 141–150. DOI: 10.1023/A:1007580809588

Antwi-Agyei, P., & Nyantakyi-Frimpong, H. (2021). Evidence of climate change coping and adaptation practices by smallholder farmers in northern Ghana. *Sustainability (Basel)*, 13(3), 1308. DOI: 10.3390/su13031308

Arcanjo, M. (2019). Eco-anxiety: mental health impacts of environmental disasters and climate change. *A Climate Institute Publication. Washington.*

Arif, S., Taweekun, J., Ali, H. M., Yanjun, D. A. I., & Ahmed, A. (2021). Feasibility study and economic analysis of grid connected solar powered net zero energy building (NZEB) of shopping mall for two different climates of Pakistan and Thailand. *Case Studies in Thermal Engineering*, 26, 101049. DOI: 10.1016/j.csite.2021.101049

Arora-Jonsson, S. (2011). Virtue and vulnerability: Discourses on women, gender, and climate change. *Global Environmental Change*, 21(2), 744–751. DOI: 10.1016/j.gloenvcha.2011.01.005

Aryankhesal, A., Pakjouei, S., & Kamali, M. (2018). Safety Needs of People with Disabilities During Earthquakes. *Disaster Medicine and Public Health Preparedness*, 12(5), 615–621. DOI: 10.1017/dmp.2017.121 PMID: 28965529

Ashraf, S. N., & Singh, A. K. (2022). Implications of appropriate technology and farm inputs in the agricultural sector of Gujarat: Empirical analysis based on primary data. *Agricultural Economics Research Review*, 35(2), 1–18. DOI: 10.5958/0974-0279.2022.00031.3

Asian Development Bank, United Nations, and World Bank. (2005). Maldives Tsunami: impact and recovery. Joint needs Assessment by World Bank–ADB–UN System, Male.

Asim, N., Badiei, M., Mohammad, M., Razali, H., Rajabi, A., Chin Haw, L., & Jameelah Ghazali, M. (2022). Sustainability of heating, ventilation and air-conditioning (HVAC) systems in buildings—An overview. *International Journal of Environmental Research and Public Health*, 19(2), 1016. DOI: 10.3390/ijerph19021016 PMID: 35055838

Aslan, M., & Yıldız, A. (2019). How blameless are hospitals in climate change? An example of a province in Turkey. *Environ Socio-economic Stud*, 7(4), 45–53. DOI: 10.2478/environ-2019-0023

Ayejoto, D. A., Agbasi, J. C., Nwazelibe, V. E., Egbueri, J. C., & Alao, J. O. (2023). Understanding the connections between climate change, air pollution, and human health in Africa: Insights from a literature review. *Journal of Environmental Science and Health. Part C, Environmental Carcinogenesis & Ecotoxicology Reviews*, 41(3-4), 77–120. PMID: 37880976

Aziz, N., Adman, M. A., Suhaimi, N. S., Misbari, S., Alias, A. R., Abd Aziz, A., & Khan, M. M. H. (2023, February). Indoor air quality (IAQ) and related risk factors for sick building syndrome (SBS) at the office and home: A systematic review. [). IOP Publishing.]. *IOP Conference Series. Earth and Environmental Science*, 1140(1), 012007. DOI: 10.1088/1755-1315/1140/1/012007

Bahroun, Z., Anane, C., Ahmed, V., & Zacca, A. (2023). Transforming education: A comprehensive review of generative artificial intelligence in educational settings through bibliometric and content analysis. *Sustainability (Basel)*, 15(17), 12983. DOI: 10.3390/su151712983

Balarajan, Y., Selvaraj, S., & Subramanian, S. V. (2021). Health care and equity in India. *Lancet*, 377(9764), 505–515. DOI: 10.1016/S0140-6736(10)61894-6 PMID: 21227492

Bandi, A., Adapa, P. V. S. R., & Kuchi, Y. E. V. P. K. (2023). The power of generative ai: A review of requirements, models, input–output formats, evaluation metrics, and challenges. *Future Internet*, 15(8), 260. DOI: 10.3390/fi15080260

Bandla, S., Nappinnai, N. R., & Gopalasamy, S. (2019). Psychiatric morbidity in December 2015 flood-affected population in Tamil Nadu, India. *The International Journal of Social Psychiatry*, 65(4), 338–344. DOI: 10.1177/0020764019846166 PMID: 31068043

Banholzer, S., Kossin, J., & Donner, S. (2014). The impact of climate change on natural disasters. *Reducing Disaster: Early Warning Systems for Climate Change*, In *Springer eBooks* (pp. 21–49). DOI: 10.1007/978-94-017-8598-3_2

Barasa, E. W., Cloete, K., & Gilson, L. (2017). From Bouncing Back, To Nurturing Emergence: Reframing The Concept Of Resilience In Health Systems Strengthening. *Health Policy and Planning*, 32(suppl_3), 91–94. DOI: 10.1093/heapol/czx118 PMID: 29149319

Barkin, J. L., Philipsborn, R. P., Curry, C. L., Upadhyay, S., Geller, P. A., Pardon, M., Dimmock, J., Bridges, C. C., Sikes, C. A., Kondracki, A. J., & Buoli, M. (2024). Climate change is an emerging threat to perinatal mental health. *Journal of the American Psychiatric Nurses Association*, 30(3), 683–689. DOI: 10.1177/10783903221139831 PMID: 36482670

Barreca, A., & Schaller, J. (2019). The impact of high ambient temperatures on delivery timing and gestational lengths. *Nature Climate Change*, 10(1), 77–82. DOI: 10.1038/s41558-019-0632-4

Barrett, C. B. (2010). Measuring Food Insecurity. *Science*, 327(5967), 825–828. DOI: 10.1126/science.1182768 PMID: 20150491

Bastami, F., Mohammadi, R., Piri, Z. A., Valipour, E., Ahmadi, P., Almasian, M., & Motlagh, S. N. (2024). prevalence of post-traumatic stress disorder and its relationship with coping strategies among flood victims: Evidence from Iran. *Journal of Affective Disorders Reports*, 15(1), 10–17. DOI: 10.1016/j.jadr.2023.100704

Basu, R., Sarovar, V., & Malig, B. J. (2016). Association between high ambient temperature and risk of stillbirth in California. *American Journal of Epidemiology*, 183(10), 894–901. DOI: 10.1093/aje/kwv295 PMID: 27037268

Batalden, M., Batalden, P., Margolis, P., Seid, M., Armstrong, G., Opipari-Arrigan, L., & Hartung, H. (2023). Coproduction of healthcare service. *BMJ Quality & Safety*. PMID: 26376674

Batterbury, S. P. J. (2008). Anthropology and global warming: The need for environmental engagement. *The Australian Journal of Anthropology*, 19(1), 62–65. DOI: 10.1111/j.1835-9310.2008.tb00108.x

Baudon, P., & Jachens, L. (2021). A scoping review of interventions for the treatment of eco-anxiety. *International Journal of Environmental Research and Public Health*, 18(18), 9636. DOI: 10.3390/ijerph18189636 PMID: 34574564

Bazga, B. (2015). Food security component of sustainable development—Prospects and challenges in the future. *Procedia Economics and Finance*, 32(1), 1075–1082. DOI: 10.1016/S2212-5671(15)01570-1

Bekkar, B., Pacheco, S., Basu, R., Basu, R., & Denicola, N. (2020). Association of air pollution and heat exposure with preterm birth, low birth weight, and stillbirth in the US: A systematic review. *JAMA Network Open*, 3(6), e208243. Advance online publication. DOI: 10.1001/jamanetworkopen.2020.8243 PMID: 32556259

Bell, J. E., Brown, C. L., Conlon, K., Herring, S., Kunkel, K. E., Lawrimore, J., Luber, G., Schreck, C., Smith, A., & Uejio, C. (2018). Changes in extreme events and the potential impacts on human health. *Journal of the Air & Waste Management Association*, 68(4), 265–287. DOI: 10.1080/10962247.2017.1401017 PMID: 29186670

Béné, C., Wood, R. G., Newsham, A., & Davies, M. (2012). Resilience: New Utopia or New Tyranny? Reflection about the Potentials And Limits Of The Concept Of Resilience in Relation To Vulnerability Reduction Programmes. *IDS Working Paper*, 2012(405), 1–61. DOI: 10.1111/j.2040-0209.2012.00405.x

Benevolenza, M. A., & DeRigne, L. (2018). The impact of climate change and natural disasters on vulnerable populations: A systematic review of literature. *Journal of Human Behavior in the Social Environment*, 29(2), 266–281. DOI: 10.1080/10911359.2018.1527739

Bergwerk, M., Gonen, T., Lustig, Y., Amit, S., Lipsitch, M., Cohen, C., Mandelboim, M., Levin, E. G., Rubin, C., Indenbaum, V., Tal, I., Zavitan, M., Zuckerman, N., Bar-Chaim, A., Kreiss, Y., & Regev-Yochay, G. (2021). Covid-19 breakthrough infections in vaccinated health care workers. *The New England Journal of Medicine*, 385(16), 1474–1484. DOI: 10.1056/NEJMoa2109072 PMID: 34320281

Berkes, F., & Jolly, D. (2001). Adapting to Climate Change: Social-Ecological Resilience in a Canadian Western Arctic Community. *Conservation Ecology*, 5(2), art18. DOI: 10.5751/ES-00342-050218

Berry, H. L., Bowen, K., & Kjellstrom, T. (2010). Climate change and mental health: A causal pathways framework. *International Journal of Public Health*, 55(2), 123–132. DOI: 10.1007/s00038-009-0112-0 PMID: 20033251

Berry, H. L., Kelly, B. J., Hanigan, I. C., Coates, J. H., McMichael, A. J., Welsh, J. A., & Kjellstrom, T. (2008). *Rural mental health impacts of climate change. Commissioned report for the Garnaut Climate Change Review*. The Australian National University.

Berry, H. L., Waite, T., Dear, K., Capon, A., & Murray, V. (2018). The case for systems thinking about climate change and mental health. *Nature Climate Change*, 8(4), 282–290. DOI: 10.1038/s41558-018-0102-4

Besir, A. B., & Cuce, E. (2018). Green roofs and facades: A comprehensive review. *Renewable & Sustainable Energy Reviews*, 82, 915–939. DOI: 10.1016/j.rser.2017.09.106

Bhowmik, J., Irfanullah, H. M., & Selim, S. A. (2021). Empirical evidence from Bangladesh of assessing climate hazard-related loss and damage and the state of adaptive capacity to address them. *Climate Risk Management*, 31, 100273. DOI: 10.1016/j.crm.2021.100273

Blanchet, K., Nam, S. L., Ramalingam, B., & Pozo-Martin, F. (2017). Governance And Capacity To Manage Resilience Of Health Systems: Towards A New Conceptual Framework. *International Journal of Health Policy and Management*, 6(8), 431–435. DOI: 10.15171/ijhpm.2017.36 PMID: 28812842

Bodeker, G. (2023). *WHO global atlas of traditional, complementary, and alternative medicine* (Vol. 1). World Health Organization.

Bodeker, G., & Kronenberg, F. (2022). A public health agenda for traditional, complementary, and alternative medicine. *American Journal of Public Health*, 92(10), 1582–1591. DOI: 10.2105/AJPH.92.10.1582 PMID: 12356597

Bodin, Ö., & Crona, B. I. (2009). The role of social networks in natural resource governance: What relational patterns make a difference? *Global Environmental Change*, 19(3), 366–374. DOI: 10.1016/j.gloenvcha.2009.05.002

Bodin, P., & Wiman, B. (2004). Resilience And Other Stability Concepts In Ecology: Notes On Their Origin, Validity, And Usefulness. *Ess Bulletin*, 2, 33–43.

Boluda-Verdú, I., Senent-Valero, M., Casas-Escolano, M., Matijasevich, A., & Pastor-Valero, M. (2022). Fear for the future: Eco-anxiety and health implications, a systematic review. *Journal of Environmental Psychology*, 84, 101904. DOI: 10.1016/j.jenvp.2022.101904

Bonti-Ankomah, S. (2001). *Addressing food insecurity in South Africa, The National Institute for Economic Policy.* Paper presented at the SARPN conference on Land Reform and Poverty Alleviation in Southern Africa. Pretoria.

Bourque, F., & Cunsolo, W.A. (2014). Climate change: The next challenge for public mental health? *Int Rev Psychiatry.* 19;26(4):415–22.

Bowman, S. (2023). Impact of electronic health record systems on information integrity: Quality and safety implications. *Perspectives in Health Information Management*, 10(Fall). PMID: 24159271

Božić, J., Ilić, P., & Ilić, S. (2019). Indoor air quality in the hospital: The influence of heating, ventilating and conditioning systems. *Brazilian Archives of Biology and Technology*, 62, e19180295. DOI: 10.1590/1678-4324-2019180295

Brewer, E., Brueggemann, B., Hetrick, N., & Yergeau, M. (2012). Introduction, background, and history. In Brueggemann, B. J. (Ed.), *Arts and Humanities* (pp. 1–62). SAGE Publications. DOI: 10.4135/9781452218335.n1

Brisenden, S. (1986). Independent Living and the Medical Model of Disability. *Disability, Handicap & Society*, 1(2), 173–178. DOI: 10.1080/02674648666780171

Brosse, M., Benateau, S., Gaudard, A., Stamm, C., & Altermatt, F. (2022). The importance of indirect effects of climate change adaptations on alpine and pre-alpine freshwater systems. *Ecological Solutions and Evidence*, 3(1), e12127. DOI: 10.1002/2688-8319.12127

Bruinsma, J. (Ed.). (2003). *World agriculture: towards 2015/2030.* FAO: Earthscan.

Bryan-Kinns, N., Banar, B., Ford, C., Reed, C. N., Zhang, Y., Colton, S., & Armitage, J. (2023). Exploring XAI for the arts: Explaining latent space in generative music. *arXiv preprint arXiv:2308.05496*.

Bryant, R. A., Waters, E., Gibbs, L., Gallagher, H. C., Pattison, P., Lusher, D., MacDougall, C., Harms, L., Block, K., Snowdon, E., Sinnott, V., Ireton, G., Richardson, J., & Forbes, D. (2014). Psychological outcomes following the Victorian Black Saturday bushfires. *The Australian and New Zealand Journal of Psychiatry*, 48(7), 634–643. DOI: 10.1177/0004867414534476 PMID: 24852323

Budhwar, P., Chowdhury, S., Wood, G., Aguinis, H., Bamber, G. J., Beltran, J. R., Boselie, P., Lee Cooke, F., Decker, S., DeNisi, A., Dey, P. K., Guest, D., Knoblich, A. J., Malik, A., Paauwe, J., Papagiannidis, S., Patel, C., Pereira, V., Ren, S., & Varma, A. (2023). Human resource management in the age of generative artificial intelligence: Perspectives and research directions on ChatGPT. *Human Resource Management Journal*, 33(3), 606–659. DOI: 10.1111/1748-8583.12524

Buliva, E., Elhakim, M., Tran Minh, N. N., Elkholy, A., Mala, P., Abubakar, A., & Malik, S. M. M. R. (2017). Emerging and reemerging diseases in the World Health Organization (WHO) Eastern Mediterranean Region—Progress, challenges, and WHO initiatives. *Frontiers in Public Health*, 5, 276. DOI: 10.3389/fpubh.2017.00276 PMID: 29098145

Burke, M., González, F., Baylis, P., Heft-Neal, S., Baysan, C., Basu, S., & Hsiang, S. (2018). Higher temperatures increase suicide rates in the United States and Mexico. *Nature Climate Change*, 8(8), 723–729. DOI: 10.1038/s41558-018-0222-x

Burroughs, H. E., & Hansen, S. J. (2020). *Managing indoor air quality*. River Publishers. DOI: 10.1201/9781003151654

Burton, I., Huq, S., Lim, B., Pilifosova, O., & Schipper, E. L. (2002). From Impacts Assessment To Adaptation Priorities: The Shaping Of Adaptation Policy. *Climate Policy*, 2(2-3), 145–159. DOI: 10.3763/cpol.2002.0217

Byrne, M. (2023). The disruptive impacts of next generation generative artificial intelligence. *CIN: Computers, Informatics. Nursing*, 41(7), 479–481. PMID: 37417716

Callaghan, W. M., Rasmussen, S. A., Jamieson, D. J., Ventura, S. J., Farr, S. L., Sutton, P. D., & Posner, S. F. (2007). Health concerns of women and infants in times of natural disasters: Lessons learned from Hurricane Katrina. *Maternal and Child Health Journal*, 11(4), 307–311. DOI: 10.1007/s10995-007-0177-4 PMID: 17253147

Callaway, M. V., Connor, S. R., & Foley, K. M. (2018). World Health Organization public health model: A roadmap for palliative care development. *Journal of Pain and Symptom Management*, 55(2), S6–S13. DOI: 10.1016/j.jpainsymman.2017.03.030 PMID: 28801003

Cammarano, D., Valdivia, R. O., Beletse, Y. G., Durand, W., Crespo, O., Tesfuhuney, W. A., Jones, M. R., Walker, S., Mpuisang, T. N., Nhemachena, C., Ruane, A. C., Mutter, C., Rosenzweig, C., & Antle, J. (2020). Integrated assessment of climate change impacts on crop productivity and income of commercial maize farms in northeast South Africa. *Food Security*, 12(3), 659–678. DOI: 10.1007/s12571-020-01023-0

Canelón, S. P., & Boland, M. R. (2020). A systematic literature review of factors affecting the timing of Menarche: The potential for climate change to impact women's health. *International Journal of Environmental Research and Public Health*, 17(5), 1703. DOI: 10.3390/ijerph17051703 PMID: 32150950

Carleton, T. A. (2017). Crop-damaging temperatures increase suicide rates in India. *Proceedings of the National Academy of Sciences of the United States of America*, 114(33), 8746–8751. DOI: 10.1073/pnas.1701354114 PMID: 28760983

Cénat, J. M., & Derivois, D. (2014). Assessment of prevalence and determinants of posttraumatic stress disorder and depression symptoms in adults survivors of earthquake in Haiti after 30 months. *Journal of Affective Disorders*, 159(1), 111–117. DOI: 10.1016/j.jad.2014.02.025 PMID: 24679398

Cengiz, M., Kayri, İ., & Aydın, H. (2024). A collated overview on the evaporative cooling applications for photovoltaic modules. *Renewable & Sustainable Energy Reviews*, 197, 114393. DOI: 10.1016/j.rser.2024.114393

Center for Women's Mood Disorders. (2022). Perinatal mood and anxiety disorders. https://www.med.unc.edu/psych/ wmd/resources/mood-disorders/perinatal/

Centers for Disease Control and Prevention. (2020). Disability and Health Related Conditions. www.cdc.gov

Centre for Disease Control and Prevention (CDC). (2021). *Climate Effects on Health*.www.cdc.gov

Cerrato, G., & Fumo, N. (2024). Performance of photocatalytic oxidizing air cleaners in different experimental setups–a review. *Energy Reports*, 11, 1412–1419. DOI: 10.1016/j.egyr.2024.01.009

Chafe, Z. A., Brauer, M., Klimont, Z., Van Dingenen, R., Mehta, S., Rao, S., Riahi, K., Dentener, F., & Smith, K. R. (2014). Household cooking with solid fuels contributes to ambient PM2.5 air pollution and the burden of disease. *Environmental Health Perspectives*, 122(12), 1314–1320. DOI: 10.1289/ehp.1206340 PMID: 25192243

Chapin, F. S., Peterson, G., Berkes, F., Callaghan, T. V., Anglestam, P., & Apps, M.. (2004). Resilience and vulnerability of northern regions to social and environmental change. *Ambio*, 33(6), 344–349. DOI: 10.1579/0044-7447-33.6.344 PMID: 15387072

Chartier, Y. (Ed.). (2024). *Safe management of wastes from health-care activities*. World Health Organization.

Charuvastra, A., & Cloitre, M. (2008). Social bonds and posttraumatic stress disorder. *Annual Review of Psychology*, 59(1), 301–328. DOI: 10.1146/annurev.psych.58.110405.085650 PMID: 17883334

Chase, S., Kimberly, F., & Christian, M. (2019). Climate Change and Food Security: A Test of U.S leadership in a Fragile world. Center for strategic and international studies. Brief Report). https://www.csis.org/analysis/climate

Chaudhary, M. T., & Piracha, A. (2021). Natural Disasters—Origins, Impacts, management. *Encyclopedia*, 1(4), 1101–1131. DOI: 10.3390/encyclopedia1040084

Chaudhury, A. S., Thornton, T. F., Helfgott, A., Ventresca, M. J., & Sova, C. (2017). Ties that bind: Local networks, communities and adaptive capacity in rural Ghana. *Journal of Rural Studies*, 53, 214–228. DOI: 10.1016/j.jrurstud.2017.05.010

Cheng, Z., Lee, D., & Tambe, P. (2022). Innovae: Generative ai for understanding patents and innovation. *Available atSSRN* 3868599. DOI: 10.2139/ssrn.3868599

Chen, J., Li, X., Martel, J. L., Brissette, F. P., Zhang, X. J., & Frei, A. (2021). The relative importance of internal climate variability versus anthropogenic climate change in global climate change. *Journal of Climate*, 34(2), 465–478. DOI: 10.1175/JCLI-D-20-0424.1

Chepkoech, W., Mungai, N. W., Stöber, S., & Lotze-Campen, H. (2020). Understanding the adaptive capacity of smallholder African Indigenous vegetable farmers to climate change in Kenya. *Climate Risk Management*, 27, 100204. DOI: 10.1016/j.crm.2019.100204

Chersich, M. F., Gray, G., Fairlie, L., Eichbaum, Q., Mayhew, S., Allwood, B., English, R., Scorgie, F., Luchters, S., Simpson, G., Haghighi, M. M., Pham, M. D., & Rees, H. (2020). COVID-19 in Africa: Care and protection for frontline healthcare workers. *Globalization and Health*, 16(1), 1–6. DOI: 10.1186/s12992-020-00574-3 PMID: 32414379

Chiabai, A., Balakrishnan, S., Sarangi, G., & Nischal, S. (2010). Human Health. In *Costing Adaptation: Preparing for Climate Change in India*. A. Markandya., & A. Mishra (Eds.). New Delhi: TERI Press.

Chui, M., Hazan, E., Roberts, R., Singla, A., & Smaje, K. (2023). The economic potential of generative AI.

Cianconi, P., Betrò, S., & Janiri, L. (2020). The impact of climate change on mental health: A systematic descriptive review. *Frontiers in Psychiatry*, 11(1), 74. DOI: 10.3389/fpsyt.2020.00074 PMID: 32210846

Clayton, S. (2020). Climate anxiety: Psychological responses to climate change. *Journal of Anxiety Disorders*, 74, 102263. DOI: 10.1016/j.janxdis.2020.102263 PMID: 32623280

Clayton, S. (2021). Climate Change and Mental Health. *Current Environmental Health Reports*, 8(1), 1–6. DOI: 10.1007/s40572-020-00303-3 PMID: 33389625

Clayton, S., Manning, C. M., Krygsman, K., & Speiser, M. (2017). *Mental Health and Our Changing Climate: Impacts, Implications, and Guidance*. American Psychological Association, and ecoAmerica.

Coêlho, A. E. L. (2004). Psychological responses to drought in Northeastern Brazil. *Interamerican Journal of Psychology*, 38(1), 95–103. DOI: 10.30849/rip/ijp.v38i1.620

Coffey, Y., Bhullar, N., Durkin, J., Islam, M. S., & Usher, K. (2021). Understanding Eco-anxiety: A systematic scoping review of current literature and identified knowledge gaps. *The Journal of Climate Change and Health*, 3, 10037. DOI: 10.1016/j.joclim.2021.100047

Connolly-Boutin, L., & Smit, B. (2016). Climate change, food security, and livelihoods in sub-Saharan Africa. *Regional Environmental Change*, 16(2), 385–399. DOI: 10.1007/s10113-015-0761-x

Conte, H. R., Weiner, M. B., & Plutchik, R. (1982). Measuring death anxiety: Conceptual, psychometric, and factor-analytic aspects. *Journal of Personality and Social Psychology*, 43(4), 775–785. DOI: 10.1037/0022-3514.43.4.775 PMID: 7175677

Convention on the Rights of Persons with Disabilities. (2006). Convention on the Rights of Persons with Disabilities. www.ohchr.org

Conway, F., Portela, A., Filippi, V., Chou, D., & Kovats, S. (2024). Climate change, air pollution and maternal and newborn health: An overview of reviews of health outcomes. *Journal of Global Health*, 14, 14. DOI: 10.7189/jogh.14.04128 PMID: 38785109

Coppola, I. G. (2021). Eco-Anxiety in "the Climate Generation": Is Action an Antidote?.

Costello, A. (2018). The Lancet Countdown on Health and Climate Change: From 25 Years of Inaction to a Global Transformation for Public Health. *Lancet*, 391(10120), 581–630. DOI: 10.1016/S0140-6736(17)32464-9 PMID: 29096948

Costello, A., Abbas, M., Allen, A., Ball, S., Bell, S., Bellamy, R., Friel, S., Groce, N., Johnson, A., Kett, M., Lee, M., Levy, C., Maslin, M., McCoy, D., McGuire, B., Montgomery, H., Napier, D., Pagel, C., Patel, J., & Patterson, C. (2009). Managing the health effects of climate change. *Lancet*, 373(9676), 1693–1733. DOI: 10.1016/S0140-6736(09)60935-1 PMID: 19447250

Crimmins, A. R., Balbus, J., Gamble, J. L., ... (2016). The Impacts of Climate Change on Human Health in the United States: A Scientific Assessment. U.S. Global Change Research Program. USA: Washington DC.

Crimmins, A., Balbus, J., Gamble, J. L., Beard, C. B., Bell, J. E., Dodgen, D., ... (2016). The impacts of climate change on human health in the United States: A scientific assessment (p. 312). Washington, DC: US Global change research program. DOI: 10.7930/J0R49NQX

Crush, J. (2013). Migration, Urbanization and Food Security in Cities of the Global South: 26–27 November 2012, Cape Town, South Africa.

Cueto, M., Brown, T. M., & Fee, E. (2019). *The world health organization: A history*. Cambridge University Press. DOI: 10.1017/9781108692878

Cunsolo Willox, A., Harper, S. L., Ford, J. D., Edge, V. L., Landman, K., Houle, K., Blake, S., & Wolfrey, C. (2013). Climate change and mental health: An exploratory case study from Rigolet, Nunatsiavut, Canada. *Climatic Change*, 121(2), 255–270. DOI: 10.1007/s10584-013-0875-4

Cunsolo, W. A., Harper, S., Ford, J., Landman, K., Houle, K., & Edge, V.Rigolet Inuit Community Government. (2012). From this place and of this place: Climate change, health, and place in Rigolet, Nunatsiavut, Canada. *Social Science & Medicine*, 75(3), 538–547. DOI: 10.1016/j.socscimed.2012.03.043 PMID: 22595069

Cusack, L., de Crespigny, C., & Athanasos, P. (2011). Heatwaves and their impact on people with alcohol, drug and mental health conditions: A discussion paper on clinical practice considerations. *Journal of Advanced Nursing*, 67(4), 915–922. DOI: 10.1111/j.1365-2648.2010.05551.x PMID: 21214621

Cutter, S. L., Barnes, L., Berry, M., Burton, C., Evans, E., Tate, E., & Webb, J. (2008). A Place-Based Model For Understanding Community Resilience To Natural Disasters. *Global Environmental Change*, 18(4), 598–606. DOI: 10.1016/j.gloenvcha.2008.07.013

D'alessandro, D., Gola, M., Appolloni, L., Dettori, M., Fara, G. M., Rebecchi, A., & Capolongo, S. (2020). COVID-19 and living space challenge. Well-being and public health recommendations for a healthy, safe, and sustainable housing. *Acta Biomedica*, 91(9-S), 61. PMID: 32701918

Dai, W., Chen, L., Lai, Z., Li, Y., Wang, J., & Liu, A. (2016). The incidence of post-traumatic stress disorder among survivors after earthquakes:a systematic review and meta-analysis. *BMC Psychiatry*, 16(1), 188. DOI: 10.1186/s12888-016-0891-9 PMID: 27267874

Dankelman, I., Alam, K., Ahmed, W. B., Gueye, Y. D., Fatema, N., & Mensah-Kutin, R. (2008). Gender, climate change, and human security: Lessons from Bangladesh, Ghana, and Senegal. WEDO with ABANTU for Development in Ghana, Action Aid Bangladesh, and ENDA Senegal. http://www.wedo.org/wp-content/uploads/hsn-study-final-may-20-2008.pdf

Dari, L., Addo, A., & Dzisi, K. A. (2016). Pesticide use in the production of Tomato (Solanum lycopersicum L.) in some areas of Northern Ghana. *African Journal of Agricultural Research*, 11(5), 352–355. DOI: 10.5897/AJAR2015.10325

Davydzenka, T., Tahmasebi, P., & Shokri, N. (2024). Unveiling the global extent of land subsidence: The sinking crisis. *Geophysical Research Letters*, 51(1), 1–11. DOI: 10.1029/2023GL104497

de Almeida, B. A., & Mostafavi, A. (2016). Resilience of infrastructure systems to sea-level rise in coastal areas: Impacts, adaptation measures, and implementation challenges. *Sustainability (Basel)*, 8(11), 11–15. DOI: 10.3390/su8111115

de Lima, C. Z., Buzan, J. R., Moore, F. C., Uris, L. C., Baldos, M. H., & Hertel, T. W. (2021). Heat stress on agricultural workers exacerbates crop impacts of climate change. *Environmental Research Letters*, 16(4), 044020. DOI: 10.1088/1748-9326/abeb9f

De Waal, A. (1989). *Famine That Kills: Darfur, Sudan, 1984–1985*. Clarendon Press.

Dehghani, M. H., Karri, R. R., Vera, T., & Hassan, S. K. M. (Eds.). (2024). *Health Effects of Indoor Air Pollution: Volume 2: Air Pollution, Human Health, and the Environment*. Elsevier.

Deitchman, S. D., Kirsch, T. D., Auerbach, P. S., & Hill, A. C. (2021). Climate Resilience: It is Time for a National Approach. Health Security. *Health Security*, 19(6), 10–24. DOI: 10.1089/hs.2021.0108 PMID: 34698530

Dercon, S. (2004). Growth and shocks: Evidence from rural Ethiopia. *Journal of Development Economics*, 74(2), 309–329. DOI: 10.1016/j.jdeveco.2004.01.001

Desai, B. H., & Mandal, M. (2021). Role of climate change in exacerbating sexual and gender-based violence against women: A new challenge for international law. *Environmental Policy and Law*, 51(3), 137–157. DOI: 10.3233/EPL-210055

Dessler, A. E., & Parson, E. A. (2019). *The science and politics of global climate change: A guide to the debate*. Cambridge University Press. DOI: 10.1017/9781316832158

Dewa, C. S., Loong, D., Bonato, S., & Trojanowski, L. (2017). The relationship between physician burnout and quality of healthcare in terms of safety and acceptability: A systematic review. *BMJ Open*, 7(6), e015141. DOI: 10.1136/bmjopen-2016-015141 PMID: 28637730

Dhoni, P., & Kumar, R. (2023). Synergizing generative ai and cybersecurity: Roles of generative ai entities, companies, agencies, and government in enhancing cybersecurity. *Authorea Preprints*.

Dimitrov, B. G. (2019). Effects of climate change on women. *Res. Rev. Int. J. Multidis*, 4(5), 201–215.

Dodds, J. (2021). The psychology of climate anxiety. *BJPsych Bulletin*, 45(4), 222–226. DOI: 10.1192/bjb.2021.18 PMID: 34006345

Doherty, T. J., & Clayton, S. (2011). The psychological impacts of global climate change. *The American Psychologist*, 66(4), 265–276. DOI: 10.1037/a0023141 PMID: 21553952

Doshi, A. R., & Hauser, O. (2023). Generative artificial intelligence enhances creativity. *Available atSSRN*. DOI: 10.2139/ssrn.4535536

Drager, N., & Sunderland, L. (2024). Public health in a globalising world: The perspective from the World Health Organization. In *Governing Global Health* (pp. 67–78). Routledge.

Dwivedi, Y. K., Kshetri, N., Hughes, L., Slade, E. L., Jeyaraj, A., Kar, A. K., Baabdullah, A. M., Koohang, A., Raghavan, V., Ahuja, M., Albanna, H., Albashrawi, M. A., Al-Busaidi, A. S., Balakrishnan, J., Barlette, Y., Basu, S., Bose, I., Brooks, L., Buhalis, D., & Wright, R. (2023). "So what if ChatGPT wrote it?" Multidisciplinary perspectives on opportunities, challenges and implications of generative conversational AI for research, practice and policy. *International Journal of Information Management*, 71, 102642. DOI: 10.1016/j.ijinfomgt.2023.102642

Dzanja, J. L., Christie, M., Fazey, I., & Hyde, A. G. (2013). The role of social capital on rural food security: The case study of Dowa and Lilongwe Districts in Central Malawi. *Access International Journal of Agricultural Sciences*, 1(4), 46–56.

Ebi, K. L., Boyer, C., Ogden, N., Paz, S., Berry, P., Campbell-Lendrum, D., Hess, J. J., & Woodward, A. (2021). Burning embers: Synthesis of the health risks of climate change. *Environmental Research Letters*, 16(4), 044042. DOI: 10.1088/1748-9326/abeadd

Eger, H., Chacko, S., El-Gamal, S., Gerlinger, T., Kaasch, A., Meudec, M., Munshi, S., Naghipour, A., Rhule, E., Sandhya, Y. K., & Uribe, O. L. (2024). Towards a Feminist Global Health Policy: Power, intersectionality, and transformation. *PLOS Global Public Health*, 4(3), e0002959. DOI: 10.1371/journal.pgph.0002959 PMID: 38451969

Enarson, E. (2000). *Gender and Natural Disasters. IPCRR Working Paper no. 1.* International Labour Organization., Retrieved from https://www.ilo.org/publications/gender-and-natural-disasters

Erlangga, D., Suhrcke, M., Ali, S., & Bloor, K. (2019). The impact of public health insurance on health care utilisation, financial protection and health status in low-and middle-income countries: A systematic review. *PLoS One*, 14(8), e0219731. DOI: 10.1371/journal.pone.0219731 PMID: 31461458

Esbjörn-Hargens, S., & Zimmerman, M. E. (2009). *Integral ecology: Uniting multiple perspectives on the natural world.* Random House/ Integral Books.

Esfahani, P., Mosadeghrad, A. M., & Akbarisari, A. (2018). The success of strategic planning in health care organizations of Iran. *International Journal of Health Care Quality Assurance*, 31(6), 563–574. DOI: 10.1108/IJHCQA-08-2017-0145 PMID: 29954262

Evans, S. (2021). Which countries are historically responsible for climate change? www.carbonbrief.org

Ezzati, F., Mosadeghrad, A. M., & Jaafaripooyan, E. (2023). Resiliency of the Iranian healthcare facilities against the Covid-19 pandemic: Challenges and solutions. *BMC Health Services Research*, 23(1), 1–16. DOI: 10.1186/s12913-023-09180-6 PMID: 36859284

Fadda, J. (2020). Climate change: an overview of potential health impacts associated with climate change environmental driving forces. In *Renewable Energy and Sustainable Buildings: Selected Papers from the World Renewable Energy Congress WREC 2018* (pp. 77-119). Springer International Publishing. DOI: 10.1007/978-3-030-18488-9_8

Fairbrother, N., Janssen, P., Antony, M. M., Tucker, E., & Young, A. H. (2016). Perinatal anxiety disorder prevalence and incidence. *Journal of Affective Disorders*, 20(2), 148–155. DOI: 10.1016/j.jad.2015.12.082 PMID: 27131505

Fang, H., Eggleston, K., Hanson, K., & Wu, M. (2019). Enhancing financial protection under China's social health insurance to achieve universal health coverage. *bmj, 365*.

Fann, N., Brennan, T., Dolwick, P., Gamble, J. L., Ilacqua, V., & Kolb, L.. (2016). Air Quality Impacts. In *The impacts of climate change on human health in the United States: A scientific assessment* (pp. 69–98). Global Change Research Program., health2016.globalchange.gov/

FAO. (1983). World food security: a reappraisal of the concepts and approaches. Director General's Report. www.fao.org

FAO. (2002). The State of Food Insecurity in the World 2001. Food and Agricultural Organization, Rome. www.fao.org

FAO. (2017). Extreme Drought and Floods Threaten Food Security in Sri Lanka. Food and Agriculture Organization of the United Nations 2017. www.fao.org

FAO. (2018). The Future of Food and Agricultural Alternative Pathways to 2050. Food and Agricultural Organization of United Nations. Rome Italy. www.fao.org

Faridi, R., & Wadood, S. N. (2010). An econometric assessment of household food security in Bangladesh. *Bangladesh Development Studies*, 33(3), 97–111. https://www.jstor.org/stable/23339854

Fatema, S. R., East, L., Islam, S., & Usher, K. (2023). Gender-based vulnerabilities for women during natural disasters in Bangladesh. *Frontiers in Communication*, 8, 1–8, 1180406. DOI: 10.3389/fcomm.2023.1180406

Fergusson, D. M., Horwood, L. J., Boden, J. M., & Mulder, R. T. (2014). Impact of a major disaster on the mental health of a well-studied cohort. *JAMA Psychiatry*, 71(9), 1025–1031. DOI: 10.1001/jamapsychiatry.2014.652 PMID: 25028897

Fewtrell, L., & Kay, D. (2008). An attempt to quantify the health impacts of flooding in the UK using an urban case study. *Public Health*, 122(5), 446–451. DOI: 10.1016/j.puhe.2007.09.010 PMID: 18222511

Field, B. C. (2012). *Summary for Policymakers in Managing the Risks of Extreme Events and Disasters to Advance Climate Change Adaptation: A Special Report of Working Groups I and II of the Intergovernmental Panel on Climate Change*. Cambridge University Press. DOI: 10.1017/CBO9781139177245

Field, C. B., Barros, V. R., Dokken, D. J., Mach, K. J., & Mastrandrea, M. D. (Eds.). (2014). *Impacts, Adaptation, and Vulnerability. Part A: Global and Sectoral Aspects. Contribution of Working Group II to the Fifth Assessment Report of the Intergovernmental Panel on Climate Change*. Cambridge University Press.

First, J. M. (2024). Post-traumatic stress and depression following disaster: Examining the mediating role of disaster resilience. *Frontiers in Public Health*, 12(1), 12–19. DOI: 10.3389/fpubh.2024.1272909 PMID: 38299076

Flory, K., Hankin, B. L., Kloos, B., Cheely, C., & Turecki, G. (2009). Alcohol and cigarette use and misuse among Hurricane Katrina survivors: Psychosocial risk and protective factors. *Substance Use & Misuse*, 44(12), 1711–1724. DOI: 10.3109/10826080902962128 PMID: 19895302

Folke, C. (2006). Resilience: The Emergence of A Perspective For Social-Ecological Systems Analyses. *Global Environmental Change*, 16(3), 253–267. DOI: 10.1016/j.gloenvcha.2006.04.002

Folland, S., Goodman, A. C., & Stano, M. (2024). *The economics of health and health care: Pearson new international edition*. Routledge.

Food and Agricultural Organization. Economic Commission for Africa, & African Union. (2020). 2019 Africa Regional Overview of Food Security and Nutrition: Containing the damage of economic slowdowns and downturns to food security in Africa. www.fao.org

Forbes, B., Stammler, F., Kumula, T., Meschtyb, N., Pajunen, A., & Kaarlejärvi, E. (2009). High resilience in the Yamal-Nenets social-ecological system, West Siberian Arctic, Russia. *Proceedings of the National Academy of Sciences of the United States of America*, 106(52), 22041–22048. DOI: 10.1073/pnas.0908286106 PMID: 20007776

Fortanier, F., Kolk, A., & Pinkse, J. (2021). Harmonization in CSR reporting: MNEs and global CSR standards. *MIR. Management International Review*, 51(5), 665–696. DOI: 10.1007/s11575-011-0089-9

Frankenburger, T. R. (1992). Indicators and Data Collection Methods for Assessing Household Food Security. In Maxwell, S., & Frankenburger, T. (Eds.), *Household food security: Concepts, indicators, and measurement: A technical review*. UNICEF and IFAD.

Fraser, N. (2000). Rethinking Recognition. *New Left Review*, 3, 107–120.

Frey, C. B., & Osborne, M. (2023). Generative AI and the Future of Work: A Re-appraisal. *The Brown Journal of World Affairs*, •••, 1–12.

Friel, S. (2023). Climate change mitigation: Tackling the commercial determinants of planetary health inequity. *Lancet*, 402(10419), 2269–2271. DOI: 10.1016/S0140-6736(23)02512-6 PMID: 37977168

Fritze, J. G., Blashki, G. A., Burke, S., & Wiseman, J. (2008). Hope, despair, and transformation: Climate change and the promotion of mental health and wellbeing. *International Journal of Mental Health Systems*, 2(1), 1–10. DOI: 10.1186/1752-4458-2-13 PMID: 18799005

Fuse Brown, E. C., & Hall, M. A. (2024). Private Equity and the Corporatization of Health Care. *Stanford Law Review*, •••, 76.

Gabrielson, A. T., Odisho, A. Y., & Canes, D. (2023). Harnessing generative artificial intelligence to improve efficiency among urologists: Welcome ChatGPT. *The Journal of Urology*, 209(5), 827–829. DOI: 10.1097/JU.0000000000003383 PMID: 36795957

Galea, S., Nandi, A., & Vlahov, D. (2005). The epidemiology of posttraumatic stress disorder after disasters. *Epidemiologic Reviews*, 27(1), 78–91. DOI: 10.1093/epirev/mxi003 PMID: 15958429

Gandhi, T. K., Kaplan, G. S., Leape, L., Berwick, D. M., Edgman-Levitan, S., Edmondson, A., Meyer, G. S., Michaels, D., Morath, J. M., Vincent, C., & Wachter, R. (2018). Transforming concepts in patient safety: A progress report. *BMJ Quality & Safety*, 27(12), 1019–1026. DOI: 10.1136/bmjqs-2017-007756 PMID: 30018115

Ganesh, C., Schmeltz, M., & Smith, J. (2020). Introduction climate change and the legal, ethical, and Health issues facing Healthcare and Public Health Systems. *The Journal of Law, Medicine & Ethics*, 48(4), 636–642. DOI: 10.1177/1073110520979370 PMID: 33404343

Gan, W. H., Lim, J. W., & Koh, D. (2020). Preventing intra-hospital infection and transmission of coronavirus disease 2019 in health-care workers. *Safety and Health at Work*, 11(2), 241–243. DOI: 10.1016/j.shaw.2020.03.001 PMID: 32292622

Gebre, G. G., Amekawa, Y., Fikadu, A. A., & Rahut, D. B. (2023). Farmers' use of climate change adaptation strategies and their impacts on food security in Kenya. *Climate Risk Management*, 40, 100495. DOI: 10.1016/j.crm.2023.100495 PMID: 37283879

Gentle, P., & Maraseni, T. N. (2012). Climate change, poverty and livelihoods: Adaptation practices by rural mountain communities in Nepal. *Environmental Science & Policy*, 21, 24–34. DOI: 10.1016/j.envsci.2012.03.007

Ghiasipour, M., Mosadeghrad, A. M., Arab, M., & Jaafaripooyan, E. (2017). Leadership challenges in health care organizations: The case of Iranian hospitals. *Medical Journal of the Islamic Republic of Iran*, 31(1), 560–567. DOI: 10.14196/mjiri.31.96 PMID: 29951397

Ghimire, P., Kim, K., & Acharya, M. (2023). Generative ai in the construction industry: Opportunities & challenges. *arXiv preprint arXiv:2310.04427.*

Gibon, F., Pellarin, T., Román-Cascón, C., Alhassane, A., Traoré, S., Kerr, Y., Seen, D. L., & Baron, C. (2018). Millet yield estimates in the Sahel using satellite-derived soil moisture time series. *Agricultural and Forest Meteorology*, 262, 100–109. DOI: 10.1016/j.agrformet.2018.07.001

Gilson, L., Barasa, E., Nxumalo, N., Cleary, S., Goudge, J., Molyneux, S., Tsofa, B., & Lehmann, U. (2017a). Everyday Resilience In District Health Systems: Emerging Insights From The Front Lines In Kenya And South Africa. *BMJ Global Health*, 2(2), 2. DOI: 10.1136/bmjgh-2016-000224 PMID: 29081995

Girardi, G., & Bremer, A. A.Girardi & Bremer. (2022). Effects of Climate and Environmental Changes on Women's Reproductive Health. *Journal of Women's Health*, 31(6), 755–757. DOI: 10.1089/jwh.2021.0631 PMID: 35708571

Gitz, V., Meybeck, A., Lipper, L., Young, C. D., & Braatz, S. (2016). Climate change and food security: risks and responses. Rome: Food and Agriculture Organization of the United Nations (FAO) Report.

Godman, B., Haque, M., Islam, S., Iqbal, S., Urmi, U. L., Kamal, Z. M., Shuvo, S. A., Rahman, A., Kamal, M., Haque, M., Jahan, I., Islam, M. Z., Hossain, M. M., Munzur-E-Murshid, , Kumar, S., Charan, J., Bhatt, R., Dutta, S., Abhayanand, J. P., & Sefah, I. (2020). Rapid assessment of price instability and paucity of medicines and protection for COVID-19 across Asia: Findings and public health implications for the future. *Frontiers in Public Health*, 8, 585832. DOI: 10.3389/fpubh.2020.585832 PMID: 33381485

Goh, A. (2012). A literature review of the gender- differentiated impacts of climate change in developing countries. *CAPRi, 106*, 1–38. Retrieved from http://www.indiaenvironmentportal.org.in/files/file/a literature review of the gender-differentiated impacts of climate change.pdf

Goldmann, E., & Galea, S. (2014). Mental health consequences of disasters. *Annual Review of Public Health*, 35(1), 169–183. DOI: 10.1146/annurev-publhealth-032013-182435 PMID: 24159920

Goldsteen, R. L., Goldsteen, K., & Dwelle, T. (2024). Introduction to public health: promises and practices.

Gostin, L. O. (2022). A framework convention on global health: Health for all, justice for all. *Journal of the American Medical Association*, 307(19), 2087–2092. PMID: 22665108

Gostin, L. O., Habibi, R., & Meier, B. M. (2020). Has global health law risen to meet the COVID-19 challenge? Revisiting the International Health Regulations to prepare for future threats. *The Journal of Law, Medicine & Ethics*, 48(2), 376–381. DOI: 10.1177/1073110520935354 PMID: 32631189

Gostin, L. O., & Sridhar, D. (2024). Global health and the law. *The New England Journal of Medicine*, 370(18), 1732–1740. DOI: 10.1056/NEJMra1314094 PMID: 24785208

Gostin, L. O., Sridhar, D., & Hougendobler, D. (2024). The normative authority of the World Health Organization. *Public Health*, 129(7), 854–863. DOI: 10.1016/j.puhe.2015.05.002 PMID: 26100341

Gray, C., & Mueller, V. (2012). Drought and population mobility in rural Ethiopia. *World Development*, 40(1), 134–145. DOI: 10.1016/j.worlddev.2011.05.023 PMID: 22523447

Grech, S. (2009). Disability, Poverty, and Development: Critical reflections on the majority world debate. *Disability & Society*, 24(6), 771–784. DOI: 10.1080/09687590903160266

Grech, S. (2015). Decolonizing Eurocentric disability studies: Why colonialism matters in the disability and global South debate. *Social Identities*, 21(1), 6–21. DOI: 10.1080/13504630.2014.995347

Greenwood, D., & Loewenthal, D. (2005). The use of 'case study'in psychotherapeutic research and education. *Psychoanalytic Psychotherapy*, 19(1), 35–47. DOI: 10.1080/02668730512331341564

Greg, E. E., Anam, B. E., William, M. F., & Duru, E. J. C. (2011). Climate change, food security and agricultural productivity in African: Issues and policy directions. *International Journal of Humanities and Social Science*, 1(21), 205–223.

Grupe, D. W., & Nitschke, J. B. (2013). Uncertainty and anticipation in anxiety: An integrated neurobiological and psychological perspective. *Nature Reviews. Neuroscience*, 14(7), 488–501. DOI: 10.1038/nrn3524 PMID: 23783199

Guenther, R., & Balbus, J. (2014). *Primary Protection: Enhancing Health Care Resilience for a Changing Climate*. U.S. Department of Health and Human Services.

Gunderson, L. H., & Holling, C. S. (Eds.). (2002). *Panarchy: understanding transformations in human and natural systems*. Island Press.

Guodaar, L., Bardsley, D. K., & Suh, J. (2021). Integrating local perceptions with scientific evidence to understand climate change variability in northern Ghana: A mixed-methods approach. *Applied Geography (Sevenoaks, England)*, 130(102440), 102440. DOI: 10.1016/j.apgeog.2021.102440

Gursoy, D., Li, Y., & Song, H. (2023). ChatGPT and the hospitality and tourism industry: An overview of current trends and future research directions. *Journal of Hospitality Marketing & Management*, 32(5), 579–592. DOI: 10.1080/19368623.2023.2211993

Hadi, M. U., Qureshi, R., Shah, A., Irfan, M., Zafar, A., Shaikh, M. B., & Mirjalili, S. (2023). A survey on large language models: Applications, challenges, limitations, and practical usage. *Authorea Preprints*. DOI: 10.36227/techrxiv.23589741.v1

Haffaf, A., Lakdja, F., Abdeslam, D. O., & Meziane, R. (2021). Solar energy for air conditioning of an office building in a case study: Techno-economic feasibility assessment. *Renewable Energy Focus*, 39, 148–162. DOI: 10.1016/j.ref.2021.09.002

Hama Radha, C. (2023). Retrofitting for improving indoor air quality and energy efficiency in the hospital building. *Sustainability (Basel)*, 15(4), 3464. DOI: 10.3390/su15043464

Hanlon, C., Alem, A., Lund, C., Hailemariam, D., Assefa, E., Giorgis, T.W, et al. (2019). Moving towards universal health coverage for mental disorders in Ethiopia. *Int J Ment Health Syst.* 25;13(11).

Hansen, J., Sato, M., Ruedy, R., Lo, K., Lea, D. W., & Medina-Elizade, M. (2006). Global temperature change. *Proceedings of the National Academy of Sciences of the United States of America*, 103(39), 14288–14293. DOI: 10.1073/pnas.0606291103 PMID: 17001018

Harrer, S. (2023). Attention is not all you need: The complicated case of ethically using large language models in healthcare and medicine. *EBioMedicine*, 90, 90. DOI: 10.1016/j.ebiom.2023.104512

Harville, E. W., Xiong, X., & Buekens, P. (2010). Disasters and perinatal health: A systematic review. *Obstetrical & Gynecological Survey*, 65(11), 713–728. DOI: 10.1097/OGX.0b013e31820eddbe PMID: 21375788

Ha, S., Liu, D., Zhu, Y., Kim, S. S., Sherman, S., Grantz, K. L., & Mendola, P. (2017). Ambient temperature and stillbirth: A multi-center retrospective cohort study. *Environmental Health Perspectives*, 125(6), 067011. DOI: 10.1289/EHP945 PMID: 28650842

Hassan, I., Chisty, A., & Bui, T. (2024). Structural and Social Determinants of Health. In *Leading an Academic Medical Practice* (pp. 343–355). Springer International Publishing.

Hawkins, V. R., Marcham, C. L., Springston, J. P., Miller, J. D., Braybrooke, G., Maunder, C., ... & Kollmeyer, B. (2020). The value of IAQ: a review of the scientific evidence supporting the benefits of investing in better indoor air quality.

Hayes, K., Berry, P., & Ebi, K. L. (2019). Factors influencing the mental health consequences of climate change in Canada. *International Journal of Environmental Research and Public Health*, 16(9), 1583. DOI: 10.3390/ijerph16091583 PMID: 31064134

Hayes, K., Blashki, G., Wiseman, J., Burke, S., & Reifels, L. (2018). Climate change and mental health: Risks, impacts and priority actions. *International Journal of Mental Health Systems*, 12(1), 1–12. DOI: 10.1186/s13033-018-0210-6 PMID: 29881451

Hayes, K., & Poland, B. (2018). Addressing Mental Health in a Changing Climate: Incorporating Mental Health Indicators into Climate Change and Health Vulnerability and Adaptation Assessment. *International Journal of Environmental Research and Public Health*, 15(9), 1806. DOI: 10.3390/ijerph15091806 PMID: 30131478

Heid, M. (2022,April 22). *How to cope with Eco-Anxiety: 5 expert tips*. EverydayHealth.com. https://www.everydayhealth.com/eco-anxiety/how-to-cope-with-eco-anxiety/

Hemingway, L., & Priestley, M. (2006). Natural Hazards, Human Vulnerability and Disabling Societies: A Disaster for Disabled People? *Review of Disability Studies*, 2(3).

Herslund, L. B., Jalayer, F., Jean-Baptiste, N., Jørgensen, G., Kabisch, S., Kombe, W., Lindley, S., Nyed, P. K., Pauleit, S., Printz, A., & Vedeld, T. (2016). A multi-dimensional assessment of urban vulnerability to climate change in Sub-Saharan Africa. *Natural Hazards*, 82(2), 149–172. DOI: 10.1007/s11069-015-1856-x

Hesselberg, J., & Yaro, J. A. (2006). An assessment of the extent and causes of food insecurity in northern Ghana using a livelihood vulnerability framework. *GeoJournal*, 67(1), 41–55. DOI: 10.1007/s10708-006-9007-2

Hess, J. J., Ranadive, N., Boyer, C., Aleksandrowicz, L., Anenberg, S. C., Aunan, K., Belesova, K., Bell, M. L., Bickersteth, S., Bowen, K., Burden, M., Campbell-Lendrum, D., Carlton, E., Cissé, G., Cohen, F., Dai, H., Dangour, A. D., Dasgupta, P., Frumkin, H., & Ebi, K. L. (2020). Guidelines for modeling and reporting health effects of climate change mitigation actions. *Environmental Health Perspectives*, 128(11), 115001. DOI: 10.1289/EHP6745 PMID: 33170741

Hill, P. S. (2021). Understanding global health governance as a complex adaptive system. *Global Public Health: An International Journal for Research, Policy and Practice*, 6(6), 593–605. DOI: 10.1080/17441691003762108 PMID: 20432102

Hoelterhoff, M., & Chung, M. C. (2013). Death anxiety and well-being; coping with life-threatening events. *Traumatology*, 19(4), 280–291. DOI: 10.1177/1534765613477499

Holling, C. S. (1973). Resilience and Stability Of Ecological Systems. *Annual Review of Ecology and Systematics*, 4(1), 1–23. DOI: 10.1146/annurev.es.04.110173.000245

Hollnagel, E., Woods, D. D., & Leveson, N. (2007). *Resilience Engineering: Concepts and Precepts*. Ashgate Publishing, Ltd.

Huong, N. T. L., Bo, Y. S., & Fahad, S. (2019). The economic impact of climate change on agriculture using the Ricardian approach: A case of northwest Vietnam. *Journal of the Saudi Society of Agricultural Sciences*, 18(4), 449–457. DOI: 10.1016/j.jssas.2018.02.006

Iberdrola. (2021, April 22). *What Is Eco-Anxiety*. Iberdrola. https://www.iberdrola.com/social-commitment/what-is-ecoanxiety

IFAD. (2009). Research and Innovation for Smallholder Farmers in the Context of Climate Change. Discussion paper prepared for the Round Table organized during the Thirty-second session of IFAD's Governing Council, 18 February 2009. http://www.ifad.org

Igwaran, A., & Okoh, A. I. (2019). Human campylobacteriosis: A public health concern of global importance. *Heliyon*, 5(11), e02814. DOI: 10.1016/j.heliyon.2019.e02814 PMID: 31763476

Ingstad, B. (1999). The myth of disability in developing nations. *Lancet*, 354(9180), 757–758. DOI: 10.1016/S0140-6736(99)06049-3 PMID: 10475204

Institute of Development Studies. (2008). *Gender and climate change: Mapping the linkages*. BRIDGE, Institute of Development Studies. https://www.adequations.org/IMG/pdf/GenderAndClimateChange.pdf

Intergovernmental Panel on Climate Change (IPCC). (2018). *Global Warming of 1.5°C.* www.ipcc.ch

Intergovernmental Panel on Climate Change. (2014). Summary for policymakers. In: *Climate Change 2014: Impacts, Adaptation, and Vulnerability. Part A: Global and Sectoral Aspects.* Contribution of Working Group II to the Fifth Assessment Report of the Intergovernmental Panel on Climate Change [Field, C.B., V.R. Barros, D.J. Dokken, K.J. Mach, M.D. Mastrandrea, T.E. Bilir, M. Chatterjee, K.L. Ebi, Y.O. Estrada, R.C. Genova, B. Girma, E.S. Kissel, A.N. Levy, S. MacCracken, P.R. Mastrandrea, and L. L. White (eds.)]. Cambridge University Press, Cambridge, United Kingdom and New York, NY, USA, 1-32.

International Union for Conservation of Nature. (2015). *Gender and climate change: IUCN issues in brief.* https://www.iucn.org/resources/issues-brief/gender-and-climate-change

IPCC. (2022). Summary for Policymakers in H.-O. Pörtner, D.C. Roberts, M. Tignor, E.S. Poloczanska, K. Mintenbeck, A. Alegría, M. Craig, S. Langsdorf, S. Löschke, V. Möller, A. Okem, B. Rama (Eds.). *Climate Change 2022: Impacts, Adaptation, and Vulnerability. Contribution of Working Group II to the Sixth Assessment Report of the Intergovernmental Panel on Climate Change.* Cambridge University Press, Cambridge, UK and New York, NY, USA, pp. 3-33.

Islam, S. A., Rahaman, M. O., Hossain, M. F., Mridha, M. S., & Islam, M. J. (2023). *design, fabrication, and performance analysis of a solar powered air purifier* (Doctoral dissertation, Sonargoan University (SU)).

Islam, S., Cenacchi, N., Sulser, T. B., Gbegbelegbe, S., Hareau, G., Kleinwechter, U., Mason-D'Croz, D., Nedumaran, S., Robertson, R., Robinson, S., & Wiebe, K. (2016). Structural approaches to modeling the impact of climate change and adaptation technologies on crop yields and food security. *Global Food Security*, 10(September), 63–70. DOI: 10.1016/j.gfs.2016.08.003

Jacobs, P. T., Altman, M., & Hart, T. (2009). The Status of Household Food Security Targets in South Africa. *Agrekon*, 48(4), 410–433. DOI: 10.1080/03031853.2009.9523834

Jammaers, E. (2021). Embodied reflections of an able-bodied disability scholar. *Gender, Work and Organization*, 28(5), 1885–1892. DOI: 10.1111/gwao.12714

Jebril, N. (2020). World Health Organization declared a pandemic public health menace: a systematic review of the coronavirus disease 2019 "COVID-19". *Available atSSRN* 3566298. DOI: 10.2139/ssrn.3566298

Ji, X., Chun, S. A., Wei, Z., & Geller, J. (2023). Twitter sentiment classification for measuring public health concerns. *Social Network Analysis and Mining*, 5, 1–25. PMID: 32226558

Joakim, E. P., Mortsch, L., & Oulahen, G. (2021). Using vulnerability and resilience concepts to advance climate change adaptation. In *Environmental Hazards and Resilience* (pp. 13–31). Routledge. DOI: 10.4324/9781003171430-1

Jodoin, S., Lofts, K. A., & Ananthamoorthy, N. (2020). A Disability Rights Approach to Climate Governance. *SSRN*, 47(1). Advance online publication. DOI: 10.2139/ssrn.3610193

Jones, A. D., Ngure, F. M., Pelto, G., & Young, S. L. (2013). What are we assessing when we measure food security? A compendium and review of current metrics. *Advances in Nutrition*, 4(5), 481–505. DOI: 10.3945/an.113.004119 PMID: 24038241

Jones, L., & Boyd, E. (2011). Exploring social barriers to adaptation: Insights from Western Nepal. *Global Environmental Change*, 21(4), 1262–1274. DOI: 10.1016/j.gloenvcha.2011.06.002

Jonsson, U. (2010). The rise and fall of paradigms in world food and nutrition policy. *World Nutrition,* 1(3).

Joshi, G., Jain, A., Araveeti, S. R., Adhikari, S., Garg, H., & Bhandari, M. (2024). FDA-Approved Artificial Intelligence and Machine Learning (AI/ML)-Enabled Medical Devices: An Updated Landscape. *Electronics (Basel)*, 13(3), 498. DOI: 10.3390/electronics13030498

Juhola, S., Heikkinen, M., Pietilä, T., Groundstroem, F., & Käyhkö, J. (2022). Connecting climate justice and adaptation planning: An adaptation justice index. *Environmental Science & Policy*, 136, 609–619. DOI: 10.1016/j.envsci.2022.07.024

Kabir, R., Khan, H. T. A., Ball, E., & Caldwell, K. (2016). Climate change impact: The experience of the coastal areas of Bangladesh affected by Cyclones Sidr and Aila. *Journal of Environmental and Public Health*, 2016, 1–9. Advance online publication. DOI: 10.1155/2016/9654753 PMID: 27867400

Kafer, A. (2013). *Feminist, Queer, Crip*. Indiana University Press.

Kansiime, M. K., Tambo, J. A., Mugambi, I., Bundi, M., Kara, A., & Owuor, C. (2021). COVID-19 implications on household income and food security in Kenya and Uganda: Findings from a rapid assessment. *World Development*, 137, 105199. DOI: 10.1016/j.worlddev.2020.105199 PMID: 32982018

Karmakar, D. (2021). Gender and Climate Change: The Condition of Women in the Indian Sundarbans. *Scholars Journal of Arts. Humanities and Social Sciences*, 9(1), 1–5. DOI: 10.36347/sjahss.2021.v09i01.001

Kar, N., & Bastia, B. K. (2006). Post-traumatic stress disorder, depression and generalised anxiety disorder in adolescents after a natural disaster: A study of comorbidity. *Clinical Practice and Epidemiology in Mental Health*, 2(17), 1–12. DOI: 10.1186/1745-0179-2-17 PMID: 16869979

Kartiki, K. (2011). Climate change and migration: A case study from rural Bangladesh. *Gender and Development*, 19(1), 23–38. DOI: 10.1080/13552074.2011.554017

Kett, M., Sriskanthan, G., & Cole, E. (2021). *Disability and Climate Justice: A Research Project*. Open Society Foundations.

Keya, T. A., Leela, A., Habib, N., Rashid, M., & Bakthavatchalam, P. (2023). Mental health disorders due to disaster exposure: A systematic review and meta-analysis. *Cureus*, 15(4), 1–7. DOI: 10.7759/cureus.37031 PMID: 37143625

Khan, I., Lei, H., Shah, I. A., Ali, I., Khan, I., Muhammad, I., Huo, X., & Javed, T. (2020). Farm households' risk perception, attitude and adaptation strategies in dealing with climate change: Promise and perils from rural Pakistan. *Land Use Policy*, 91, 104395. DOI: 10.1016/j.landusepol.2019.104395

Kickbusch, I., Brindley, C., & World Health Organization. (2023). *Health in the post-2015 development agenda: an analysis of the UN-led thematic consultations, High-Level Panel report and sustainable development debate in the context of health*. World Health Organization.

Kiefer, M., Rodriguez-Guzman, J., Watson, J., Van Wendel De Joode, B., Mergler, D., & Da Silva, A. S. (2016). Worker health and safety and climate change in the Americas: Issues and research needs. Revista Panamericana de Salud Publica/Pan American. *Journal of Public Health*, 40(3), 192–197.

King, M.M., & Gregg, M.A. (2022). Disability and Climate Change: a critical realist model of climate justice. *Sociology Compass* 1 6(1): e12954.

King, S., Dancause, K., Turcotte-Tremblay, A. M., Veru, F., & Laplante, D. P. (2012). Using natural disasters to study the effects of prenatal maternal stress on child health and development. *Birth Defects Research. Part C, Embryo Today*, 96(4), 273–288. DOI: 10.1002/bdrc.21026 PMID: 24203917

Knagg, R., Dorey, J., Evans, R., & Hitchman, J. (2024). Sustainability in healthcare: Patient and public perspectives. *Anaesthesia*, 79(3), 278–283. DOI: 10.1111/anae.16159 PMID: 38205532

Knight, A., & McNaught, A. (Eds.). (2011). *Understanding Well-being: an introduction for students and practitioners of health and social care*. Lantern.

Korzynski, P., Mazurek, G., Altmann, A., Ejdys, J., Kazlauskaite, R., Paliszkiewicz, J., & Ziemba, E. (2023). *Generative artificial intelligence as a new context for management theories: analysis of ChatGPT*. Central European Management Journal.

Kousky, C. (2016). Impacts of natural disasters on children. *The Future of Children*, 26(1), 73–92. DOI: 10.1353/foc.2016.0004

Kruk, M. E., Ling, E. J., Bitton, A., Cammett, M., Cavanaugh, K., Chopra, M., El-Jardali, F., Macauley, R. J., Muraguri, M. K., Konuma, S., Marten, R., Martineau, F., Myers, M., Rasanathan, K., Ruelas, E., Soucat, A., Sugihantono, A., & Warnken, H. (2017). Building Resilient Health Systems: A Proposal For A Resilience Index. *BMJ (Clinical Research Ed.)*, •••, J2323. DOI: 10.1136/bmj.j2323 PMID: 28536191

Kruk, M. E., Myers, M., Varpilah, S. T., & Dahn, B. T. (2015). What Is A Resilient Health System? Lessons From Ebola. *Lancet*, 385(9980), 1910–1912. DOI: 10.1016/S0140-6736(15)60755-3 PMID: 25987159

Kshetri, N., Dwivedi, Y. K., Davenport, T. H., & Panteli, N. (2023). Generative artificial intelligence in marketing: Applications, opportunities, challenges, and research agenda. *International Journal of Information Management*, •••, 102716.

Kumar, A., Ahmad, M. M., & Sharma, P. (2017). Influence of climatic and non-climatic factors on sustainable food security in India: A statistical investigation. *International Journal of Sustainable Agricultural Management and Informatics*, 3(1), 1–100. DOI: 10.1504/IJSAMI.2017.082917

Kumar, P., & Akhai, S. (2022). Effective energy management in smart buildings using VRV/VRF systems. *Additive Manufacturing in Industry*, 4(0), 27–35. DOI: 10.1201/9781003360001-2

Kumar, P., Singh, A. B., Arora, T., Singh, S., & Singh, R. (2023). Critical review on emerging health effects associated with the indoor air quality and its sustainable management. *The Science of the Total Environment*, 872, 162163. DOI: 10.1016/j.scitotenv.2023.162163 PMID: 36781134

Kumar, S., & Singh, A. K. (2023). Modeling the effects of climate change on agricultural productivity: Evidence from Himachal Pradesh, India. *Asia-Pacific Journal of Regional Science*, 7(2), 521–548. DOI: 10.1007/s41685-023-00291-w

Kumar, V. (2016). Gender-responsive climate change strategies for sustainable development. *Productivity*, 57(2), 182.

Kuo, H. W., Wu, S. J., Ma, T. C., Chiu, M. C., & Chou, S. Y. (2007). Posttraumatic symptoms were worst among quake victims with injuries following the Chi-chi quake in Taiwan. *Journal of Psychosomatic Research*, 62(4), 495–500. DOI: 10.1016/j.jpsychores.2004.11.012 PMID: 17383502

Lachat, C., Van Camp, J., De Henauw, S., Matthys, C., Larondelle, Y., Remaut–De Winter, A. M., & Kolsteren, P. (2023). A concise overview of national nutrition action plans in the European Union Member States. *Public Health Nutrition*, 8(3), 266–274. DOI: 10.1079/PHN2004691 PMID: 15918923

Lai, L., Wang, X., Hu, E., & Ng, K. C. (2024). A vision of dew point evaporative cooling: Opportunities and challenges. *Applied Thermal Engineering*, 244, 122683. DOI: 10.1016/j.applthermaleng.2024.122683

Laneri, K., Bhadra, A., Ionides, E. L., Bouma, M., Dhiman, R. C., Yadav, R. S., & Pascual, M. (2010). Forcing versus feedback: Epidemic malaria and monsoon rains in northwest India. *PLoS Computational Biology*, 6(9), e1000898. DOI: 10.1371/journal.pcbi.1000898 PMID: 20824122

Law, R., Lin, K. J., Ye, H., & Fong, D. K. C. (2023). Artificial intelligence research in hospitality: A state-of-the-art review and future directions. *International Journal of Contemporary Hospitality Management*.

Lawrance, E. L., Thompson, R., Newberry Le Vay, J., Page, L., & Jennings, N. (2022). The impact of climate change on mental health and emotional wellbeing: A narrative review of current evidence, and its implications. *International Review of Psychiatry (Abingdon, England)*, 34(5), 443–498. DOI: 10.1080/09540261.2022.2128725 PMID: 36165756

Leal Filho, W., Simane, B., Kalangu, J., Wuta, M., Munishi, P., & Musiyiwa, K. (2017). *Climate change adaptation in Africa*. Springer International Publishing. DOI: 10.1007/978-3-319-49520-0

Leap, S. R., Soled, D. R., Sampath, V., & Nadeau, K. C. (2024). Effects of Extreme Weather on Health in Underserved Communities. *Annals of Allergy, Asthma & Immunology*.

Lebel, L., Anderies, J. M., Campbell, B., Folke, C., Hatfield-Dodds, S., Hughes, T. P., & Wilson, J. (2006). Governance And The Capacity To Manage Resilience In Regional Social-Ecological Systems. *Ecology and Society*, 11(1), 11. DOI: 10.5751/ES-01606-110119

Lee, G. O., Surkan, P. J., Zelner, J., Olórtegui, M. P., Yori, P. P., Ambikapathi, R., Caulfield, L. E., Gilman, R. H., & Kosek, M. N. (2018). Social connectedness is associated with food security among peri-urban Peruvian Amazonian communities. *SSM - Population Health*, 4, 254–262. DOI: 10.1016/j.ssmph.2018.02.004 PMID: 29854909

Leen, J. L., & Juurlink, D. N. (2019). Carfentanil: A narrative review of its pharmacology and public health concerns. *Canadian Journal of Anaesthesia*, 66(4), 414–421. DOI: 10.1007/s12630-019-01294-y PMID: 30666589

Leichenko, R., & O'Brien, K. (2019). *Climate and Society, Transforming the Future*. Polity Press.

Leisner, C. P. (2020). Climate change impacts food security-focus on perennial cropping systems and nutritional value. *Plant Science*, 293, 110412. DOI: 10.1016/j.plantsci.2020.110412 PMID: 32081261

Lemery, J., Knowlton, K., & Sorenson, C. (2021). *Global climate change and human health: from Science to Practice*. John Wiley & Sons.

Leone, D., Schiavone, F., Appio, F. P., & Chiao, B. (2021). How does artificial intelligence enable and enhance value co-creation in industrial markets? An exploratory case study in the healthcare ecosystem. *Journal of Business Research*, 129, 849–859. DOI: 10.1016/j.jbusres.2020.11.008

Lewis, D., & Ballard, K. (2012). *Disability and Climate Change: Understanding vulnerability and building resilience in a changing world*. CBM.

Lewis, M., & Mercer, E. (2024). The AI Roadmap–charting our future in healthcare communications. *Current Medical Research and Opinion*, 40(1), 5–7. DOI: 10.1080/03007995.2023.2261737 PMID: 37791596

Lindsey, L., & Saut, S. (2020). The Interconnection Between Climate Change and Mental Health. *Climate Change and Human Responses*, 115.

Li, X., Jiang, Y., Rodriguez-Andina, J. J., Luo, H., Yin, S., & Kaynak, O. (2021). When medical images meet generative adversarial network: Recent development and research opportunities. *Discover Artificial Intelligence*, 1(1), 1–20. DOI: 10.1007/s44163-021-00006-0

Li, Y., Wang, H., Jin, X. R., Li, X., Pender, M., Song, C. P., Tang, S.-L., Cao, J., Wu, H., & Wang, Y. G. (2018). Experiences and challenges in the health protection of medical teams in the Chinese Ebola treatment center, Liberia: A qualitative study. *Infectious Diseases of Poverty*, 7(1), 1–12. DOI: 10.1186/s40249-018-0468-6 PMID: 30134982

Lloyd, S. J., Kovats, R. S., & Chalabi, Z. (2011). Climate Change, Crop Yields, and Undernutrition: Development of a Model to Quantify the Impact of Climate Scenarios on Child Undernutrition. *Environmental Health Perspectives*, 119(12), 1817–1823. DOI: 10.1289/ehp.1003311 PMID: 21844000

Lopes, M. B. S. (2017). The 2017 World Health Organization classification of tumors of the pituitary gland: A summary. *Acta Neuropathologica*, 134(4), 521–535. DOI: 10.1007/s00401-017-1769-8 PMID: 28821944

Lovibond, P. F., & Lovibond, S. H. (1995). The structure of negative emotional states: Comparison of the Depression Anxiety Stress Scales (DASS) with the Beck Depression and Anxiety Inventories. *Behaviour Research and Therapy*, 33(3), 335–343. DOI: 10.1016/0005-7967(94)00075-U PMID: 7726811

Lutz, P. K., Zelenski, J. M., & Newman, D. B. (2023). Eco-anxiety in daily life: Relationships with well-being and pro-environmental behavior. *Current Research in Ecological and Social Psychology*, 4, 100110. DOI: 10.1016/j.cresp.2023.100110

Maddipoti, A. (2023). *Pathway Forward for Responsible Generative AI Implementation in Healthcare* (Doctoral dissertation).

Magnusson, R. S. (2019). Rethinking global health challenges: Towards a 'global compact' for reducing the burden of chronic disease. *Public Health*, 123(3), 265–274. DOI: 10.1016/j.puhe.2008.12.023 PMID: 19278695

Magray, M., Hussain, K., Aijax, M., Singh, P., & Nabi, A. (2017). Post-Harvest Management of Vegetable Crops: Introduction. *Journal of Pharmacology and Phytochemistry*, 12(23). Pp12-24.

Makwana, N. (2019). Disaster and its impact on mental health: A narrative review. *Journal of Family Medicine and Primary Care*, 8(10), 3090–3095. DOI: 10.4103/jfmpc.jfmpc_893_19 PMID: 31742125

Malhi, G. S., Kaur, M., & Kaushik, P. (2021). Impact of climate change on agriculture and its mitigation Strategies: A review. *Sustainability (Basel)*, 13(3), 1318. DOI: 10.3390/su13031318

Malik, S., Tyagi, A. K., & Mahajan, S. (2022). Architecture, Generative Model, and Deep Reinforcement Learning for IoT Applications: Deep Learning Perspective. *Artificial Intelligence-based Internet of Things Systems*, 243-265.

Mall, M. K., Singh, R., Gupta, A., Srinivasan, G., & Rathore, L. S. (2006). Impact of climate change on Indian agriculture: A review. *Climatic Change*, 78(1), 445–478. DOI: 10.1007/s10584-005-9042-x

Malloy, J. T., & Ashcraft, C. M. (2020). A framework for implementing socially just climate adaptation. *Climatic Change*, 160(1), 1–14. DOI: 10.1007/s10584-020-02705-6

Mamun, M. A., Huq, N., Papia, Z. F., Tasfina, S., & Gozal, D. (2019). Prevalence of depression among Bangladeshi village women subsequent to a natural disaster: A pilot study. *Psychiatry Research*, 276, 124–128. DOI: 10.1016/j.psychres.2019.05.007 PMID: 31077883

Ma, N., Aviv, D., Guo, H., & Braham, W. W. (2021). Measuring the right factors: A review of variables and models for thermal comfort and indoor air quality. *Renewable & Sustainable Energy Reviews*, 135, 110436. DOI: 10.1016/j.rser.2020.110436

Mansouri, A., Wei, W., Alessandrini, J. M., Mandin, C., & Blondeau, P. (2022). Impact of climate change on indoor air quality: A review. *International Journal of Environmental Research and Public Health*, 19(23), 15616. DOI: 10.3390/ijerph192315616 PMID: 36497689

Marcus, M., Yasamy, M. T., van Ommeren, M. V., Chisholm, D., & Saxena, S. (2012). Depression: A global public health concern.

Margaret, G. R. (2018). Climate Change and Individual. *The American Journal of Comparative Law*, 66(suppl_1), 345–378. DOI: 10.1093/ajcl/avy018

Martin, K. S., Rogers, B. L., Cook, J. T., & Joseph, H. M. (2004). Social capital is associated with a decreased risk of hunger. *Social Science & Medicine*, 58(12), 2645–2654. DOI: 10.1016/j.socscimed.2003.09.026 PMID: 15081212

Martin, M. A., Sendra, O. A., Bastos, A., Bauer, N., Bertram, C., Blenckner, T., Bowen, K., Brando, P. M., Rudolph, T. B., Büchs, M., Bustamante, M., Chen, D., Cleugh, H., Dasgupta, P., Denton, F., Donges, J. F., Donkor, F. K., Duan, H., Duarte, C. M., & Woodcock, J. (2021). Ten new insights in climate science 2021: A horizon scan. *Glob. Sustain.*, 4, E25. DOI: 10.1017/sus.2021.25

Mashallahi, A., Ardalan, A., Nejati, A., & Ostadtaghizadeh, A. (2022). Climate adaptive hospital: A systematic review of determinants and actions. *Journal of Environmental Health Science & Engineering*, 20(2), 983–1013. DOI: 10.1007/s40201-022-00810-5 PMID: 36406601

Masika, R. (2002). *Gender, development and climate change.* Oxfam GB., DOI: 10.3362/9780855987374

Mata, T. M., Martins, A. A., Calheiros, C. S., Villanueva, F., Alonso-Cuevilla, N. P., Gabriel, M. F., & Silva, G. V. (2022). Indoor air quality: A review of cleaning technologies. *Environments (Basel, Switzerland)*, 9(9), 118. DOI: 10.3390/environments9090118

Matthews, J.B., Robin, M., Vincent, van Diemen, R., Fuglestvedt, J.S., Masson-Delmotte, V., Méndez, C., Semenov, S., & Reisinger. (2021). *Annex VII. Glossary: IPCC – Intergovernmental Panel on Climate Change.* IPCC Sixth Assessment Report.

McCoy, D., Kapilashrami, A., Kumar, R., Rhule, E., & Khosla, R. (2024). Developing an agenda for the decolonization of global health. *Bulletin of the World Health Organization*, 102(2), 130–136. DOI: 10.2471/BLT.23.289949 PMID: 38313156

McLaren, K., & Kelleher, D. (2005). New horizons in organizational development. Au Courant. *Canadian Council for International Cooperation*, 13, 6–9.

McMichael, C., Barnett, J., & McMichael, A. J. (2012). An ill wind? Climate change, migration, and health. *Environmental Health Perspectives*, 120(5), 646–654. DOI: 10.1289/ehp.1104375 PMID: 22266739

Meier, B. M., Cinà, M. M., & Gostin, L. O. (2020). Advancing human rights through global health governance. *Foundations of global health and human rights.*

Metcalf, H. (2009). Pay gaps across the equality strands: a review. Equality and Human Rights Commission Research Report Series. www.equalityhumanrights.com

Miao, H., Li, C., & Wang, J. (2023). A future of smarter digital health empowered by generative pretrained transformer. *Journal of Medical Internet Research*, 25, e49963. DOI: 10.2196/49963 PMID: 37751243

Minssen, T., Vayena, E., & Cohen, I. G. (2023). The challenges for regulating medical use of ChatGPT and other large language models. *Journal of the American Medical Association*, 330(4), 315. DOI: 10.1001/jama.2023.9651 PMID: 37410482

Mirzabaev, A. (2013). Impact of Weather Variability and Climate Change on Agricultural Resources in Central Asia. *Zeitschrift für Ausländische Landwirtschaft*, 52(3), 237–252.

Mirza, M. M. Q. (2002). Global warming and changes in the probability of occurrence of floods in Bangladesh and implications. *Global Environmental Change*, 12(2), 127–138. DOI: 10.1016/S0959-3780(02)00002-X

Mohamadi, S., Mujtaba, G., Le, N., Doretto, G., & Adjeroh, D. A. (2023). ChatGPT in the age of generative AI and large language models: a concise survey. *arXiv preprint arXiv:2307.04251.*

Molotoks, A., Smith, P., & Dawson, T. P. (2021). Impacts of land use, population, and climate change on global food security. *Food and Energy Security*, 10(1), e261. DOI: 10.1002/fes3.261

Mondal, S., Das, S., & Vrana, V. G. (2023). How to bell the cat? A theoretical review of generative artificial intelligence towards digital disruption in all walks of life. *Technologies*, 11(2), 44. DOI: 10.3390/technologies11020044

Moran, M., Hillman, W., Mann, J., Kinnear, S., Taylor, D., & Gaskin, C. J. (2017). Of People with Disability: A Systematic Review. *Weather, Climate, and Society*, 9(4), 801–814. DOI: 10.1175/WCAS-D-16-0126.1

Morley, J., DeVito, N. J., & Zhang, J. (2023). Generative AI for medical research. *bmj, 382*.

Mosadeghrad, A.M., Isfahani, P., Eslambolchi, L., Zahmatkesh, M., Afshari, M. (2023). Strategies to strengthen a climate-resilient health system: a scoping review. *Global Health*, 28;19(1):62.

Mosadeghrad, A. M. (2022). A practical model for health policy making and analysis. *Payesh*, 21(1), 7–24. DOI: 10.52547/payesh.21.1.7

Mosadeghrad, A. M., & Shahsavani, A. (2023). Strengthening the resilience of Iran's health system against climate change. *Payesh*, 22(4), 479–482. DOI: 10.61186/payesh.22.4.479

Mousavi, S. M., Pouramini, Z., Babapoor, A., Binazadeh, M., Rahmanian, V., Gholami, A., Omidfar, N., Althomali, R. H., Chiang, W.-H., & Rahman, M. M. (2024). Photocatalysis air purification systems for coronavirus removal: Current technologies and future trends. *Chemosphere*, 353, 141525. DOI: 10.1016/j.chemosphere.2024.141525 PMID: 38395369

Mujib, R. B., Nausahad, K., & Inayatullah, J. (2007). Post-Harvest Losses in Tomato Crop. A case of Peshawar valley. *Sarhad Journal of Agriculture*, 23(4), 1281.

Mullins, J. T., & White, C. (2019). Temperature and Mental Health: Evidence from the Spectrum of Mental Health Outcomes. *Journal of Health Economics*, 68, 102240. DOI: 10.1016/j.jhealeco.2019.102240 PMID: 31590065

Musah, B. A., Mumuni, E., Abayomi, O., & Jibrel, M. B. (2013). Effects of floods on the livelihoods and food security of households in the Tolon/kumbumgu district of the Northern region of Ghana. *American Journal of Research Communication*, 1(8), 160–171.

Music, T. (2022). Protecting patients, protecting healthcare workers: A review of the role of influenza vaccination. *International Nursing Review*, 59(2), 161–167. DOI: 10.1111/j.1466-7657.2011.00961.x PMID: 22591085

Mycoo, M., Wairiu, M., Campbell, D., Duvat, V., Golbuu, Y., Maharaj, S., ... (2022). Small Islands. In: H.O. Pörtner., D.C. Roberts., M. Tignor., E.S. Poloczanska., K. Mintenbeck., A. Alegria., M. Craig., S. Langsdorf., S. Löschke., V. Möller., A. Okem., & B. Rama. (Eds). *Climate Change 2022: Impacts, Adaptation and Vulnerability. Contribution of Working Group II to the Sixth Assessment Report of the Intergovernmental Panel on Climate Change.* Cambridge, UK: Cambridge University Press; 2022. p. 2043–2121.

Naeema, F., Ayub, M., Masood, K., Gul, H., Farrukh, A., Saheen, A., Waheed, W., & Chaudhry, H. R. (2011). Prevalence and psychosocial risk factors of PTSD: 18 months after Kashmir earthquake in Pakistan. *Journal of Affective Disorders*, 130(1-2), 268–274. DOI: 10.1016/j.jad.2010.10.035 PMID: 21071091

Nag, P. K., & Nag, P. K. (2019). Sick building syndrome and other building-related illnesses. *Office Buildings: Health, Safety and Environment*, 53-103.

Nazril, I. S., & Winkel, J. (2017). *Climate Change and Social Inequality. Department of Economic and Social Affairs Working Paper No. 152* ST/ESA/2017/DWP/152. 2017. www.un.org

Neria, P., & Schultz, J. M. (2012). Mental health effects of Hurricane Sandy characteristics, potential aftermath, and response. *Journal of the American Medical Association*, 308(24), 2571–2572. DOI: 10.1001/jama.2012.110700 PMID: 23160987

Neria, Y., Nandi, A., & Galea, S. (2008). Post-traumatic stress disorder following disasters: A systematic review. *Psychological Medicine*, 38(4), 467–480. DOI: 10.1017/S0033291707001353 PMID: 17803838

Neufeldt, H., Christiansen, L., & Dale, T. W. (2020). *Adaptation Gap Report 2020*. United Nations Environment Programme.

Neumayer, E., & Plümper, T. (2007). The gendered nature of natural disasters: The impact of catastrophic events on the gender gap in life expectancy, 1981–2002. *Annals of the Association of American Geographers*, 97(3), 551–566. DOI: 10.1111/j.1467-8306.2007.00563.x

Niang, I., Ruppel, O. C., Abdrabo, M. A., Essel, A., Lennard, C., Padgham, J., & Urquhart, P. (2014). Africa. In Barros, V. R. (Eds.), *Climate Change 2014: Impacts, Adaptation, and Vulnerability* (pp. 1199–1265). University of Cambridge Press.

Niza, I. L., de Souza, M. P., da Luz, I. M., & Broday, E. E. (2024). Sick building syndrome and its impacts on health, well-being and productivity: A systematic literature review. *Indoor and Built Environment*, 33(2), 218–236. DOI: 10.1177/1420326X231191079

Norris, F. H., Friedman, M. J., Watson, P. J., Byrne, C. M., Diaz, E., & Kaniasty, K. (2002). 60,000 disaster victims speak: Part I. An empirical review of the empirical literature, 1981-2001. *Psychiatry*, 65(3), 207–239. DOI: 10.1521/psyc.65.3.207.20173 PMID: 12405079

North, C. S., Kawasaki, A., Spitznagel, E. L., & Hong, B. A. (2004). The course of PTSD, major depression, substance abuse, and somatization after a natural disaster. *The Journal of Nervous and Mental Disease*, 293(12), 823–829. DOI: 10.1097/01.nmd.0000146911.52616.22 PMID: 15583503

Nosratabadi, S., Khazami, N., Abdallah, M. B., Lackner, Z., Band, S., Mosavi, A., & Mako, C. (2020). Social capital contributions to food security: A comprehensive literature review. *Foods*, 9(11), 1650. DOI: 10.3390/foods9111650 PMID: 33198127

O'Brien, K., & Sygna, L. (2013). Responding to climate change: The three spheres of transformation. Proceedings of Transformation in a Changing Climate, 19-21 June 2013, Oslo, Norway: University of Oslo (pp.16-23).

O'Brien, A. J., & Elders, A. (2022). Climate anxiety. When it's good to be worried. *Journal of Psychiatric and Mental Health Nursing*, 29(3), 387–389. DOI: 10.1111/jpm.12817 PMID: 34967950

O'Brien, K. (2010). *Responding to climate change. Integral Theory in Action: Applied, Theoretical, and Constructive Perspectives on the AQAL Model*. State University of New York Press.

O'brien, K. (2012). Global Environmental Change Ii: From Adaptation to Deliberate Transformation. *Progress in Human Geography*, 36(5), 667–676. DOI: 10.1177/0309132511425767

O'Brien, K., & Hochachka, G. (2010). Integral adaptation to climate change. *Journal of Integral Theory & Practice*, 5(1), 89–102.

Obradovich, N., Migliorini, R., Paulus, M. P., & Rahwan, I. (2018). Empirical evidence of mental health risks posed by climate change. *Proceedings of the National Academy of Sciences of the United States of America*, 115(43), 10953–10958. DOI: 10.1073/pnas.1801528115 PMID: 30297424

Obrist, B., Iteba, N., Lengeler, C., Makemba, A., Mshana, C., Nathan, R., Alba, S., Dillip, A., Hetzel, M. W., Mayumana, I., Schulze, A., & Mshinda, H. (2007). Access To Health Care In Contexts Of Livelihood Insecurity: A Framework For Analysis And Action. *PLoS Medicine*, 4(10), 4. DOI: 10.1371/journal.pmed.0040308 PMID: 17958467

Office of the United Nations High Commissioner for Human Rights (UNHCHR). (2020). Analytical study on the promotion and protection of the rights of persons with disabilities in the context of climate change. www.undocs.org

Ogugua, J. O., Anyanwu, E. C., Olorunsogo, T., Maduka, C. P., & Ayo-Farai, O. (2024). Ethics and strategy in vaccination: A review of public health policies and practices. *International Journal of Science and Research Archive*, 11(1), 883–895. DOI: 10.30574/ijsra.2024.11.1.0141

Ojala, M., & Bengtsson, H. (2019). Young People's Coping Strategies Concerning Climate Change: Relations to Perceived Communication With Parents and Friends and Pro-environmental Behavior. *Environment and Behavior*, 51(8), 907–935. DOI: 10.1177/0013916518763894

Okoro, C. A., Hollis, N. D., Cyrus, A. C., & Griffin-Blake, S. (2018). Prevalence of Disabilities and Health Care Access by Disability Status and Type Among Adults — United States, 2016. *MMWR. Morbidity and Mortality Weekly Report*, 67(32), 882–887. DOI: 10.15585/mmwr.mm6732a3 PMID: 30114005

Oliver, M. (2013). The social model of disability: Thirty years on. *Disability & Society*, 28(7), 1024–1026. DOI: 10.1080/09687599.2013.818773

Ortiz, M., Itard, L., & Bluyssen, P. M. (2020). Indoor environmental quality related risk factors with energy-efficient retrofitting of housing: A literature review. *Energy and Building*, 221, 110102. DOI: 10.1016/j.enbuild.2020.110102

Ortoleva, S., & Lewis, H. (2012). *Forgotten Sisters - A Report on Violence Against Women with Disabilities: An Overview of its Nature, Scope, Causes and Consequences (No. 104–2012)*. Northeastern University School of Law Research Paper.

Osborn, D., Cutter, A., & Ullah, F. (2015). Universal Sustainable Development Goals. *Underst Transformational Chall Developed Ctries*, 2(1), 1–25.

Osborn, T. J., Jones, P. D., Lister, D. H., Morice, C. P., Simpson, I. R., Winn, J. P., Hogan, E., & Harris, I. C. (2021). Land Surface Air Temperature Variations Across the Globe Updated to 2019: The CRUTEM5 Data Set. *Journal of Geophysical Research. Atmospheres*, 126(2), e2019JD032352. DOI: 10.1029/2019JD032352

Owusu, V., Ma, W., Emuah, D., & Renwick, A. (2021). Perceptions and vulnerability of farming households to climate change in three agroecological zones of Ghana. *Journal of Cleaner Production*, 293, 12. DOI: 10.1016/j.jclepro.2021.126154

Padhy, S. K., Sarkar, S., Panigrahi, M., & Paul, S. (2015). Mental health effects of climate change. *Indian Journal of Occupational and Environmental Medicine*, 19(1), 3. DOI: 10.4103/0019-5278.156997 PMID: 26023264

Page, L. A., Hajat, S., Kovats, R. S., & Howard, L. M. (2012). Temperature-related deaths in people with psychosis, dementia, and substance misuse. *The Journal of Mental Science*, 200(6), 485–490. PMID: 22661680

Palinkas, L. A., & Wong, M. (2020). Global climate change and mental health. *Current Opinion in Psychology*, 32, 12–16. DOI: 10.1016/j.copsyc.2019.06.023 PMID: 31349129

Pandey, R. A., Chalise, P., Khadka, S., Chaulagain, B., Maharjan, B., Pandey, J., Nepal, J., & Pandey, C. (2023). Post-traumatic stress disorder and its associated factors among survivors of 2015 earthquake in Nepal. *BMC Psychiatry*, 23(1), 340. DOI: 10.1186/s12888-023-04836-3 PMID: 37189110

Parsons, L. A., Shindell, D., Tigchelaar, M., Zhang, Y., & Spector, J. T. (2021). Increased labor losses and decreased adaptation potential in a warmer world. *Nature Communications*, 12(1), 7286. DOI: 10.1038/s41467-021-27328-y PMID: 34907184

Pasquini, L., van Aardenne, L., Godsmark, C. N., Lee, J., & Jack, C. (2020). Emerging climate change-related public health challenges in Africa: A case study of the heat-health vulnerability of informal settlement residents in Dar es Salaam, Tanzania. *The Science of the Total Environment*, 747, 141355. DOI: 10.1016/j.scitotenv.2020.141355 PMID: 32777515

Patel, V., Saxena, S., Lund, C., Thornicroft, G., Baingana, F., Bolton, P., et al. (2018). *The Lancet Commission on global mental health and sustainable development.* The Lancet: Lancet Publishing Group.

Patel, S. K., Agrawal, G., Mathew, B., Patel, S., Mohanty, B., & Singh, A. (2020). Climate change and women in South Asia: A review and future policy implications. *World Journal of Science. Technology and Sustainable Development*, 17(2), 145–166. DOI: 10.1108/WJSTSD-10-2018-0059

Patel, S., Zhou, C., Rattan, S., & Flaws, J. A. (2015). Effects of endocrine-disrupting chemicals on the ovary. *Biology of Reproduction*, 93(1), 20–21. DOI: 10.1095/biolreprod.115.130336 PMID: 26063868

Paulson, D. S. (2008). Wilber's integral philosophy: A summary and critique. *Journal of Humanistic Psychology*, 48(3), 364–388. DOI: 10.1177/0022167807309748

Peek, L., & Stough, L. M. (2010). Children with Disabilities in the Context of Disaster: A Social Vulnerability Perspective. *Child Development*, 81(4), 1260–1270. DOI: 10.1111/j.1467-8624.2010.01466.x PMID: 20636694

Pereira, L. (2017). *Climate change impacts on agriculture across Africa.* Oxford Research Encyclopedia of Environmental Science.

Perrin, S., Meiser-Stedman, R., & Smith, P. (2005). The Children's Revised Impact of Event Scale (CRIES): Validity as a Screening Instrument for PTSD. *Behavioural and Cognitive Psychotherapy*, 33(4), 487–498. DOI: 10.1017/S1352465805002419

Philipsborn, R. P., Shefield, P., White, A., Osta, A., Anderson, M. S., & Bernstein, A. (2021). Climate change and the practice of medicine: Essentials for resident education. *Academic Medicine*, 96(3), 355–367. DOI: 10.1097/ACM.0000000000003719 PMID: 32910006

Pichdara, L., Monin, N., Marong, C., Sivmuy, D., & Saren, K. (2022). Challenges of targeting poor and vulnerable groups to reduce climate change vulnerability: The case of a Water and Sanitation project in Kampong Svay District, Cambodia. Financial Crises, Poverty and Environmental Sustainability: Challenges in the Context of the SDGs and Covid-19 Recovery, pp.155-169.

Pickson, R. B., & Boateng, E. (2022). Climate change: A friend or foe to food security in Africa? *Environment, Development and Sustainability*, 24(3), 1–26. DOI: 10.1007/s10668-021-01621-8 PMID: 34276245

Pihkala, P. (2018). Eco-anxiety, tragedy, and hope: Psychological and spiritual dimensions of climate change. *Zygon*, 53(2), 545–569. DOI: 10.1111/zygo.12407

Pihkala, P. (2020). Anxiety and the ecological crisis: An analysis of eco-anxiety and climate anxiety. *Sustainability (Basel)*, 12(19), 7836. DOI: 10.3390/su12197836

Pihkala, P. (2020). Eco-anxiety and environmental education. *Sustainability (Basel)*, 12(23), 10149. DOI: 10.3390/su122310149

Pollack, A.A., Weiss, B., Trung, L.T. (2016). Mental health, life functioning, and risk factors among people exposed to frequent natural disasters and chronic poverty in Vietnam. *BJPsych Open,* 2;2(3):221–32.

Pool, J., Akhlaghpour, S., Fatehi, F., & Burton-Jones, A. (2024). A systematic analysis of failures in protecting personal health data: A scoping review. *International Journal of Information Management*, 74, 102719. DOI: 10.1016/j.ijinfomgt.2023.102719

Prasad, A. S., & Francescutti, L. H. (2017). Natural disasters. *International Encyclopedia of Public Health*, 3(2), 215–222. DOI: 10.1016/B978-0-12-803678-5.00519-1

Prats, E. V. (2020). WHO guidance for climate-resilient and environmentally sustainable health care facilities. www.who.int

Priddy, A. (2019). Disability in Armed Conflict. The Geneva Academy. www.geneva-academy.ch

Prina, M., Khan, N., & Akhter Kan, S. (2024). Healthy ageing and climate change: An assessment of the impact of climate hazards on older people. *Global Journal of Health, 14.*

Proulx, K., Daelmans, B., Baltag, V., & Banati, P. (2024). Climate change impacts on child and adolescent health and wellbeing: A narrative review. *Journal of Global Health*, •••, 4. PMID: 38781568

Pun, V. C., Manjourides, J., & Suh, H. (2017). Association of ambient air pollution with depressive and anxiety symptoms in older adults: Results from the NSHAP Study. *Environmental Health Perspectives*, 125(3), 342–348. DOI: 10.1289/EHP494 PMID: 27517877

Rahaman, M. A., Rahman, M. M., & Rahman, S. H. (2019). Pathways of climate-resilient health systems in Bangladesh. Confronting Climate Change in Bangladesh. *Adaptation and Resilience*, 3, 119–143.

Rahmani, M., Muzwagi, A., & Pumariega, A. J. (2022). Cultural factors in disaster response among diverse children and youth around the world. *Current Psychiatry Reports*. *Current Psychiatry Reports*, 24(10), 481–491. DOI: 10.1007/s11920-022-01356-x PMID: 35953637

Raile, P. (2023). Psychodrama and eco-anxiety—A case study. *Environment and Social Psychology, 9*(1).

Rajasekhar, B., Tushar, W., Lork, C., Zhou, Y., Yuen, C., Pindoriya, N. M., & Wood, K. L. (2020). A survey of computational intelligence techniques for air-conditioners energy management. *IEEE Transactions on Emerging Topics in Computational Intelligence*, 4(4), 555–570. DOI: 10.1109/TETCI.2020.2991728

Raj, R., Singh, A., Kumar, V., & Verma, P. (2023). Analyzing the potential benefits and use cases of ChatGPT as a tool for improving the efficiency and effectiveness of business operations. *BenchCouncil Transactions on Benchmarks. Standards and Evaluations*, 3(3), 100140.

Raju, S., Siddharthan, T., & McCormack, M. C. (2020). Indoor air pollution and respiratory health. *Clinics in Chest Medicine*, 41(4), 825–843. DOI: 10.1016/j.ccm.2020.08.014 PMID: 33153698

Rane, N. L. (2023). Multidisciplinary collaboration: key players in successful implementation of ChatGPT and similar generative artificial intelligence in manufacturing, finance, retail, transportation, and construction industry.

Rane, N., Choudhary, S., & Rane, J. (2023). Integrating Building Information Modelling (BIM) with ChatGPT, Bard, and similar generative artificial intelligence in the architecture, engineering, and construction industry: applications, a novel framework, challenges, and future scope. *Bard, and similar generative artificial intelligence in the architecture, engineering, and construction industry: applications, a novel framework, challenges, and future scope (November 22, 2023).*

Rane, N. (2023). Roles and challenges of ChatGPT and similar generative artificial intelligence for achieving the sustainable development goals (SDGs). *Available atSSRN* 4603244. DOI: 10.2139/ssrn.4603244

Ratten, V., & Jones, P. (2023). Generative artificial intelligence (ChatGPT): Implications for management educators. *International Journal of Management Education*, 21(3), 100857. DOI: 10.1016/j.ijme.2023.100857

Ray, D.K., West, P.C., Clark, M., Gerber, S.S., Alexander, P.V., & Chattejee, S. (2019). Climate Change has likely Affected Global Food Production. *Plose One* 14(5) e 0217148.

Ray, S. J., & Sibara, J. (Eds.). (2017). *Disability Studies and the Environmental Humanities: Toward an Eco-Crip Theory*. University of Nebraska Press. DOI: 10.2307/j.ctt1p6jht5

Renzaho, A. M., & Mellor, D. (2010). Food security measurement in cultural pluralism: Missing the point or conceptual misunderstanding? *Nutrition (Burbank, Los Angeles County, Calif.)*, 26(1), 1–9. DOI: 10.1016/j.nut.2009.05.001 PMID: 19804955

Rofi, A., Doocy, S., & Robinson, C. (2006). Tsunami mortality and displacement in Aceh province, Indonesia. *Disasters*, 30(3), 340–350. DOI: 10.1111/j.0361-3666.2005.00324.x PMID: 16911432

Romanello, M., di Napoli, C., Green, C., Kennard, H., Lampard, P., & Scamman, D. (2023). The 2023 report of the Lancet Countdown on health and climate change: The imperative for a health-centred response in a world facing irreversible harms. *Lancet*, 402(10419), 2346–2394. DOI: 10.1016/S0140-6736(23)01859-7 PMID: 37977174

Romanello, M., McGushin, A., Di Napoli, C., Drummond, P., Hughes, N., Jamart, L., Kennard, H., Lampard, P., Solano Rodriguez, B., Arnell, N., Ayeb-Karlsson, S., Belesova, K., Cai, W., Campbell-Lendrum, D., Capstick, S., Chambers, J., Chu, L., Ciampi, L., Dalin, C., & Hamilton, I. (2021). The 2021 Report of the Lancet Countdown on Health and Climate Change: Code Red for a Healthy Future. *Lancet*, 398(10311), 1619–1662. DOI: 10.1016/S0140-6736(21)01787-6 PMID: 34687662

Rose, W. M. (2023). Climate Anxiety: The Emotional Impact of Climate Change.

Ross, A., Mack, E. A., Marcantonio, R., Miller-Graff, L., Pearson, A. L., Smith, A. C., & Zimmer, A. (2024). A mediation analysis of the linkages between climate variability, water insecurity, and interpersonal violence. *Climate and Development*, 16(2), 134–148. DOI: 10.1080/17565529.2023.2186746

Roush, S. E., & Sharby, N. (2011). Disability reconsidered: The paradox of physical therapy. *Physical Therapy*, 91(12), 1715–1727. DOI: 10.2522/ptj.20100389 PMID: 22003161

Runciman, B., Merry, A., & Walton, M. (2017). *Safety and ethics in healthcare: a guide to getting it right*. CRC Press.

Russ, S., Regalado, M., & Halfon, N. (2008). Healthcare. *Encyclopaedia of Infant and Early Childhood Development,* pp. 66-80.

Russell, M. (2019). *Capitalism and disability: Selected writings by Marta Russell.* Haymarket Books.

Sadrizadeh, S., Yao, R., Yuan, F., Awbi, H., Bahnfleth, W., Bi, Y., Cao, G., Croitoru, C., de Dear, R., Haghighat, F., Kumar, P., Malayeri, M., Nasiri, F., Ruud, M., Sadeghian, P., Wargocki, P., Xiong, J., Yu, W., & Li, B. (2022). Indoor air quality and health in schools: A critical review for developing the roadmap for the future school environment. *Journal of Building Engineering*, 57, 104908. DOI: 10.1016/j.jobe.2022.104908

Saeed, S. A., & Gargano, S. P. (2022). Natural disasters and mental health. *International Review of Psychiatry (Abingdon, England)*, 34(1), 16–25. DOI: 10.1080/09540261.2022.2037524 PMID: 35584023

Safi, M. (2017). Suicides of nearly 60,000 Indian farmers linked to climate change, study claims. *The Guardian, 31.*

Saini, J., Dutta, M., Marques, G., Saini, J., Dutta, M., & Marques, G. (2021). Indoor Air Quality: Impact on Public Health. *Internet of Things for Indoor Air Quality Monitoring*, 1-14.

Saini, J., Dutta, M., & Marques, G. (2020). Indoor air quality prediction systems for smart environments: A systematic review. *Journal of Ambient Intelligence and Smart Environments*, 12(5), 433–453. DOI: 10.3233/AIS-200574

SAMHSA. (2017). SAMHSA Disaster Technical assistance center supplemental research bulletin greater impact: How disasters affect people of low socioeconomic status. SAMHSA. https://www.samhsa.gov/sites/default/files/dtac/srb-low-ses_2.pdf

Sanyaolu, A., Okorie, C., Qi, X., Locke, J., & Rehman, S. (2019). Childhood and adolescent obesity in the United States: a public health concern. *Global pediatric health, 6*, 2333794X19891305.

Saran, S., Gurjar, M., Baronia, A., Sivapurapu, V., Ghosh, P. S., Raju, G. M., & Maurya, I. (2020). Heating, ventilation and air conditioning (HVAC) in intensive care unit. *Critical Care*, 24(1), 1–11. DOI: 10.1186/s13054-020-02907-5 PMID: 32375844

Sasaki, N., Rosenberg, M., Shin, J. H., Kunisawa, S., & Imanaka, Y. (2024). Hidden Populations for Healthcare Financial Protection in the Super-Aging Society: Closing the Gap Between Policy and Practice. *Clinical Social Work Journal*, 52(3), 1–12. DOI: 10.1007/s10615-023-00914-x PMID: 39188583

Schipper, E. L. F. (2020). Maladaptation: When adaptation to climate change goes very wrong. *One Earth*, 3(4), 409–414. DOI: 10.1016/j.oneear.2020.09.014

Schlosberg, D. (2013). Theorizing environmental justice: The expanding sphere of a discourse. *Environmental Politics*, 22(1), 37–55. DOI: 10.1080/09644016.2013.755387

Schmidhuber, J., & Tubiello, F. N. (2007). Global food security under climate change. *National Academy of Sciences*, 104(50).

Schneider, M. J. (2020). *Introduction to public health.* Jones & Bartlett Learning.

Schnitter, R., Verret, M., Berry, P., Chung, T., Fook, T., Hales, S., Lal, A., & Edwards, S. (2019). An assessment of climate change and health vulnerability and adaptation in Dominica. *International Journal of Environmental Research and Public Health*, 16(1), 70. DOI: 10.3390/ijerph16010070 PMID: 30597870

Sen, A. (1982). *Poverty and famines: an essay on entitlement and deprivation.* Oxford University Press.

Sen, B., Dhimal, M., Latheef, A. T., & Ghosh, U. (2017). Climate change: Health effects and response in South Asia. *BMJ (Clinical Research Ed.)*, 5, 359. DOI: 10.1136/bmj.j5117 PMID: 29146578

Seneviratne, S. I., Nicholls, N., Easterling, D., Goodess, C. M., Kanae, S., Kossin, J., Luo, Y., Marengo, J., McInnes, K., Rahimi, M., Reichstein, M., Sorteberg, A., Vera, C., & Zhang, X. (2012). Changes in climate extremes and their impacts on the natural physical environment. In Field, C. B., V. Barros, T.F. Stocker, D. Qin, D.J. Dokken, K.L. Ebi, M.D. Mastrandrea, K.J. Mach, G.-K. Plattner, S.K. Allen, M. Tignor, and P.M. Midgley (eds.), *Managing the Risks of Extreme Events and Disasters to Advance Climate Change Adaptation. A Special Report of Working Groups I and II of the Intergovernmental Panel on Climate Change (IPCC)* (pp. 109-230). Cambridge University Press, Cambridge, UK, and New York, NY, USA.) DOI: 10.1017/CBO9781139177245.006

Sertse, S. F., Khan, N. A., Shah, A. A., Liu, Y., & Naqvi, S. A. A. (2021). Farm households' perceptions and adaptation strategies to climate change risks and their determinants: Evidence from Raya Azebo district, Ethiopia. *International Journal of Disaster Risk Reduction*, 60, 102255. DOI: 10.1016/j.ijdrr.2021.102255

Sezgin, A. U., & Punamäki, R. L. (2016). Perceived changes in social relations after earthquake trauma among eastern Anatolian women: Associated factors and mental health consequences. *Stress and Health*, 32(4), 355–366. DOI: 10.1002/smi.2629 PMID: 25516368

Sezgin, E. (2023). Artificial intelligence in healthcare: Complementing, not replacing, doctors and healthcare providers. *Digital Health*, 9, 20552076231186520. DOI: 10.1177/20552076231186520 PMID: 37426593

Shakespeare, T. (2013). *Disability Rights and Wrongs Revisited* (second). New York: Routledge.

Sharma, A., & Singh, B. (2022). Measuring Impact of E-commerce on Small Scale Business: A Systematic Review. *Journal of Corporate Governance and International Business Law*, 5(1).

Shaw, F. E., Asomugha, C. N., Conway, P. H., & Rein, A. S. (2024). The Patient Protection and Affordable Care Act: Opportunities for prevention and public health. *Lancet*, 384(9937), 75–82. DOI: 10.1016/S0140-6736(14)60259-2 PMID: 24993913

Shivanna, K. R. (2022). Climate change and its impact on biodiversity and human welfare. *Proceedings of the Indian National Science Academy. Part A, Physical Sciences*, 88(2), 160–171. DOI: 10.1007/s43538-022-00073-6

Shooshtari, S., Abedi, M., Bahrami, M., & Samouei, R. (2018). The mental health needs of women in natural disasters: A qualitative study with a preventive approach. *Journal of Family Medicine and Primary Care*, 7(4), 678. DOI: 10.4103/jfmpc.jfmpc_333_17 PMID: 30234037

Shultz, J. M., Neria, Y., Allen, A., & Espinel, Z. (2013). Psychological impacts of natural disasters. In Bobrowsky P. T. (Edt.), *Encyclopaedia of Natural Hazards. Netherlands*, pp. 779-791. Springer. DOI: 10.1007/978-1-4020-4399-4_279

Sikka, R., Morath, J. M., & Leape, L. (2020). The quadruple aim: Care, health, cost and meaning in work. *BMJ Quality & Safety*, 24(10), 608–610. DOI: 10.1136/bmjqs-2015-004160 PMID: 26038586

Singh, B. (2023). Blockchain Technology in Renovating Healthcare: Legal and Future Perspectives. In *Revolutionizing Healthcare Through Artificial Intelligence and Internet of Things Applications* (pp. 177-186). IGI Global.

Singh, A. K., Ashraf, S. N., Sharma, S. K., & Kumar, S. (2023a). Factors affecting farm income and climate adaptation strategies in the farming sector of Gujarat state of India. In Sadhik, B. J., Adewole, J., Yeneneh, A., & Abidoye, L. (Eds.), *Opportunities and Challenges in Climate-Friendly Clean Water and Energy Technologies* (pp. 122–147). IGI Global., DOI: 10.4018/978-1-6684-7303-0.ch006

Singh, A. K., Jyoti, B., & Mishra, V. K. (2023c). Policy implications from selected countries for a sustainable future of the society through green entrepreneurship and economic development. In Abdelli, M., Ul-Durar, S., & Wasef Hattab, H. (Eds.), *Entrepreneurship Ecosystems and Their Opportunities and Challenges* (pp. 129–162). IGI Global., DOI: 10.4018/978-1-6684-7140-1.ch006

Singh, A. K., Jyoti, B., & Sankaranarayanan, K. G. (2024). A review for analyzing the impact of climate change on agriculture and food security in India. In Samanta, D., & Garg, M. (Eds.), *The Climate Change Crisis and Its Impact on Mental Health* (pp. 227–251). IGI Global., DOI: 10.4018/979-8-3693-3272-6.ch017

Singh, A. K., Kumar, S., & Jyoti, B. (2022). Influence of climate change on agricultural sustainability in India: A state-wise panel data analysis. *Asian Journal of Agriculture*, 6(1). Advance online publication. DOI: 10.13057/asianjagric/g060103

Singh, B. (2019). Affordability of Medicines, Public Health and TRIPS Regime: A Comparative Analysis. *Indian Journal of Health and Medical Law*, 2(1), 1–7.

Singh, B. (2023). Blockchain Technology in Renovating Healthcare: Legal and Future Perspectives. In Kaushik, K., Dahiya, S., Aggarwal, S., & Dwivedi, A. (Eds.), *Revolutionizing Healthcare Through Artificial Intelligence and Internet of Things Applications* (pp. 177–186). IGI Global., DOI: 10.4018/978-1-6684-5422-0.ch012

Singh, B. (2023). Federated Learning for Envision Future Trajectory Smart Transport System for Climate Preservation and Smart Green Planet: Insights into Global Governance and SDG-9 (Industry, Innovation and Infrastructure). *National Journal of Environmental Law*, 6(2), 6–17.

Singh, B. (2023). Tele-Health Monitoring Lensing Deep Neural Learning Structure: Ambient Patient Wellness via Wearable Devices for Real-Time Alerts and Interventions. *Indian Journal of Health and Medical Law*, 6(2), 12–16.

Singh, B. (2023). Unleashing Alternative Dispute Resolution (ADR) in Resolving Complex Legal-Technical Issues Arising in Cyberspace Lensing E-Commerce and Intellectual Property: Proliferation of E-Commerce Digital Economy. *Revista Brasileira de Alternative Dispute Resolution-Brazilian Journal of Alternative Dispute Resolution-RBADR*, 5(10), 81–105. DOI: 10.52028/rbadr.v5i10.ART04.Ind

Singh, B. (2024). Cherish Growth, Advancement and Tax Structure: Addressing Social and Economic Prospects. *Journal of Taxation and Regulatory Framework*, 7(1), 7–10.

Singh, B. (2024). Evolutionary Global Neuroscience for Cognition and Brain Health: Strengthening Innovation in Brain Science. In Prabhakar, P. (Ed.), *Biomedical Research Developments for Improved Healthcare* (pp. 246–272). IGI Global., DOI: 10.4018/979-8-3693-1922-2.ch012

Singh, B. (2024). Featuring Consumer Choices of Consumable Products for Health Benefits: Evolving Issues from Tort and Product Liabilities. *Journal of Law of Torts and Consumer Protection Law*, 7(1).

Singh, B. (2024). Green Infrastructure in Real Estate Landscapes: Pillars of Sustainable Development and Vision for Tomorrow. *National Journal of Real Estate Law*, 7(1), 4–8.

Singh, B. (2024). Legal Dynamics Lensing Metaverse Crafted for Videogame Industry and E-Sports: Phenomenological Exploration Catalyst Complexity and Future. *Journal of Intellectual Property Rights Law*, 7(1), 8–14.

Singh, B. (2024). Lensing Legal Dynamics for Examining Responsibility and Deliberation of Generative AI-Tethered Technological Privacy Concerns: Infringements and Use of Personal Data by Nefarious Actors. In Ara, A., & Ara, A. (Eds.), *Exploring the Ethical Implications of Generative AI* (pp. 146–167). IGI Global., DOI: 10.4018/979-8-3693-1565-1.ch009

Singh, B. (2024). Social Cognition of Incarcerated Women and Children: Addressing Exposure to Infectious Diseases and Legal Outcomes. In Reddy, K. (Ed.), *Principles and Clinical Interventions in Social Cognition* (pp. 236–251). IGI Global., DOI: 10.4018/979-8-3693-1265-0.ch014

Singh, B., Jain, V., Kaunert, C., Dutta, P. K., & Singh, G. (2024). Privacy Matters: Espousing Blockchain and Artificial Intelligence (AI) for Consumer Data Protection on E-Commerce Platforms in Ethical Marketing. In Saluja, S., Nayyar, V., Rojhe, K., & Sharma, S. (Eds.), *Ethical Marketing Through Data Governance Standards and Effective Technology* (pp. 167–184). IGI Global., DOI: 10.4018/979-8-3693-2215-4.ch015

Singh, B., & Kaunert, C. (2024). Computational Thinking for Innovative Solutions and Problem-Solving Techniques: Transforming Conventional Education to Futuristic Interdisciplinary Higher Education. In Fonkam, M., & Vajjhala, N. (Eds.), *Revolutionizing Curricula Through Computational Thinking, Logic, and Problem Solving* (pp. 60–82). IGI Global., DOI: 10.4018/979-8-3693-1974-1.ch004

Sirba, H. Y., & Chimdessa, T. B. (2021). Review of Impact of Climate Change on Food Security in Africa. *Int. J. Res. Innov. Earth Sci*, 8, 2394–1375.

Slaunwhite, A., Min, J. E., Palis, H., Urbanoski, K., Pauly, B., Barker, B., ... & Nosyk, B. (2024). Effect of Risk Mitigation Guidance opioid and stimulant dispensations on mortality and acute care visits during dual public health emergencies: retrospective cohort study. *bmj, 384*.

Smith, K. R., Woodward, A., Campbell-Lendrum, D., Chadee, D. D., & Honda, Y. (2014). Human Health: Impacts, Adaptation, and Co-Benefits. In C.B, Field, V.R, Barros., D.J, Dokken., K.J. Mach., & M.D. Mastrandrea (Eds.). *Climate Change 2014: Impacts, Adaptation, and Vulnerability; Part A: Global and Sectoral Aspects. Contribution of Working Group II to the Fifth Assessment Report of the Intergovernmental Panel on Climate Change.* New York: Cambridge University Press.

Solomon, S., Qin, D., Manning, M., Averyt, K., & Marquis, M. (2007). [-*the physical science basis: Working group I contribution to the fourth assessment report of the IPCC.* UK: Cambridge university press.]. *Climatic Change*, •••, 2007.

Soni, V. (2023). Impact of Generative AI on Small and Medium Enterprises' Revenue Growth: The Moderating Role of Human, Technological, and Market Factors. *Reviews of Contemporary Business Analytics*, 6(1), 133–153.

Sorensen, C., Saunik, S., Sehgal, M., Tewary, A., Govindan, M., Lemery, J., & Balbus, J. (2018). Climate change and women's health: Impacts and opportunities in India. *GeoHealth*, 2(10), 283–297. DOI: 10.1029/2018GH000163 PMID: 32159002

Stafoggia, M., Forastiere, F., Agostini, D., Biggeri, A., Bisanti, L., & Cadum, E.. (2006). Vulnerability to Heat-Related Mortality: A Multicity, Population-Based, Case-Crossover Analysis. *Epidemiology (Cambridge, Mass.)*, 17(3), 315–323. DOI: 10.1097/01.ede.0000208477.36665.34 PMID: 16570026

Stanley, S. K., Hogg, T. L., Leviston, Z., & Walker, I. (2021). From anger to action: Differential impacts of eco-anxiety, eco-depression, and eco-anger on climate action and wellbeing. *The Journal of Climate Change and Health*, 1, 100003. DOI: 10.1016/j.joclim.2021.100003

Stasevych, M., & Zvarych, V. (2023). Innovative robotic technologies and artificial intelligence in pharmacy and medicine: Paving the way for the future of health care—a review. *Big Data and Cognitive Computing*, 7(3), 147. DOI: 10.3390/bdcc7030147

Stella, E., Mari, L., Gabrieli, J., Barbante, C., & Bertuzzo, E. (2020). Permafrost dynamics and the risk of anthrax transmission: A modeling study. *Scientific Reports*, 10(1), 16460. DOI: 10.1038/s41598-020-72440-6 PMID: 33028874

Stephane, H., Mook, B., Laura, B., Marianne, F., Tamaro, K., Ulf, N., Julie, R., David, T., & Adrien, V. (2015). *Shock Waves.* World Bank Publications.

Stevanovic, M., Popp, A., Lotze-canpen, H., & Dietrich, J. (2016). The Impact of High-end Climate Change on Agricultural Welfare. *Science Advances*, 12(14), 16–34. DOI: 10.1126/sciadv.1501452 PMID: 27574700

Stork, A., Travis, C., & Halle, S. (2015). Gender-Sensitivity in natural resource management in Côte d'Ivoire and Sudan. *Peace Review*, 27(2), 147–155. DOI: 10.1080/10402659.2015.1037617

Strand, L. B., Barnett, A. G., & Tong, S. (2011). The influence of season and ambient temperature on birth outcomes: A review of the epidemiological literature. *Environmental Research*, 111(3), 451–462. DOI: 10.1016/j.envres.2011.01.023 PMID: 21333980

Sundblad, E.-L., Biel, A., & Gärling, T. (2007). Cognitive and affective risk judgements related to climate change. *Journal of Environmental Psychology*, 27(2), 97–106. DOI: 10.1016/j.jenvp.2007.01.003

Swamy, G. S. N. V. K. S. N. (2021). Development of an indoor air purification system to improve ventilation and air quality. *Heliyon*, 7(10), e08153. DOI: 10.1016/j.heliyon.2021.e08153 PMID: 34703923

Syed, M., Folz, R. J., & Ali, U. (2023). Environmental factors and their impact on airway diseases: Exploring air pollution, indoor and outdoor allergens, and climate change. *Current Pulmonology Reports*, 12(3), 162–170. DOI: 10.1007/s13665-023-00319-8

Talebi Bezmin Abadi, A., Rizvanov, A. A., Haertlé, T., & Blatt, N. L. (2019). World Health Organization report: Current crisis of antibiotic resistance. *BioNanoScience*, 9(4), 778–788. DOI: 10.1007/s12668-019-00658-4

Tam, K. P., Chan, H. W., & Clayton, S. (2023). Climate change anxiety in China, India, Japan, and the United States. *Journal of Environmental Psychology*, 87, 101991. DOI: 10.1016/j.jenvp.2023.101991

Tanwar, N., & Akhai, S. (2017). Survey Analysis for Quality Control Comfort Management in Air Conditioned Classroom. *Journal of Advanced Research in Civil and Environmental Engineering*, 4(1&2), 20–23. DOI: 10.24321/2393.8307.201702

Tan, Y., Yang, R., Zhao, J., Cao, Z., Chen, Y., & Zhang, B. (2017). The associations between air pollution and adverse pregnancy outcomes in China. *Advances in Experimental Medicine and Biology*, 1017, 181–214. DOI: 10.1007/978-981-10-5657-4_8 PMID: 29177963

Tarvainen, M. (2019). Ableism and the life stories of people with disabilities. *Scandinavian Journal of Disability Research*, 21(1), 291–299. DOI: 10.16993/sjdr.632

Thacker, S., Adshead, D., Fantini, C., Palmer, R., Ghosal, R., & Adeoti, T.. (2021). *Infrastructure for climate action*. United Nations Office for Project Services. Denmark.

Thamaga-Chitja, J. M., & Tamako, N. (2017). Does social capital play a role in climate change adaptation among smallholder farmers for improving food security and livelihoods? *Journal of Consumer Sciences*, 2, 16–27.

The Lancet. (2015). *Health Professionals: Be Prepared for Heatwaves.*www.thelancet.org

Thirunavukarasu, A. J., Ting, D. S. J., Elangovan, K., Gutierrez, L., Tan, T. F., & Ting, D. S. W. (2023). Large language models in medicine. *Nature Medicine*, 29(8), 1930–1940. DOI: 10.1038/s41591-023-02448-8 PMID: 37460753

Thomas, C. (1999). *Female Forms: Experiencing and understanding disability. Buckingham*. Open University Press.

Thomas, C. (2004). Developing the social-relational in the social model of disability: a theoretical agenda. In Barnes, C., & Mercer, G. (Eds.), *Implementing the Social Model of Disability: Theory and Research* (pp. 32–47). Disability Press.

Thompson, N. J., Fiorillo, D., Rothbaum, B. O., Ressler, K. J., & Michopoulos, V. (2018). Coping strategies as mediators in relation to resilience and posttraumatic stress disorder. *Journal of Affective Disorders*, 225(1), 153–159. DOI: 10.1016/j.jad.2017.08.049 PMID: 28837948

Tien, P. W., Wei, S., Darkwa, J., Wood, C., & Calautit, J. K. (2022). Machine learning and deep learning methods for enhancing building energy efficiency and indoor environmental quality–a review. *Energy and AI*, 10, 100198. DOI: 10.1016/j.egyai.2022.100198

Tigchelaar, M., Cheung, W. W. L., Mohammed, E. Y., Phillips, M. J., Payne, H. J., Selig, E. R., Wabnitz, C. C. C., Oyinlola, M. A., Frölicher, T. L., Gephart, J. A., Golden, C. D., Allison, E. H., Bennett, A., Cao, L., Fanzo, J., Halpern, B. S., Lam, V. W. Y., Micheli, F., Naylor, R. L., & Troell, M. (2021). Compound climate risks threaten aquatic food system benefits. *Nature Food*, 2(9), 673–682. DOI: 10.1038/s43016-021-00368-9 PMID: 37117477

Tiotiu, A. I., Novakova, P., Nedeva, D., Chong-Neto, H. J., Novakova, S., Steiropoulos, P., & Kowal, K. (2020). Impact of air pollution on asthma outcomes. *International Journal of Environmental Research and Public Health*, 17(17), 6212. Advance online publication. DOI: 10.3390/ijerph17176212 PMID: 32867076

Tobin-Gurley, J., Peek, L., & Loomis, J. (2010). Displaced single mothers in the aftermath of Hurricane Katrina: Resource needs and resource acquisition. *International Journal of Mass Emergencies and Disasters*, 28(2), 170–206. DOI: 10.1177/028072701002800202

Tochkin, J., Richmond, J., & Hertelendy, A. (2022). Healthcare system leadership and climate change: Five lessons for improving health systems resiliency. *BMJ Leader*, 4(7), 21–32. PMID: 37013877

Tomas, M. E., Kundrapu, S., Thota, P., Sunkesula, V. C., Cadnum, J. L., Mana, T. S. C., Jencson, A., O'Donnell, M., Zabarsky, T. F., Hecker, M. T., Ray, A. J., Wilson, B. M., & Donskey, C. J. (2023). Contamination of health care personnel during removal of personal protective equipment. *JAMA Internal Medicine*, 175(12), 1904–1910. DOI: 10.1001/jamainternmed.2015.4535 PMID: 26457544

Ton, K. T., & Adamson, C. (2020). Disaster justice for people with disabilities. *Disaster Prevention and Management*, 30(2), 125–136. DOI: 10.1108/DPM-08-2020-0248

Tran, V. V., Park, D., & Lee, Y. C. (2020). Indoor air pollution, related human diseases, and recent trends in the control and improvement of indoor air quality. *International Journal of Environmental Research and Public Health*, 17(8), 2927. DOI: 10.3390/ijerph17082927 PMID: 32340311

Trautmann, S., Rehm, J., & Wittchen, H. (2016). The economic costs of mental disorders. *EMBO Reports*, 17(9), 1245–1249. DOI: 10.15252/embr.201642951 PMID: 27491723

Tulenko, K., Mgedal, S., Afzal, M. M., Frymus, D., Oshin, A., Pate, M., & Zodpey, S. (2023). Community health workers for universal health-care coverage: From fragmentation to synergy. *Bulletin of the World Health Organization*, 91(11), 847–852. DOI: 10.2471/BLT.13.118745 PMID: 24347709

Tunstall, S., Tapsell, S., Green, C., Floyd, P., & George, C. (2006). The health effects of flooding: Social research results from England and Wales. *Journal of Water and Health*, 4(3), 365–380. DOI: 10.2166/wh.2006.031 PMID: 17036844

U.S. Global Change Research Program (USGCRP). (2016). *The Impacts of Climate Change on Human Health in the United States: A Scientific Assessment.*

Uddin, M. S., Haque, C. E., Khan, M. N., Doberstein, B., & Cox, R. S. (2021). Disasters threaten livelihoods, and people cope, adapt and make transformational changes: Community resilience and livelihoods reconstruction in coastal communities of Bangladesh. *International Journal of Disaster Risk Reduction*, 63(1), 23–34. DOI: 10.1016/j.ijdrr.2021.102444

UN General Assembly. (2007). *Convention on the Rights of Persons with Disabilities: resolution / adopted by the General Assembly, A/RES/61/106.* www.refworld.org

UNICEF. (2022). *UNICEF Fact Sheet Children with Disabilities.* www.unicef.org

Union of the Physically Impaired Against Segregation (UPIAS). (1976). *Fundamental Principles of Disability*. UPIAS.

United Nations Framework Convention on Climate Change. (2011). *Reducing Vulnerability to Climate Change, Climate Variability and Extremes, Land Degradation and Loss of Biodiversity: Environmental and Developmental Challenges and Opportunities*.https://unfccc.int/resource/docs/publications/pub_nwp_reducing_vulnerability.pdf

United Nations General Assembly. (2015). *Transforming our World: The 2030 Agenda for Sustainable Development*. United Nations General Assembly.

United Nations High Commissioner for Refugees. (2003). *Sexual and Gender-Based Violence against Refugees, Returnees, and Internally Displaced Persons: Guidance for Prevention and Response*.https://www.unhcr.org/sites/default/files/legacy-pdf/3f696bcc4.pdf

United Nations International Strategy for Disaster Reduction. (2008). *Gender perspectives: Integrating disaster risk reduction into climate change adaptation*. United Nations International Strategy for Disaster Reduction. https://www.unisdr.org/files/3391_GenderPerspectivesIntegrating DRRCC Adaptation.pdf

United Nations. (2015). *General Comment on equal recognition before the law*. www.ohchr.org

United Nations. (2015). *Paris Agreement*. United Nations.

United Nations. (2018). *Persons with disabilities: Breaking down barriers*.www.un.org

United Nations. (2022). *The human right to a clean, healthy and sustainable environment: resolution/adopted by the General Assembly*.www.un.org

Usher, K., Durkin, J., & Bhullar, N. (2019). Eco-anxiety: How thinking about climate change-related environmental decline is affecting our mental health. *International Journal of Mental Health Nursing*, 28(6), 1233–1234. DOI: 10.1111/inm.12673 PMID: 31724833

Van Den Berg, B., Grievink, L., Yzermans, J., & Lebret, E. (2005). Medically unexplained physical symptoms in the aftermath of disasters. *Epidemiologic Reviews*, 27(1), 92–106. DOI: 10.1093/epirev/mxi001 PMID: 15958430

Van Ittersum, M. K., Van Bussel, L. G., Wolf, J., Grassini, P., Van Wart, J., Guilpart, N., ... & Cassman, K. G. (2016). Can sub-Saharan Africa feed itself?. *National Academy of Sciences,* 113(52), 14964-14969.

Van Lerberghe, W. (2018). [*primary health care: now more than ever.* World Health Organization.]. *World Health Report*, •••, 2008.

Van Niel, M. S., & Payne, J. L. (2020). Perinatal depression: A review. *Cleveland Clinic Journal of Medicine*, 87(5), 273–277. DOI: 10.3949/ccjm.87a.19054 PMID: 32357982

Van Zutphen, A. R., Lin, S., Fletcher, B. A., & Hwang, S. A. (2012). A population-based case–control study of extreme summer temperature and birth defects. *Environmental Health Perspectives*, 120(10), 1443–1449. DOI: 10.1289/ehp.1104671 PMID: 23031822

Vandentorren, S., Bretin, P., Zeghnoun, A., Mandereau-Bruno, L., Croisier, A., Cochet, C., & Ledrans, M. (2006, December). August 2003 heat wave in France: Risk factors for death of elderly people living at home. *European Journal of Public Health*, 16(6), 583–591. Advance online publication. DOI: 10.1093/eurpub/ckl063 PMID: 17028103

Vardoulakis, S., Giagloglou, E., Steinle, S., Davis, A., Sleeuwenhoek, A., Galea, K. S., Dixon, K., & Crawford, J. O. (2020). Indoor exposure to selected air pollutants in the home environment: A systematic review. *International Journal of Environmental Research and Public Health*, 17(23), 8972. DOI: 10.3390/ijerph17238972 PMID: 33276576

Vigo, D., Thornicroft, G., & Atun, R. (2016). Estimating the true global burden of mental illness. *The Lancet. Psychiatry*, 3(2), 171–178. DOI: 10.1016/S2215-0366(15)00505-2 PMID: 26851330

Viguié, V., Lemonsu, A., Hallegatte, S., Beaulant, A. L., Marchadier, C., Masson, V., Pigeon, G., & Salagnac, J. L. (2020). Early adaptation to heat waves and future reduction of air-conditioning energy use in Paris. *Environmental Research Letters*, 15(7), 075006. DOI: 10.1088/1748-9326/ab6a24

Vornanen-Winqvist, C., Järvi, K., Andersson, M. A., Duchaine, C., Létourneau, V., Kedves, O., Kredics, L., Mikkola, R., Kurnitski, J., & Salonen, H. (2020). Exposure to indoor air contaminants in school buildings with and without reported indoor air quality problems. *Environment International*, 141, 105781. DOI: 10.1016/j.envint.2020.105781 PMID: 32417615

Vyas, B., & Rajendran, R. M. (2023). Generative Adversarial Networks for Anomaly Detection in Medical Images. *International Journal of Multidisciplinary Innovation and Research Methodology, ISSN: 2960-2068,* 2(4), 52-58.

Wach, K., Duong, C. D., Ejdys, J., Kazlauskaitė, R., Korzynski, P., Mazurek, G., Paliszkiewicz, J., & Ziemba, E. (2023). The dark side of generative artificial intelligence: A critical analysis of controversies and risks of ChatGPT. *Entrepreneurial Business and Economics Review*, 11(2), 7–30. DOI: 10.15678/EBER.2023.110201

Wade, T. (2022). Health risks associated with sea level rise. National Collaborating Centre for Environmental Health (NCCEH). https://ncceh.ca/resources/evidence-reviews/health-risks-associated-sea-level-rise

Wahid, S. S., Raza, W. A., Mahmud, I., & Kohrt, B. A. (2023). Climate-related shocks and other stressors associated with depression and anxiety in Bangladesh: A nationally representative panel study. *The Lancet. Planetary Health*, 7(2), e137–e146. DOI: 10.1016/S2542-5196(22)00315-1 PMID: 36754469

Walker, B., Holling, C. S., Carpenter, S. R., & Kinzig, A. (2004). Resilience, Adaptability, and Transformability In Social–Ecological Systems. *Ecology and Society*, 9(6), 15–28. DOI: 10.5751/ES-00650-090205

Walter, L., Filho, U. M., & Azeiteiro, F. A. (2016). *Climate Change and Health*. Springer.

Wamba, S. F., Queiroz, M. M., Jabbour, C. J. C., & Shi, C. V. (2023). Are both generative AI and ChatGPT game changers for 21st-Century operations and supply chain excellence? *International Journal of Production Economics*, 265, 109015. DOI: 10.1016/j.ijpe.2023.109015

Wang, F., Fang, Y., Deng, H., & Wei, F. (2022). How community medical facilities can promote resilient community constructions under the background of pandemics. *Indoor and Built Environment*, 31(4), 1018–1027. DOI: 10.1177/1420326X211048537 PMID: 35431637

Wang, H., Zhao, L., Cao, R., & Zeng, W. (2021). Refrigerant alternative and optimization under the constraint of the greenhouse gas emissions reduction target. *Journal of Cleaner Production*, 296, 126580. DOI: 10.1016/j.jclepro.2021.126580

Watts, N., Amann, M., Ayeb-Karlsson, S., Belesova, K., Bouley, T., Boykoff, M., et al. (2018). The Lancet Countdown on health and climate change: from 25 years of inaction to a global transformation for public health. *Lancet*, 10;391(10120):581–630.

Watts, N., Amann, M., Ayeb-Karlsson, S., Belesova, K., Bouley, T., Boykoff, M., et al. (2018). The Lancet Countdown on health and climate change: from 25 years of inaction to a Global Transformation for Public Health. *The Lancet,* 10;391 (10120):581– 630.

Watts, N., Adger, W. N., Agnolucci, P., Blackstock, J., Byass, P., Cai, W., Chaytor, S., Colbourn, T., Collins, M., Cooper, A., Cox, P. M., Depledge, J., Drummond, P., Ekins, P., Galaz, V., Grace, D., Graham, H., Grubb, M., Haines, A., & Costello, A. (2015). Lancet Commission: Health and Climate Change: Policy Responses to Protect Public Health. *Lancet*, 386(10006), 1861–1914. DOI: 10.1016/S0140-6736(15)60854-6 PMID: 26111439

Watts, N., Amann, M., Arnell, N., Ayeb-Karlsson, S., Belesova, K., Boykoff, M., Byass, P., Cai, W., Campbell-Lendrum, D., Capstick, S., Chambers, J., Dalin, C., Daly, M., Dasandi, N., Davies, M., Drummond, P., Dubrow, R., Ebi, K. L., Eckelman, M., & Montgomery, H. (2019). The 2019 report of The Lancet Countdown on health and climate change: Ensuring that the health of a child born today is not defined by a changing climate. *Lancet*, 394(10211), 1836–1878. DOI: 10.1016/S0140-6736(19)32596-6 PMID: 31733928

Webb, P., Coates, J., Frongillo, E. A., Rogers, B., Swindale, A., & Bilinsky, P. (2006). Measuring Household Food Insecurity: Why It's So Important and Yet So Difficult to Do. *The Journal of Nutrition*, 136(5), 1404S–1408S. DOI: 10.1093/jn/136.5.1404S PMID: 16614437

Weintrobe, S. (2013). *Engaging with climate change: psychoanalytic and interdisciplinary perspectives*. Routledge.

WHO. (2002). *Towards a common language for functioning, disability and health*. www.who.org

WHO. (2004). *Promoting Mental Health: Concepts, Emerging Evidence, Practice*. www.who.int

WHO. (2007). *Everybody's business–strengthening Health Systems to Improve Health Outcomes: WHO's Framework for Action*.www.who.int

WHO. (2015). *Operational framework for building climate resilient health systems*. www.who.int

WHO. (2022). *Health and climate change: country profile 2022: Czechia*.www.who.int

Willets, E., & Haines, A. (2023). *Managing the Health Risks of Climate Change*. www.cfr.org

Williams, L. (2018). Climate change, colonialism, and women's well-being in Canada: What is to be done? *Canadian Journal of Public Health*, 109(2), 268–271. DOI: 10.17269/s41997-018-0031-z PMID: 29981032

Williams, S. J. (1999). Is anybody there? Critical realism, chronic illness and the disability debate. *Sociology of Health & Illness*, 21(6), 797–819. DOI: 10.1111/1467-9566.00184

Wing, I. S., De Cian, E., & Mistry, M. N. (2021). Global vulnerability of crop yields to climate change. *Journal of Environmental Economics and Management*, 109, 102462. DOI: 10.1016/j.jeem.2021.102462

Wisner, K. L., Sit, D. K. Y., McShea, M. C., Rizzo, D. M., Zoretich, R. A., Hughes, C. L., & Hanusa, B. H. (2013). Onset Timing, Thoughts of Self-harm, and Diagnoses in Postpartum Women with Screen-Positive Depression Findings. *JAMA Psychiatry*, 70(5), 490. DOI: 10.1001/jamapsychiatry.2013.87 PMID: 23487258

Wlokas, H. L. (2008b). The impacts of climate change on food security and health in Southern Africa. *Journal of Energy in Southern Africa*, 19(4), 12–20. DOI: 10.17159/2413-3051/2008/v19i4a3334

Wolbring, G. (2009). A Culture of Neglect: Climate Discourse and Disabled People. *M/C Journal,* 12(4).

Wolbring, G. (2011)... *People with Disabilities and Social Determinants of Health Discourses.*, 102(4), 317–319.

Wolfson, J. A., & Leung, C. W. (2020). Food insecurity and COVID-19: Disparities in early effects for US adults. *Nutrients*, 12(6), 1648. DOI: 10.3390/nu12061648 PMID: 32498323

Wong, I. A., Lian, Q. L., & Sun, D. (2023). Autonomous travel decision-making: An early glimpse into ChatGPT and generative AI. *Journal of Hospitality and Tourism Management*, 56, 253–263. DOI: 10.1016/j.jhtm.2023.06.022

World Bank Group. (2018). Sustainable energy for all database. https://databank.worldbank.org/source/sustainable-energy-for-all/preview/on

World Bank Group. (2023). What is food security? In World Bank. www.worldbank.org

World Bank Group. (2024). *Health and Climate Change.* www.worldbank.org

World Bank. (2024). *Health and Climate Change.* www.worldbank.org

World Health Organisation. (2024). *Climate Change.* www.who.int

World Health Organization (WHO). (2001). *International classification of functioning, disability and health.* www.who.org

World Health Organization (WHO). (2018). *Climate change and Health.* www.who.int

World Health Organization (WHO). (2018). *COP24 special report: health and climate change*. www.who.int

World Health Organization. (1946). Preamble to the Constitution of the World Health Organization as adopted by the International Health Conference. *Official Records of the World Health Organization*, 2, 100. https://apps.who.int/gb/bd/PDF/bd47/EN/constitution-en.pdf

World Health Organization. (2001). World Health Organization. *Gender and women's mental health. Gender disparities and mental health: The Facts.*https://www.who.int/mental_health/media/en/242.pdf

World Health Organization. (2008). *The Right to Health.* www.who.int

World Health Organization. (2014). *Gender, climate change and health.*http://apps.who.int/iris/bitstream/10665/144781/1/ 9789241508186_eng.pdf

World Health Organization. (2017). *World Health Statistics 2017: Monitoring Health for the SDGs, Sustainable Development Goals.* https://www.who.int/gho/publications/world_health_statistics/2017/en/

World Health Organization. (2022). *Mental Health and Climate Change: Policy Brief.*www.who.int

World Health Organization. (WHO) (2017). *Mental health.* 2017. www.who.int

World Meteorological Organisation (WMO). (2021). *State of the Global Climate in 2020.* World Meteorological Organization.

Wudil, A. H., Usman, M., Rosak-Szyrocka, J., Pilař, L., & Boye, M. (2022). Reversing Years for Global Food Security: A Review of the Food Security Situation in Sub-Saharan Africa (SSA). *International Journal of Environmental Research and Public Health*, 19(22), 14836. DOI: 10.3390/ijerph192214836 PMID: 36429555

Xu, M., Du, H., Niyato, D., Kang, J., Xiong, Z., Mao, S., Han, Z., Jamalipour, A., Kim, D. I., Shen, X., Leung, V. C. M., & Poor, H. V. (2024). Unleashing the power of edge-cloud generative ai in mobile networks: A survey of aigc services. *IEEE Communications Surveys and Tutorials*, 26(2), 1127–1170. DOI: 10.1109/COMST.2024.3353265

Yadav, S. S., & Lal, R. (2018). Vulnerability of women to climate change in arid and semi-arid regions: The case of India and South Asia. *Journal of Arid Environments*, 149, 4–17. DOI: 10.1016/j.jaridenv.2017.08.001

Yehuda, R., Hoge, C. W., McFarlane, A. C., Vermetten, E., Lanius, R. A., Nievergelt, C. M., Hobfoll, S. E., Koenen, K. C., Neylan, T. C., & Hyman, S. E. (2015). Post-traumatic stress disorder. *Nature Reviews. Disease Primers*, 1(1), 15–30. DOI: 10.1038/nrdp.2015.57 PMID: 27189040

Yorulmaz, D. S., & Karadeniz, H. (2021). Effects of disasters on mental health. *Doğ Afet Çev Derg*, 7(2), 392–398. DOI: 10.21324/dacd.786048

Zander, K. K., Baggen, H. S., & Garnett, S. T. (2023). Topic modeling the mobility response to heat and drought. *Climatic Change*, 176(4), 42. DOI: 10.1007/s10584-023-03524-1

Zeleke, T., Beyene, F., Deressa, T., Yousuf, J., & Kebede, T. (2021). Vulnerability of Smallholder Farmers to Climate Change-Induced Shocks in East Hararghe Zone, Ethiopia. *Sustainability (Basel)*, 13(4), 2162. DOI: 10.3390/su13042162

Zhang, Y., Bi, P., & Hiller, J. E. (2007). Climate change and disability-adjusted life years. *Journal of Environmental Health*, 70(3), 32–36. PMID: 17941401

Zhao, C. W., Yang, J., & Li, J. (2021). Generation of hospital emergency department layouts based on generative adversarial networks. *Journal of Building Engineering*, 43, 102539. DOI: 10.1016/j.jobe.2021.102539

Zhao, Q., Guo, Y., Ye, T., Gasparrini, A., Tong, S., Overcenco, A., Urban, A., Schneider, A., Entezari, A., Vicedo-Cabrera, A. M., Zanobetti, A., Analitis, A., Zeka, A., Tobias, A., Nunes, B., Alahmad, B., Armstrong, B., Forsberg, B., Pan, S. C., & Li, S. (2021). Global, regional, and national burden of mortality associated with non-optimal ambient temperatures from 2000 to 2019: A three-stage modelling study. *The Lancet. Planetary Health*, 5(7), e415–e425. DOI: 10.1016/S2542-5196(21)00081-4 PMID: 34245712

Zhou, X., Kang, L., Sun, X., Song, H., Mao, W., Huang, X., Zhang, Y., & Li, J. (2013). Prevalence and risk factors of post-traumatic stress disorder among adult survivors six months after the Wenchuan earthquake. *Comprehensive Psychiatry*, 54(5), 493–499. DOI: 10.1016/j.comppsych.2012.12.010 PMID: 23337407

About the Contributors

Jeffrey Kurebwa is a Senior Lecturer in the Department of Peace and Governance at Bindura University of Science Education, Bindura, Zimbabwe. He holds a PhD in Public Administration from Nelson Mandela University, South Africa; Master of Public Administration, and Bachelor of Science Honours in Administration Degree from the University of Zimbabwe. Jeffrey Kurebwa teaches at undergraduate, Masters and Doctoral level courses in Gender Studies, Local Governance, Research Methods, Public Policy and Management, and Community Development. He was a Visiting Lecturer for the Peace building and Post Conflict Recovery course for the PhD programme in the College of Business, Peace Leadership and Governance at Africa University, Zimbabwe (2021). Jeffrey is a Visiting Lecturer at Africa Leadership and Management Academy (ALMA) an affiliate of the National University of Science and Technology (NUST) Graduate Business School (Zimbabwe) where he teaches Management Theory and Practice, and Strategic Management. Dr Kurebwa has supervised and examined dissertations at undergraduate, masters, and doctoral level. He has examined Masters and Doctoral theses for Africa University, Women's University in Africa, Zimbabwe Open University, and Bindura University of Science Education (Zimbabwe), Durban University of Technology, University of Kwazulu Natal, and Cape Peninsula University of Technology (South Africa). Jeffrey has over 80 publications including books, book chapters, and peer reviewed journal articles in the areas of gender, local governance, youth development, natural resource governance, cyber security, and public policy. Kurebwa has edited two books on Participation of Young People in Governance Processes in Africa (2019), and Understanding Gender in the African Context (2020). Dr Kurebwa is a Programme Peer Reviewer for the Zimbabwe Council for Higher Education (ZIMCHE). Kurebwa has peer reviewed programmes for Women's University in Africa, Zimbabwe Open University, UNICAF, University of Zimbabwe, Arrupe Jesuit University, Great Zimbabwe University, and Midlands State University. Dr Kurebwa has done consultancy work for UN Women, National AIDS Council (NAC), and National

Peace and Reconciliation Commission (NPRC). Kurebwa is also an Advisory Board member for Shamva Agricultural College. He has been instrumental in designing and teaching short courses in Projects Monitoring and Evaluation; Project Planning and Management; and Gender and Transformative Studies.

David Makwerere is an experienced Leadership, Governance and Peace practitioner with over thirteen years lecturing experience in different universities in Zimbabwe. Program management specialist with demonstrated experience in overall leadership, management, supervision, and strategic vision for the implementation of systems capacity strengthening. He possess strong knowledge and experience in governance, systems strengthening, civil society capacity strengthening, peacebuilding/social cohesion and civic engagement. David is adept in managing staff and resources to ensure that programs meet targets and deliverables on-time, within budget, and at a high level of quality. Experience in ensuring governance processes are more equitable and effective, empowering communities to tackle the underlying causes of the world's toughest challenges, working collaboratively to build secure, productive and just communities. David also has experience in organizational development and monitoring and evaluation. He has an impressive profile of consulting in both the public sector and the voluntary sector, particularly civil society.

***

Shalom Akhai, Department of Mechanical Engineering, Maharishi Markandeshwar Engineering College, Maharishi Markandeshwar (Deemed to be University), Mullana, Ambala, Haryana – 133207, India.

Manmeet Kaur Arora is a Research Scholar at the School of Law, Sharda University, Greater Noida. She is a graduate of Law and has completed her Master's in Law from Amity University, Noida. Manmeet Kaur is a qualified Company Secretary Executive Programme. She has over 8+ years of Industry Experience and 1 year of experience as an Associate at a reputed IPR Law Firm. Manmeet Kaur has attended 20+ National & International Conferences, Webinars, Seminars, Symposiums, and Paper Presentations and participated in 5+ Faculty Development programs. She has over 50+ publications in various esteemed Journals and Books. She is active at the Family Dispute Resolution Clinic, Meditation Centre, Greater Noida. Manmeet Kaur has a keen interest in the areas of Corporate Law, Intellectual Property Rights, Insolvency, Artificial Intelligence, Machine Learning, and Blockchain Technology.

Anchal Bhaskaran is currently pursuing her bachelors degree from Pes University, Bangalore. Her interests include Social Psychology, Environmental Psychology, Philosophy and Statistics.

Christian Kaunert is Professor of International Security at Dublin City University, Ireland. He is also Professor of Policing and Security, as well as Director of the International Centre for Policing and Security at the University of South Wales. In addition, he is Jean Monnet Chair, Director of the Jean Monnet Centre of Excellence and Director of the Jean Monnet Network on EU Counter-Terrorism (www.eucter.net).

Nyasha Kurebwa is currently studying towards a Master's degree in Child sensitive social policies with Women's University in Africa. Works for Oxfam International as Southern Africa Region Safeguarding Adviser. She has vast experience working in mainstreaming adult and child safeguarding in organisation's programs, policies and procedures, prevention of sexual exploitation, abuse and sexual harassment in work place, gender programming and child rights advocacy. She is a holder of Master of Science Degree in Development Studies.

Sahil Lal is a Research Scholar at School of Law, Sharda University. He pursued his graduation from National Law University, Lucknow and completed his Master's in Law from Gujarat National Law University, Gandhinagar. Sahil has completed a certificate course on 'AI & Law' from Lund University, Sweden and on 'Cyber Law' from Indian Law Institute, New Delhi. Sahil Lal is a qualified UGC-NET & JRF candidate. He has a 10+ year Industry Experience and is an active member of Family Dispute Resolution Clinic, Mediation Center, Greater Noida. Sahil has attended 20+ International & National Conferences along with Paper Presentation and has also participated in more than 5+ Faculty Development Programme. Mr. Sahil has more than 20+ publications in various esteemed Journals & Books. Mr. Sahil has a keen interest in area of Artificial Intelligence, Machine Learning, Robotics, Block-Chain Technology and Intellectual Property Rights.

Srishti Muralidharan is a UGC NET qualified postgraduate in Clinical Psychology and a Faculty of Psychology in PES University, RR Nagar Bangalore. She is currently pursuing a Masters in Philosophy from IGNOU. Her interests include Clinical Psychology, Indian Psychology, Environmental Psychology, Philosophy, Gender and Sexuality, Social Psychology and Neuropsychology.

Sudhanshu Negi has completed his masters of Psychology from Department of Humanities and Social Science, Graphic Era (Deemed to be University) Dehradun. As Postgraduate student, my research interest began in the field of environmental psychology, community psychology, and applied social psychology so that I can understand complex human behaviour. I am learning new skills to serve as a psychologist for intervention and rehabilitation of the marginalized communities. Currently I am member of American Psychological Association as Student affiliate.

Shreya Pharswan is currently pursuing a Master of Psychology at the Banaras Hindu University, Varanasi. As a PG student, she started to do research in pursuit of knowledge of the psychological field. Her research started with exploring various subjects and analyzing data from behavioral science. She is keen to address issues such as gender, domestic violence, and social justice.

Urvashi Pipaliya is currently in 2nd year of bachelor's in psychology (Hons) program which she is pursuing from the Graphic Era (Deemed to be University). Her journey for research began in her search to expand her knowledge and understanding of the psychological sciences. She aims that one day she would be able to contribute her perspectives to this field.

Kamalesh Ravesangar, highly committed academician with over ten years of experience in teaching Certificate, Diploma, Degree, and supervision for MBA(coursework mode) students from various social and cultural backgrounds at private colleges and universities. Currently, she works as Assistant Professor at Tunku Abdul Rahman University of Management & Technology in Malaysia. Moreover, she has a total of three years of experience in industries related to educational, manufacturing, recruitment, and engineering. She is also a Professional Member of Malaysian Institute of Human Resource Management (MIHRM). A subject matter expert in human resource management, human resource development, management and organizational behavior studies. She has published articles and reviewed articles in online journal publications. She holds a Phd (doctorate) and a Master's degree in Organizational Behavior and Human Resource Management. She is also peer-reviewer for a few journals in the field of Business, Management, HRM. In addition, she has been awarded " Certificate of Recognition & Excellence in Reviewing by Asian Journal of Economics, Business & Accounting and Journal of Global Economics, Management and Business Research & Institute of Industry and Academic Research Incorporated (IIARI).

Ajay K. Singh is working as an Associate Professor (Research) in the Department of Humanities and Social Science, Graphic Era (Deemed to be University) Dehradun. He has worked as Assistant Professor (Economics) in the School of Liberal Arts & Management, DIT University Dehradun for 6 years. He did Post-Doctorate Research with EDI of India, Ahmedabad, Gujarat (India). He received MPhil (Economics) from DAVV Indore (India), and PhD (Economics) from IIT Indore (India. He has published several research papers in the diversified area such as climate change, agricultural productivity, assessment of food security, estimation of GFSI, development of environmental sustainability index (ESI) and its association with socio-economic indicators; measurement and determinants of entrepreneurship ecosystem, and dimension of sustainable development and its interlinkages with economic development.

Bhupinder Singh working as Professor at Sharda University, India. Also, Honorary Professor in University of South Wales UK and Santo Tomas University Tunja, Colombia. His areas of publications as Smart Healthcare, Medicines, fuzzy logics, artificial intelligence, robotics, machine learning, deep learning, federated learning, IoT, PV Glasses, metaverse and many more. He has 3 books, 139 paper publications, 163 paper presentations in international/national conferences and seminars, participated in more than 40 workshops/FDP's/QIP's, 25 courses from international universities of repute, organized more than 59 events with international and national academicians and industry people's, editor-in-chief and co-editor in journals, developed new courses. He has given talks at international universities, resource person in international conferences such as in Nanyang Technological University Singapore, Tashkent State University of Law Uzbekistan; KIMEP University Kazakhstan, All'ah meh Tabatabi University Iran, the Iranian Association of International Criminal law, Iran and Hague Center for International Law and Investment, The Netherlands, Northumbria University Newcastle UK,

Ravindra Singh has completed his Ph.D. from Banaras Hindu University (Institute of Eminence). Currently he is working as an Assistant Professor Psychology Graphic Era Deemed to be University, Dehradun. Earlier he was Indian Youth Ambassador at BRICS. Beijing China 2017 during Indian delegation. Dr. Singh has also worked as Junior Project Fellow, NCERT funded project, New Delhi in 2018. He worked as the Former Teaching Assistant at IIT, BHU. Recently he was the Youth Member at G20 Youth for India's G-20 Presidency. He served as an Independent Reviewer in Frontiers in Psychology Journal and Member at APA, International Affiliate, Washington DC and also the member of prestigious National Academy of Psychology, (NAOP) India. He has 7 Years of teaching experience and supervising 3 PhD candidate along with many posts' graduate thesis supervision.

Index

F

G

H

I

M

N

P

R

S

V

W

www.ingramcontent.com/pod-product-compliance
Lightning Source LLC
LaVergne TN
LVHW080640011026
849160LV00014B/1274
* 9 7 9 8 3 3 7 3 0 6 1 8 6 *